Communications in Computer and Information Science

2744

Series Editors

Gang Li, *School of Information Technology, Deakin University, Burwood, VIC, Australia*
Joaquim Filipe, *Polytechnic Institute of Setúbal, Setúbal, Portugal*
Zhiwei Xu, *Chinese Academy of Sciences, Beijing, China*

Rationale

The CCIS series is devoted to the publication of proceedings of computer science conferences. Its aim is to efficiently disseminate original research results in informatics in printed and electronic form. While the focus is on publication of peer-reviewed full papers presenting mature work, inclusion of reviewed short papers reporting on work in progress is welcome, too. Besides globally relevant meetings with internationally representative program committees guaranteeing a strict peer-reviewing and paper selection process, conferences run by societies or of high regional or national relevance are also considered for publication.

Topics

The topical scope of CCIS spans the entire spectrum of informatics ranging from foundational topics in the theory of computing to information and communications science and technology and a broad variety of interdisciplinary application fields.

Information for Volume Editors and Authors

Publication in CCIS is free of charge. No royalties are paid, however, we offer registered conference participants temporary free access to the online version of the conference proceedings on SpringerLink (http://link.springer.com) by means of an http referrer from the conference website and/or a number of complimentary printed copies, as specified in the official acceptance email of the event.

CCIS proceedings can be published in time for distribution at conferences or as post-proceedings, and delivered in the form of printed books and/or electronically as USBs and/or e-content licenses for accessing proceedings at SpringerLink. Furthermore, CCIS proceedings are included in the CCIS electronic book series hosted in the SpringerLink digital library at http://link.springer.com/bookseries/7899. Conferences publishing in CCIS are allowed to use our online conference service (Meteor) for managing the whole proceedings lifecycle (from submission and reviewing to preparing for publication) free of charge.

Publication process

The language of publication is exclusively English. Authors publishing in CCIS have to sign the Springer CCIS copyright transfer form, however, they are free to use their material published in CCIS for substantially changed, more elaborate subsequent publications elsewhere. For the preparation of the camera-ready papers/files, authors have to strictly adhere to the Springer CCIS Authors' Instructions and are strongly encouraged to use the CCIS LaTeX style files or templates.

Abstracting/Indexing

CCIS is abstracted/indexed in DBLP, Google Scholar, EI-Compendex, Mathematical Reviews, SCImago, Scopus. CCIS volumes are also submitted for the inclusion in ISI Proceedings.

How to start

To start the evaluation of your proposal for inclusion in the CCIS series, please send an e-mail to ccis@springer.com

Frédéric Barbaresco · François Gerin
Editors

Quantum Engineering Sciences and Technologies for Industry and Services

First International Conference, QUEST-IS 2025
Paris, France, December 1–4, 2025
Proceedings, Part II

 Springer

Editors
Frédéric Barbaresco
Thales
Vélizy-Villacoublay, France

François Gerin
SEE
Paris, France

ISSN 1865-0929 ISSN 1865-0937 (electronic)
Communications in Computer and Information Science
ISBN 978-3-032-13854-5 ISBN 978-3-032-13855-2 (eBook)
https://doi.org/10.1007/978-3-032-13855-2

This Springer imprint is published by the registered company Springer Nature Switzerland AG
The registered company address is: Gewerbestrasse 11, 6330 Cham, Switzerland

If disposing of this product, please recycle the paper.

Preface

From Quantum Engineering to Applications for Citizens

At the heart of French deep tech, academic and industrial, QUEST-IS 2025, the 1st International Quantum Engineering conference and exhibition, gathered many companies working in the quantum ecosystem from various countries.

It thus addressed technical matters such as quantum computing and algorithms, quantum sensors, quantum communications, crypto and the Internet, including enabling technologies, and obviously quantum engineering targeting applications for citizens.

This multidisciplinary event let attendees from various backgrounds discuss the technical challenges of applied quantum technologies and business opportunities.

We present in this proceedings peer-reviewed papers addressing challenges on quantum engineering.

Quantum Engineering Challenges and Issues for Quantum Computing, Quantum Sensors and Quantum Communication

Quantum engineering denotes the systematic endeavour to transform the abstract formalism of quantum mechanics into practical technologies. It occupies a unique position at the confluence of physics, materials science, control theory, computer science, and systems engineering. The field encompasses three principal technological domains, quantum computing, quantum sensing, and quantum communications, each facing distinct, though deeply interconnected, challenges. Despite their differences, these disciplines share a common difficulty: the engineering of systems that rely upon fragile quantum states, whose coherence can be destroyed by the slightest interaction with their environment.

Quantum computing aspires to create fault-tolerant logical qubits but is constrained by decoherence and the immense overheads of error correction. Quantum sensing seeks ultimate precision, yet must balance coherence preservation with environmental coupling. Quantum communications endeavour to distribute entanglement securely over global distances, but remain limited by photon loss, imperfect memories, and the absence of scalable network architectures.

The future of quantum engineering lies in the emergence of a new discipline, quantum systems engineering, which formalises the design, verification, and integration of quantum devices in the same way that aerospace or microelectronics engineering did for their respective domains. The goal is not merely to understand quantum mechanics but to harness it reliably, reproducibly, and at scale, thereby ushering in the true technological era of the quantum world.

Quantum Computing

In quantum computing, the overarching objective is to construct machines capable of performing computations by manipulating coherent quantum states, superpositions and entanglements, across large numbers of qubits. The central challenge lies in the extreme delicacy of such states. Decoherence, resulting from unwanted coupling between a quantum system and its surroundings, rapidly erases the information encoded in superposition and entanglement. Extending coherence times whilst maintaining precise operational control constitutes one of the most formidable undertakings in modern engineering. Achieving this requires ultra-pure materials, cryogenic operation at temperatures frequently below tens of millikelvin, and stringent isolation from electromagnetic and vibrational disturbances. Even the smallest imperfections in fabrication or fluctuations in temperature and electromagnetic fields may induce decoherence and computational error.

Given that all physical qubits are inherently noisy, large-scale computation depends critically upon quantum error correction (QEC). Unlike in classical systems, redundancy cannot be implemented by straightforward duplication, since quantum states cannot be cloned. Instead, information must be encoded across entangled collections of physical qubits to form logical qubits that are resilient to noise. Implementing QEC on this scale is immensely demanding: thousands of physical qubits may be required to realise a single fault-tolerant logical qubit. Quantum dissipation engineering is emerging based on quantum feedback for bosonic qubits, to reduce drastically the number of physical qubits and by regulating qubit errors through thermodynamics dissipation (the Lindblad equation). Each qubit must be precisely calibrated, and errors must be identified and corrected in real time, often within microseconds. This imposes stringent requirements upon low-latency control electronics and cryogenic data processing.

A further source of difficulty arises from qubit control and crosstalk. Quantum gates must be executed with extremely high fidelity, typically exceeding 99.9%, if fault-tolerance is to be achieved. The generation of microwave or optical control pulses must therefore be exact, and must not disturb neighbouring qubits. As systems grow in size, the problems of frequency crowding, wiring density, and thermal load increase substantially. One promising avenue involves the development of cryogenic CMOS electronics placed in close proximity to the qubits, thereby reducing latency and electrical noise.

The issue of scalability is equally fundamental. Moving from devices containing a few dozen qubits to those comprising millions necessitates entirely new architectural paradigms. The wiring, signal routing, and thermal management of large cryogenic systems present severe engineering constraints. Modular architectures, wherein several small processors are interconnected by photonic links, may offer a practical path to scale. Another promising approach lies in the integration of heterogeneous qubit technologies, such as superconducting circuits, trapped ions, spin qubits, and photonic systems, into hybrid platforms. Ultimately, progress in this field depends upon a holistic approach known as quantum systems co-design, in which hardware, control electronics, and software are developed in concert, optimising the entire computational stack from materials to algorithms.

Quantum Sensing

Quantum sensing exploits the sensitivity of quantum systems to external perturbations in order to achieve measurements of unprecedented precision. These sensors can detect magnetic and electric fields, gravitational gradients, accelerations, or time intervals with sensitivities far beyond classical limits. Yet the very sensitivity that grants them such power also renders them vulnerable to environmental noise. The fundamental challenge is to preserve coherence long enough to perform a measurement, despite constant interaction with the environment.

Quantum sensors often employ nitrogen–vacancy centres in diamond, trapped ions, or ensembles of cold atoms in interferometric configurations. In every case, the sensor must be sufficiently coupled to the quantity of interest to register a measurable signal, while simultaneously being isolated from unwanted noise. Achieving this delicate equilibrium requires careful design of electromagnetic shielding, active feedback, and control of the working point so that the sensor operates at regions of reduced susceptibility to fluctuations.

As these technologies mature, miniaturisation and integration have become pressing concerns. Laboratory systems, typically large and fragile, must be transformed into compact, robust devices suitable for deployment in the field. This entails the integration of optical excitation, detection, and control subsystems within constrained physical volumes, as well as effective thermal management. Cold-atom systems, in particular, must balance vacuum integrity, temperature stability, and portability. Long-term calibration and drift correction present further engineering challenges, since even small instabilities in control parameters may degrade sensitivity over time. Autonomous recalibration and machine-learning-based feedback loops are therefore being investigated to ensure sustained performance.

Quantum Communications

Quantum communications seek to employ the principles of quantum mechanics, particularly entanglement and the no-cloning theorem, to enable the secure transmission of information. The best-known application, quantum key distribution (QKD), allows two parties to share cryptographic keys with absolute theoretical security. Beyond this, researchers envisage a quantum internet, capable of distributing entanglement between distant nodes for both secure communication and distributed computation.

The primary obstacle to such a network is photon loss and decoherence during transmission. Photons propagating through optical fibres are absorbed and scattered, restricting the direct distribution of entanglement to a few hundred kilometres. Free-space and satellite channels mitigate this limitation but suffer from atmospheric turbulence, beam divergence, and pointing errors. To overcome these difficulties, the community is developing quantum repeaters, which rely on intermediate nodes equipped with quantum memories to store, purify, and retransmit entanglement. This, however, requires quantum memories with long coherence times, high efficiency, and rapid read–write cycles, capabilities that remain technically elusive.

The development of reliable quantum memories is thus one of the foremost engineering challenges in the field. Candidate systems include ensembles of cold atoms, rare-earth-doped crystals, and spin systems in solid-state matrices. Each exhibits trade-offs between coherence time, storage efficiency, and compatibility with telecommunication wavelengths. The interface between photonic and matter qubits is similarly critical: efficient, low-noise conversion between optical photons and stationary qubits is essential for scalable networking. Approaches such as cavity quantum electrodynamics, optomechanical transduction, and electro-optic conversion are under active investigation, yet none has achieved the required combination of efficiency, fidelity, and scalability.

At a higher level, network synchronisation and control present formidable obstacles. Entanglement distribution and quantum teleportation protocols require extremely precise timing and phase stability across multiple nodes, often at the nanosecond scale. Large-scale quantum networks must integrate quantum channels with classical communication layers for coordination, routing, and error management. The absence of universally accepted standards and protocols compounds these difficulties. Different experimental platforms employ incompatible encodings, operating wavelengths, and modulation schemes. Achieving interoperability across heterogeneous systems will be crucial to realising a practical quantum internet.

Enabling Technologies and System Integration

Across quantum computing, sensing, and communications, certain engineering challenges are universal. Materials science remains a decisive factor: impurities, defects, and surface roughness all shorten coherence times and reduce gate fidelity. Cryogenics presents another formidable constraint; maintaining stable temperatures at the millikelvin level demands complex and expensive refrigeration systems that are difficult to scale industrially. Quantum control theory is assuming an increasingly central role, providing the mathematical framework for designing optimal pulse sequences and feedback algorithms that stabilise quantum dynamics in the presence of noise.

System integration poses further difficulties, since quantum subsystems must interact seamlessly with classical electronics responsible for control, measurement, and feedback. Timing, thermal management, and signal integrity must all be maintained under extreme conditions. Even the task of verification and validation is non-trivial, for measurement inevitably disturbs the quantum state being tested. Benchmarking and certification of quantum devices thus require sophisticated statistical and tomographic techniques. Lastly, the issue of scalability and manufacturability looms large. Many existing devices are artisanal laboratory constructs, far removed from the robust, reproducible systems required for industrial deployment. The transition to mass-manufacturable quantum technologies will necessitate advances in fabrication processes, design automation, and quality assurance.

Reviewing, Support

Again, we present in this proceedings peer-reviewed papers addressing challenges on quantum engineering. We thank the 102 reviewers from 17 countries who provided an average of 3.5 reviews per paper for the submitted 120 reviewed papers, using a single-blind process. Of the reviewed papers, 68 were selected papers for inclusion in these proceedings.

Support for the event came from Belgium, Canada, Denmark, Finland, France, Germany, Greece, Italy, Japan, The Netherlands, Saudi Arabia, Singapore, South Korea, Spain, Sweden, Switzerland, Taiwan, UAE, and the UK, also from the EU, CERN, and Quantum Flagship, and from our Platinum sponsors, Agence Innovation Défense, EDF, IBM, and Thales, and our Gold sponsors, Laboratoire National de Métrologie et d'Essais, CEA, NATO, Alice & Bob, and Quandela.

October 2025   François Gerin
                                                Frédéric Barbaresco

Organization

General Chairs

Frédéric Barbaresco   Thales, France
François Gerin   SEE, France

Local Organizing Chair

Loïc Cantu   SEE, France

Program Committee Chairs

Frédéric Barbaresco   Thales, France
François Gerin   SEE, France

Program Committee

Neil Abroug	Inria, France
Khulud Almutairi	King Saud University, Saudi Arabia
Yasutaka Amagai	AIST, Japan
Nina Amini	CentraleSupélec, France
Dimitris Angelakis	TU Crete and AngelQ Quantum, Greece and CQT, Singapore
Thomas Antoni	CentraleSupélec, France
Sofiane Bahbah	SEDI.ATI, France
Bhashyam Balaji	DRDC Ottawa, Canada
Quentin Barbeau	Danish Embassy, Paris, France
Pascale Bendotti	EDF R&D, France
Mathieu Bertrand	Thales Alenia Space, France
Quentin Bodart	Crystal Q. Computing, France
Nadia Boutabba	IAT, Abu Dhabi, UAE
Philippe Bouyer	Quantum Delta, Netherlands
Harry Buhram	Quantinuum, UK
Andrea Busch	Quobly, France
Marie-Elisabeth Campo	EM Armée air & espace, France

Joan Camps	Riverlane, UK
Victor Canivell	Qilimanjaro, Spain
Patrick Carribault	CEA DAM hpc, France
Adam Connolly	Quantinuum, UK
Franck Correia	DGA, France
Giacomo Corrielli	Ephos/IFN-CNR, Italy
Stefan Creemers	UC Louvain, Belgium
Simon Crispel	GIFAS ALAT, France
Nicolas Daval	Quobly, France
Merouane Debbah	Khalifa University, Abu Dhabi, UAE
Fabrice Debbasch	Sorbonne Université, France
Ivo Pietro Degiovanni	INRIM, Italy
Guillaume De Giovanni	Viqthor, France
Philippe Deniel	CEA, France
Amer Delilbasic	FZ Jülich, Germany
Matthieu Desjardins	C12, France
Alain Dessertaine	La Poste, France
Amanda Diez	CERN, Switzerland
Daniel Dolfi	Thales, France
Christophe Domain	EDF R&D, France
Jens Eisert	Freie Universität Berlin, Germany
Andreas Elben	PSI, Switzerland
Jaap Essing	TNO, Netherlands
Olivier Ezratty	Independent Researcher, France
Nicolas Fabre	Télécom Paris, IPP, QuantEduFrance, France
Thomas Fauvel	Choose Paris Region, France
Benjamin Frisch	CERN, Switzerland
Jacques-Henri Gagnon	Canadian Embassy, Paris, France
Luc Gérardin	University of Sussex, UK
James A. Grieve	TII Abu Dhabi, UAE
Fabrice Guillemin	Orange Labs, France
Stefan Haeussler	Hensoldt, Germany
Pascal Halffmann	ITWM Fraunhofer, Germany
Mélanie Hardman	NPL, UK
David Harvey	Thales, France
Winfried Hensiger	University of Sussex, UK
Mohamed Hibti	EDF R&D, France
Israel Hinostroza	CentraleSupélec/SONDRA, France
Magnus Hoijer	FOI, Sweden
Min-Hsiu Hsieh	Foxconn, Australia
Daniel Huerga	ONERA, France
Toshiyasu Ichioka	RIKEN, Japan

Samy Jousset	Région Ile de France, France
Mathieu Juan	Université de Sherbrooke, Canada
Marc Kaplan	Veriqloud, France
Jian Feng Kong	PIC IHPC, Singapore
Naoya Kono	Secretariat of Science, Technology and Innovation Policy, Japan
Romain Kukla	Naval Group, France
Philippe Lacomme	Université Clermont-Auvergne, France
Jacques-Charles Lafoucrière	CEA, France
Catherine Lambert	Académie des Technologies, France
Aolita Leandro	6G RC Abu Dhabi, UAE
Marc Leconte	F2S, France
Megan Lee	Quantum City, University of Calgary, Canada
Jeremy Leow Kwang Siong	FSTD, Singapore
François-Marie Le Régent	Pasqal, France
Tobias Lindstrom	NPL, UK
Craig Lloyd	NPL, UK
Tony Maindron	CEA, France
Joseph Mikael	EDF R&D, France
Swapan Mandal	Visva Bharati Institute, India
Sabrina Maniscalco	Algorithmiq, Finland
Ulrich Mans	Quantum Delta, Netherlands
Luigi Martiradonna	Riverlane, UK
Francesco Mauro	Sannio University, Italy
Mathias Möller	TU Delft, Netherlands
Mikko Möttönen	Aalto University, Finland
Stéphanie Molin	Thales, France
Liran Naaman	Quantum Delta, Netherlands
Yuichi Nakamura	NEC Corporation, Japan
Hiro Nakata	Jij, Japan
Frank Nielsen	Sony Computer Science Laboratories, Japan
Pierre-Elie Normand	Dassault Aviation, France
Myriam Nouvel	Thales, France
Yasser Omar	PQI (Portuguese Quantum Institute), Portugal
Dimitrios Papadimitriou	Université libre de Bruxelles, Belgium
Cécile Perrault	Alice & Bob, France
Frank Phillipson	TNO NL, OTAN Group, Netherlands
Nico Piatkowski	IAIS Fraunhofer, Germany
Jonathan Pisane	Thales, France
Mathilde Portais	Naval Group, France
Tony Quertier	Orange, France
Setra Rakotomavo	La Poste, France

Matthieu Ratiéville   AID and DGA, France
René Reimann   TII Abu Dhabi, UAE
Florentin Reiter   IAF Fraunhofer, Germany
Hervé Ribot   Minalogic, France
Stéphane Saillant   ONERA/SONDRA, France
Mitsuhisa Sato   RIKEN R-CCS, Japan
Tanguy Sassolas   CEA, France
Hisham Sati   CQTS, NYU, Dubai, UAE
Félicien Schopfer   LNE, France
Jan Schorer   Hensoldt, Germany
André Schrottenloher   Inria, France
Sylvain Schwartz   GIFAS-ONERA, France
Jean Sénellart   Quandela, France
Xavier Sénémaud   SNCF Group, France
Mohammed Serhir   Centrale Supélec/SONDRA, France
Najwa Sidqi   National Quantum Computing Center, UK
Priyank Singh   Aalto University, Finland
Kako Sugiyama   AIST, Japan
Dunlin Tan   Thales Asia, Singapore
Mathieu Taupin   LNE, France
Hiroaki Tezuka   AIST, Japan
Guido Torrese   Thales, Belgium
Malak Trabelsi Loeb   Vernewell Academy, Belgium
Pierfrancesco Ulpiani   Leonardo, Italy
Mathias van den Bossche   Thales, France
Christophe Vasse   Thales, France
Emmanuelle Vergnaud   Teratec, France
Benoît Vermersch   University of Grenoble-Alpes (UGA), and
                               Quobly, France

Daniel Vert   Systematic, France
Stéphane Vialle   CentraleSupélec, Université Paris-Saclay, France
Loïc Viguerie   Cryoconcept, Air Liquide, France
Mark Webber   Universal Quantum, UK
Göran Wendin   RISE, Sweden
Ronin Wu   QunaSys, France
Yu Yamashiro   Jij and Q-STAR, Japan
Jun Ye   A*STAR, Singapore
Su Yi   ED IHPC
Ugo Zanforlin   Leonardo, Italy
Gilles Zalamansky   GIFAS-Dassault Aviation, France
Dominik Zumbuhl   University of Basel, Switzerland

Supporting Countries

Sponsors

Keynote Talks

The Two Quantum Revolutions: From Concepts to Applications

Alain Aspect

Institut d'Optique - Université Paris-Saclay, France

The first quantum revolution, based on wave–particle duality, led to the society of information and communication. The second quantum revolution is based on entanglement. Will its applications lead to a new societal revolution?

Quantum Error Correction and Feedback

Pierre Rouchon

Center Automatic and Systems, Mines-Paris, University PSL, Member of Académie
des Sciences, France

Quantum error correction relies on a feedback loop. This feedback generally corresponds
to a classical controller. Quantum error correction can also exploit the dissipation associated with the phenomenon of decoherence. Called autonomous correction by physicists,
it then uses feedback where the controller is a dissipative quantum auxiliary system.
This talk focuses on the development of such quantum controllers to stabilize logical qubits encoded in harmonic oscillators (bosonic code). Two types of encoding will
be considered: cat-qubit encoded in two coherent states of opposite phases for which
bit-flip errors induced by usual noises can be experimentally almost suppressed; and
GKP-qubit encoded in finite energy grid-states approximating position/impulsion Dirac
combs where, in principle, both bit-flips and phase-flips could be almost suppressed.

The Interplay Between Quantum Engineering and Quantum Science

Olivier Ezratty
Freelance quantum engineer, mostly known for "Understanding Quantum Technologies"

Quantum engineering is a relatively new discipline that takes shape as quantum technologies are maturing and turning into commercial products. But what is it exactly? How are science and engineering intermingled in this innovation process? Is the science done, and we are just left with engineering and technology development? What is the engineering scope required for the development of complex quantum systems, particularly fault-tolerant quantum computers? How does quantum engineering connect the dots between the software and hardware stacks? Is the environmental footprint of these emerging technologies integrated in vendors' engineering goals? Are there quantum engineers? How are they trained and how will they be trained?

Quantum Enabling Technologies from Science to Engineering

Richard Versluis

Quantum Enabling Technologies Engineering, TNO/TU Delft, Netherlands

In this talk Richard will share an overview of the most critical engineering challenges in Quantum Computing that we will face in the upcoming years. He will delve into the transition from academic research to practical engineering, emphasising the complexities and hurdles that need to be overcome. Aspects like scalability, reliability and modularity will be quantified and related to practical design choices and design requirements for future quantum computers and their constituent components.

Experimental Quantum Photonics: From Testing Foundations of Quantum Mechanics to Building Practical Quantum Networks

Djeylan Aktas

Experimental Quantum Communications, Slovak Academy of Sciences, Slovakia

In this talk we will explore a few seminal experiments in which quantum photonics made it possible to interrogate nature itself by studying the very foundations of quantum theory. From there, we will see how technological progress allowed for scientists to "surf the wave" of the second quantum revolution to actually create practical quantum technologies that can be useful for society by bringing some quantum advantages like enhanced security for modern communication networks, or better sensing capabilities in various domains of applications.

The European Commission's Vision for Quantum Engineering: Challenges and Opportunities in EU-Funded Projects

Oscar Diez

European Commission Representative, DG CNECT

In this keynote, I will present the European Commission's strategy for quantum technologies and its role in building a globally competitive and resilient European quantum ecosystem. The talk will highlight how EU programmes, including Horizon Europe, the Quantum Flagship, the Chips JU and the EuroHPC Joint Undertaking, support the full quantum value chain, from foundational research and engineering to infrastructure, skills and industrial deployment. It will also address current challenges, such as scaling up quantum platforms, strengthening supply chains and fostering innovation across member states, and will outline upcoming opportunities for industry and research communities to engage with EU initiatives in quantum technologies.

Quantum Sensors

Marco Genovese
Director, Quantum Optics Research, INRIM, Italy

Quantum state coherence and entanglement are very sensitive to interaction with the environment. On the one hand, this represents a problem for developing quantum technologies such as quantum computation or quantum computation, but, on the other hand, it allows for a very high sensitivity to various parameters, leading to the possibility of developing quantum sensors largely surpassing traditional classical sensors. In this talk, I will present this exciting new field, discussing the potentialities of some of the most interesting techniques.

Fair Benchmarking of Quantum Optimisation Applications

Frank Phillipson

Quantum Computing Applications, TNO, Netherlands

Quantum computing is advancing rapidly, and optimisation is one of its most promising application areas. But measuring progress is not as simple as comparing runtimes or claiming "quantum advantage". In this keynote, I will explore why traditional benchmarking approaches – developed for CPUs and GPUs – fall short when applied to quantum and hybrid systems, and how misleading comparisons risk slowing genuine progress. Building on lessons from classical supercomputing, I will introduce principles for fair benchmarking that emphasise transparency, end-to-end workflows, solution quality, and reproducibility. I will also highlight recent initiatives, from Q-Score to TAQOS, and show how energy use and application-driven metrics are shaping the next generation of evaluation standards. The goal is to provide the community with a practical framework for assessing quantum optimisation responsibly – one that enables researchers, industry, and policymakers to interpret results with confidence and set realistic expectations for the future.

Telecom Integrated QKD Networks: The Madrid QCI Example

Vicente Martin
Quantum Engineering, CSS/Univ. Politécnica de Madrid, Spain

In the talk I will revise the types of QKD networks deployed in the field, the problems and different choices, concentrating on the solutions used in the Madrid network. This is a multidomain and highly heterogeneous network currently spanning more than 700 km, with 29 nodes hosting 23 QKD pairs from 10 different vendors. Finally several of the use cases implemented will be shown.

Contents

Session: 8 Quantum Algorithms, Computing; Simulation – Physics and Engineering Applications

Session: 9 Quantum Algorithms, Computing; Simulation – Simulation and Quantum Chemistry A

Session: 10 Quantum Algorithms, Computing; Simulation – Simulation and Quantum Chemistry B

**Session: 11 Quantum Algorithms, Computing; Simulation – Hardware
and Architectures**

**Session: 12 Quantum Algorithms, Computing; Simulation –
Specialized Applications and Security**

Session: 4 Quantum Algorithms, Computing; Simulation – Quantum Algorithms For Finance and Industry

Quantum Approaches for the Unit Commitment Problem - a Literature Survey

Frank Phillipson[1([✉])] and Sven Muller[2]

[1] Maastricht University, Maastricht, The Netherlands
`f.phillipson@maastrichtuniversity.nl`
[2] TNO, The Hague, The Netherlands

Abstract. This paper investigates the application of quantum computing to the Unit Commitment (UC) problem, a fundamental optimisation challenge in power system operations. From the literature, we can learn about various quantum approaches, including the Quantum Approximate optimisation Algorithm (QAOA), Quantum Annealing (QA), and hybrid quantum-classical methods. The review shows that in the literature it is believed that quantum computing can potentially solve the UC problem more efficiently than classical methods. QAOA shows promise in handling binary decision variables, while hybrid methods enhance computational efficiency and scalability. Quantum Annealing is effective for smaller UC instances, with larger problems requiring partitioning. Despite current hardware limitations, advancements in quantum algorithms and hybrid methods provide a strong foundation for future research. This study highlights the transformative potential of quantum computing in optimising power systems, emphasising the need for continued innovation in quantum hardware and error mitigation techniques.

Keywords: Quantum Computing · Energy Model · Unit Commitment · Optimisation

1 Introduction

The Unit Commitment (UC) problem is a critical optimisation task in electrical power systems, where the objective is to schedule the operation of generating units over a planning horizon to meet electricity demand at minimal cost while ensuring system reliability. This problem arises because large-scale energy storage is impractical, requiring real-time alignment of generation and consumption. UC is challenging due to the diversity of generation units, operational constraints such as minimum up/down times and ramp limits, and the need for advance scheduling to accommodate slow-starting units. Furthermore,

---

This work was supported by the Dutch National Growth Fund (NGF), as part of the Quantum Delta NL programme.

F. Barbaresco and F. Gerin (Eds.): QUEST-IS 2025, CCIS 2744, pp. 3–13, 2026.
https://doi.org/10.1007/978-3-032-13855-2_1

the problem must account for network constraints, reliability requirements, and market rules, adding layers of complexity to the optimisation process.

Classical approaches to UC have evolved significantly. Since 2005, power system operators have transitioned from heuristic methods based on Lagrangian relaxation to Mixed-Integer Programming (MIP) formulations [13]. These models, solved using commercial solvers like CPLEX and Gurobi, have achieved substantial cost savings—estimated at \$5 billion annually in the United States. Despite these advances, solving real-world UC problems remains computationally demanding, often involving hundreds of generators, network constraints, and strict time limits for solution generation. Recent research focuses on improving MIP formulations to enhance computational efficiency, but challenges persist.

Quantum computing offers a promising alternative for addressing these computational bottlenecks. By leveraging superposition and entanglement, quantum algorithms could accelerate the exploration of vast combinatorial search spaces, potentially enabling faster and more scalable UC solutions. Although quantum computing for UC is in its early stages, it represents a promising step toward real-time optimisation for large-scale power systems.

This paper provides an overview of existing quantum approaches to the UC problem. Section 2 introduces key concepts in quantum computing. Section 3 outlines the UC problem and its mathematical formulation. Section 4 reviews existing work on quantum methods for UC. Finally, Sect. 5 summarizes findings and identifies directions for future research.

2   Quantum Computing

Quantum computing is a revolutionary approach to computation that leverages the principles of quantum mechanics to solve problems that are challenging or even intractable for classical computers. Unlike classical systems that use bits to represent information as either 0 or 1, quantum computers utilise quantum bits or qubits, which can exist in a superposition of states, representing both 0 and 1 simultaneously. This property, combined with quantum entanglement and interference, enables quantum computers to process a vast number of possibilities in parallel [26].

Two primary paradigms define the current landscape of quantum computing: gate-based quantum computing and quantum annealing (QA). Gate-based systems operate similarly to classical computers, employing quantum logic gates to manipulate qubits in complex sequences. These systems are programmable and are often visualised through circuit diagrams, enabling the implementation of sophisticated quantum algorithms.

In contrast, quantum annealers are specialised machines tailored for optimisation problems. They function by encoding a mathematical problem into the physical states of qubits and evolving the system to find a configuration that minimises the problem's energy. Quantum annealing has its roots in the work of Kadowaki and Nishimori [12] and is particularly effective for problems formulated as Quadratic Unconstrained Binary optimisation (QUBO) tasks.

The promise of quantum computing lies in its potential to revolutionise fields requiring extensive computation, including optimisation, cryptography, and material science. For optimisation challenges such as the UC Problem, quantum computing offers innovative methods that could transcend the limitations of classical algorithms, paving the way for breakthroughs in efficiency and scalability [11]. In the quantum computing approaches in literature, we will see a number of techniques used, which will be introduced in the remainder of this section.

Quantum Annealing is a specialised approach to quantum computation designed to solve optimisation problems by finding the minimum of a specific energy function. This is achieved by initialising qubits in a superposition of states and allowing them to evolve under the influence of a problem-specific Hamiltonian, which encodes the optimisation objective. Unlike classical simulated annealing, QA leverages quantum phenomena such as tunnelling to explore and converge to optimal or near-optimal solutions. It is particularly effective for problems expressed in QUBO form and is implemented in devices like those developed by D-Wave Systems.

The **Quantum Approximate Optimisation Algorithm.** (QAOA) [6] is a hybrid quantum-classical approach for solving combinatorial optimisation problems on gate-based quantum computers. Also this approach often uses the QUBO formulation of the problem as input. QAOA approximates the process of adiabatic quantum evolution by alternately applying two parametrised quantum operators to a system of qubits. By iteratively tuning the parameters through classical optimisation, QAOA seeks to find a quantum state that minimises the problem's objective function. Its adaptability, noise resistance, and relatively low qubit requirements make it a promising candidate for near-term quantum optimisation applications.

There are several approaches within **Quantum Machine Learning** (QML). Quantum Neural Networks (QNNs) [28] combine quantum computing principles with neural network architectures, leveraging qubits and quantum gates to represent and process high-dimensional data efficiently. They exploit quantum superposition and entanglement for richer representations and potential computational speedups, with applications in classification, pattern recognition, and generative modeling. Similarly, Quantum Reinforcement Learning (QRL) [4] integrates quantum algorithms with reinforcement learning to accelerate policy optimization and improve exploration. By encoding states and actions in quantum systems, QRL aims to achieve faster convergence and better performance in domains such as robotics, game theory, and decision-making under uncertainty.

Benders Decomposition. [25] is a mathematical optimisation technique that divides a large-scale problem into smaller, more manageable sub-problems. This approach is particularly useful for mixed-integer programming and problems involving complex constraints. The main problem is split into a master problem, which involves high-level decision variables, and sub-problems, which handle detailed constraints and computations. Iteratively solving these components allows for efficient handling of computationally intensive problems. The method is widely applied in fields like energy systems, logistics, and supply chain optimisation and can be used to create sub-problems that are better solvable by quantum annealing or QAOA.

3   Unit Commitment Problem

The UC Problem is a critical optimisation challenge in power system operations, determining the most efficient schedule for turning power generation units on and off while meeting electricity demand and system constraints. It aims to minimise operational costs while adhering to technical constraints such as generator ramp rates, minimum up and down times, and reserve requirements for system reliability. Given its combinatorial nature and large-scale complexity, the UC problem is computationally demanding, especially in systems with high integration of renewable energy sources. Advanced computational approaches, including quantum computing, are increasingly explored to address the limitations of classical methods in solving this problem efficiently.

Knueven et al. [13] give a general UC problem formulation:

$$\min \sum_{g \in G} \sum_{t \in T} c_g(t), \tag{1}$$

$$\text{s.t.} \sum_{g \in G} A_g(p_g, \overline{p_g}, u_g) + N(s) = L, \tag{2}$$

$$(p_g, \overline{p_g}, u_g, c_g) \in \Pi_g \quad \forall g \in G. \tag{3}$$

Here, c_g is the cost vector associated with $p_g, \overline{p_g}, u_g$, such that the objective function, system operation cost, is minimised. The vectors $p_g, \overline{p_g}, u_g$ represent the feasible generation schedule, maximum power available, and the on/off status for generator g, respectively. The matrix $A_g(p_g, \overline{p_g}, u_g)$ determines how the generator interacts with the system requirements, which are written in matrix form.

A specific example formulation can be found in [7]. The objective of the UC problem is to minimise the total generation cost:

$$P_{i,t} = \operatorname{argmin} \sum_{i=1}^{G} \sum_{t=1}^{T} (a_i P_{i,t}^2 + b_i P_{i,t} + c_i)$$

where $P_{i,t}$ is the generation of unit i at time t, a_i, b_i and c_i are coefficients of unit i-th cost function, T is the number of operation periods, and G is the number of units. The minimisation is subject to the following constraints:

- System Demand Constraints: The total generated power should meet the demand D_t at time t:

$$\sum_{i=1}^{G} P_{i,t} = D_t$$

- Generation Capacity Constraints: Generation $P_{i,t}$ of online units are constrained between minimum P_i^{min} and maximum P_i^{max}:

$$x_{i,t}P_i^{min} \leq P_{i,t} \leq x_{i,t}P_i^{max}$$

where $x_{i,t}$ is the on/off decision variable of unit i at time t.
- Ramp-Rate Constraints: Ramp-rate constraints ensure that the change of power generation levels between two consecutive time periods does not exceed ramp rates:

$$P_{i,t} - P_{i,t-1} \leq R_i x_{i,t-1} + V_i(1 - x_{i,t-1})$$
$$P_{i,t-1} - P_{i,t} \leq R_i x_{i,t} + V_i(1 - x_{i,t})$$

where R_i is the ramp rate of unit i and V_i is the startup/shut-down ramp rate.

4   Literature Overview

In this section, we delve into the various quantum computing approaches that have been explored for addressing the UC problem.

QAOA. The literature on quantum approaches to the Unit Commitment (UC) problem presents a range of hybrid and quantum-enhanced frameworks, primarily leveraging the Quantum Approximate Optimization Algorithm (QAOA) and QUBO formulations. A core trend involves decomposing the UC problem into sub-problems solvable via quantum and classical methods, often within distributed or iterative schemes such as ADMM.

Magar, Nikmehr et al. [17,20,21,31] reformulate the UC problem as a QUBO for QAOA and develop a distributed ADMM framework for large-scale systems. Similarly, Fontana et al. [8] address single-period UC problems using hybrid solvers and demonstrate fast performance with D-Wave's platform. Halffmann et al. [9] focus on compact QUBO formulations optimized for current quantum hardware, minimizing qubit and connectivity requirements, and later expand their work [10] to robust optimisation under uncertainty using QAOA and quantum annealing (QA).

Several papers propose hybrid ADMM frameworks. Mahroo et al. [18] decompose UC into three sub-problems with QAOA handling the QUBO part, showing convergence and improved accuracy via parameter updates. Yang et al. [30] integrate Tabu Search with ADMM and QAOA (H-ADMM-TS) to escape local optima. Salgado et al. [27] refine this hybridization by applying QAOA to binary variables while using classical solvers for continuous decisions, achieving high accuracy and scalability, especially when warm-starting QAOA.

Feng et al. [7] introduce in their paper the Quantum-Surrogate Lagrangian Relaxation (QSLR) approach for large-scale, distributed UC, emphasizing scalability, data privacy, and robustness in noisy environments. Stein et al. [29] propose the QuSO and LinQuSO frameworks, exploiting quantum simulation-based speedups using QAOA and singular value transformation. These show promise for exponential gains but rely on future advancements in quantum hardware and error correction.

Koretsky et al. [14] propose a hybrid method where QAOA handles binary decision variables, and classical optimisers address continuous power output. Reformulating UC as a QUBO problem enables efficient handling of discrete variables by quantum algorithms while classical methods ensure continuous variable optimisation. Simulations on systems with up to 10 units show improved probability of finding near-optimal solutions with increased QAOA depth. Despite current quantum hardware limitations, the study emphasises hybrid approaches' scalability advantages for larger systems.

Across the literature, QUBO reformulations and QAOA remain central, often within hybrid schemes to offset current quantum hardware limitations. Techniques such as warm-starting, robust optimisation, and integration with meta-heuristics (e.g., Tabu Search) enhance solution quality and convergence. While most studies acknowledge scalability and noise challenges, results consistently demonstrate quantum methods' potential to complement or outperform classical counterparts in UC problems, especially as quantum technology matures.

Quantum Annealing. Ajagekar et al. [1] demonstrates the implementation of the UC problem on D-Wave's quantum annealer by discretising continuous variables into equally spaced grid points and reformulating the problem as a QUBO model. Their results show that smaller instances could be efficiently solved on the quantum processor, achieving solutions close to the global optimum. However, larger instances required partitioning into sub-problems, which impacted solution quality. They emphasised the trade-off between solution precision and the likelihood of obtaining global optima, concluding that advancements in quantum hardware and algorithms are necessary for solving larger and more complex problems effectively.

Building on this, Quinton et al. [24] benchmarked D-Wave's hybrid quantum annealing solver against classical solvers such as CPLEX, Gurobi, and IPOPT across several optimisation problems, including the UC problem. They confirmed that while quantum annealers show promise for small-scale UC problem instances, larger problems necessitate hybrid approaches that combine classical and quantum methods. The study underscores the potential of quantum annealing for optimisation but calls for further technological advancements to fully leverage its benefits for industrial applications.

Braun et al. [2] extended the exploration of quantum computing for the UC problem by introducing a QUBO formulation capable of addressing uncertainties in renewable energy supply, power demand, and machine failures. This formulation integrates probabilistic modelling of future scenarios, enabling optimisation that considers both cost and variance minimisation. Results showed that smaller UC problem instances could be effectively solved using D-Wave's quantum annealer, while larger problems faced the same partitioning challenges observed in previous studies. The authors highlighted the need for continued innovation in quantum hardware and algorithms to manage the complexity of large-scale, real-world UC problems effectively.

Müller [19] focuses on balancing intermittent generation and compares two methods for encoding inequality constraints in QUBO formulations: the traditional slack-variable approach and the unbalanced penalization technique. Computational experiments show that unbalanced penalization reduces the number of variables and outperforms the slack-variable method, but its uneven penalty on feasible solutions limits scalability.

Quantum Machine Learning. Liu et al. [16] introduce the Partially Connected Quantum Neural Network (PCQNN), a knowledge-based quantum-classical model designed to efficiently solve the UC problem. By reducing network connectivity based on domain knowledge, PCQNN improves precision and reduces circuit depth compared to traditional QNNs. Tested on 5- and 10-unit systems, it achieves near-optimal and exact solutions with fewer quantum layers, making it suitable for distributed quantum environments and scalable to larger problems using a divide-and-conquer strategy.

Kruse et al. [15] propose a Hamiltonian-based Quantum Reinforcement Learning (QRL) method, combining quantum computing with Neural Combinatorial Optimisation. By directly modeling the UC problem's Hamiltonian in a variational quantum circuit, the approach improves trainability and general applicability beyond graph-based problems, offering a versatile framework for complex combinatorial optimisation tasks.

Hybrid Approach Based on Benders Decomposition. Chang et al. [3] propose a hybrid algorithm combining quantum computing and classical computing through Benders decomposition. Their method splits MIPs into binary programming problems, handled by a so-called noisy intermediate-scale quantum (NISQ) [23] processor, and linear programming problems, solved on a classical processor. Tested on the D-Wave 2000Q, the approach demonstrated promise for small-scale cases and power system-inspired problems but faced challenges with quantum annealing hardware limitations and the reformulation of inequality constraints into QUBOs.

Paterakis et al. [22] introduce a hybrid framework for MILPs in power system optimisation using a multi-cut Benders decomposition scheme. This method applies QC to generate multiple valid cuts, accelerating convergence for problems like Unit Commitment. While Quantum Annealers showed faster computation for small-scale problems, noise and scalability remain barriers for larger problems. The study underscores the need for error mitigation and improved scalability, offering a Python tutorial to aid further research.

Ellinas et al. [5] expand on the hybrid paradigm for power system optimisation, addressing Optimal Transmission Switching and Neural Network verification for DC Optimal Power Flow. Their results align with previous findings, demonstrating QC's potential to enhance computation times for small-scale problems but encountering noise and scalability constraints for larger instances.

The study reiterates the importance of developing error mitigation techniques and scaling strategies while providing community resources to advance the field.

Table 1. Overview of the literature. The papers are clusters into their main topic: QAOA, Simulated Annelling, Benders Decomposition and Quantum Machine Learning. For each paper is indicated which of the other paper it cites.

Topic	Article	1	2	3	4	5	6	7	8	9	10	11	12	13	14	15	16	17	18	19	20	21
QAOA	1 Koretsky 2021 [14]													x								
QAOA	2 Nikmehr 2022a [20]																					
QAOA	3 Feng 2022 [7]		x																			
QAOA	4 Nikmehr 2022b [21]		x																			
QAOA	5 Mahroo 2022 [18]													x								
QAOA	6 Stein 2023 [29]																					
QAOA	7 Yang 2023 [30]																					
QAOA	8 Salgado 2024 [27]	x	x	x		x	x							x								
QAOA	9 Fontana 2024 [8]	x				x								x								
QAOA	10 Magar 2024 [17]	x			x			x						x								
QAOA/SA	11 Halffmann 2022 [9]													x								
QAOA/SA	12 Halffmann 2024 [10]	x										x										
SA	13 Ajagekar 2020 [1]																					
SA	14 Braun 2023 [2]											x										
SA	15 Quinton 2024 [24]																					
SA	16 Mller 2025 [19]	x	x				x	x	x					x					x	x		x
BDC/SA	17 Chang 2020 [3]																					
BDC/SA	18 Paterakis 2023 [22]	x		x	x									x								
BDC/SA	19 Ellinas 2024 [5]		x		x														x			
QML	20 Liu 2024 [16]	x			x		x													x		
QML	21 Kruse 2024 [15]	x																				

5 Conclusions

In this paper, we have explored the application of quantum computing to the Unit Commitment problem, a critical optimisation challenge in power system operations. Our review of the literature, which is summerised in Table 1, highlights several promising quantum approaches, including the Quantum Approximate Optimisation Algorithm, Quantum Annealing (QA), and hybrid quantum-classical methods.

The findings suggest that quantum computing holds significant potential for solving the UC problem more efficiently than classical methods in the future.

QAOA, in particular, has shown promise in handling binary decision variables, while classical optimisers address continuous power output. Hybrid approaches that combine quantum and classical techniques have demonstrated improved computational efficiency and scalability, making them suitable for large-scale UC scenarios. Quantum Annealing has also been effective for smaller UC instances, although larger problems require partitioning into sub-problems. The integration of quantum algorithms with robust optimisation techniques has shown potential in managing uncertainties in power systems, such as renewable energy supply and power demand fluctuations.

Despite the current limitations of quantum hardware, such as noise and scalability issues, the advancements in quantum algorithms and hybrid methods provide a strong foundation for future research. Continued innovation in quantum hardware and error mitigation techniques will be crucial in realising the full potential of quantum computing for the UC problem.

In conclusion, quantum computing offers a transformative approach to solving the UC problem, with the potential to significantly enhance the efficiency and scalability of power system optimisation. Future research should focus on further developing quantum algorithms, improving hardware capabilities, and exploring new hybrid methods to fully leverage the benefits of quantum computing in this domain.

References

1. Ajagekar, A., You, F.: Quantum computing for energy systems optimization: challenges and opportunities. Energy **179**, 76–89 (2019)
2. Braun, M., Decker, T., Hegemann, N., Kerstan, S., Lorenz, F.: Towards optimization under uncertainty for fundamental models in energy markets using quantum computers. arXiv preprint arXiv:2301.01108 (2023)
3. Chang, C.Y., Jones, E., Yao, Y., Graf, P., Jain, R.: On hybrid quantum and classical computing algorithms for mixed-integer programming. arXiv preprint arXiv:2010.07852 (2020)
4. Dong, D., Chen, C., Li, H., Tarn, T.J.: Quantum reinforcement learning. IEEE Trans. Syst. Man Cybern. Part B (Cybernetics) **38**(5), 1207–1220 (2008)
5. Ellinas, P., Chevalier, S., Chatzivasileiadis, S.: A hybrid quantum-classical algorithm for mixed-integer optimization in power systems. Electric Power Syst. Res. **235**, 110835 (2024)
6. Farhi, E., Goldstone, J., Gutmann, S.: A quantum approximate optimization algorithm. arXiv preprint arXiv:1411.4028 (2014)
7. Feng, F., Zhang, P., Bragin, M.A., Zhou, Y.: Novel resolution of unit commitment problems through quantum surrogate Lagrangian relaxation. IEEE Trans. Power Syst. **38**(3), 2460–2471 (2022)
8. Fontana, C.: Energy Management Systems and Potential Applications of Quantum Computing in the Energy Sector. Ph.D. thesis, University of Palermo (2024)
9. Halffmann, P., Holzer, P., Plociennik, K., Trebing, M.: A quantum computing approach for the unit commitment problem. In: International Conference on Operations Research, pp. 113–120. Springer (2022)

10. Halffmann, P., Trebing, M., Lenk, S.: Harnessing inferior solutions for superior outcomes: Obtaining robust solutions from quantum algorithms. In: Proceedings of the Genetic and Evolutionary Computation Conference Companion, pp. 1950–1953 (2024)
11. Horowitz, M., Grumbling, E.: Quantum computing: progress and prospects. National Academies Press (2019)
12. Kadowaki, T., Nishimori, H.: Quantum annealing in the transverse Ising model. Phys. Rev. E **58**(5), 5355 (1998)
13. Knueven, B., Ostrowski, J., Watson, J.P.: On mixed-integer programming formulations for the unit commitment problem. INFORMS J. Comput. **32**(4), 857–876 (2020)
14. Koretsky, S., et al.: Adapting quantum approximation optimization algorithm (qaoa) for unit commitment. In: 2021 IEEE International Conference on Quantum Computing and Engineering (QCE), pp. 181–187. IEEE (2021)
15. Kruse, G., Coelho, R., Rosskopf, A., Wille, R., Lorenz, J.M.: Hamiltonian-based quantum reinforcement learning for neural combinatorial optimization. In: 2024 IEEE International Conference on Quantum Computing and Engineering (QCE). vol. 1, pp. 1617–1627. IEEE (2024)
16. Liu, J., Zhou, X., Zhou, Z., Luo, L.: Exact quantum algorithm for unit commitment optimization based on partially connected quantum neural networks. arXiv preprint arXiv:2411.11369 (2024)
17. Magar, M.R., Pandit, D., Nguyen, D., Nguyen, N.: DC optimal power flow in unit commitment using quantum computing: An admm approach. In: 2024 56th North American Power Symposium (NAPS), pp. 1–6. IEEE (2024)
18. Mahroo, R., Kargarian, A.: Hybrid quantum-classical unit commitment. In: 2022 IEEE Texas Power and Energy Conference (TPEC), pp. 1–5. IEEE (2022)
19. Müller, S., Dukalski, M., Phillipson, F.: Quantum annealing for optimizing unit scheduling in renewable energy systems: Formulation and evaluation. IEEE Trans. Power Syst. (2025)
20. Nikmehr, N., Zhang, P., Bragin, M.A.: Quantum distributed unit commitment: an application in microgrids. IEEE Trans. Power Syst. **37**(5), 3592–3603 (2022)
21. Nikmehr, N., Zhang, P., Bragin, M.A.: Quantum-enabled distributed unit commitment. In: 2022 IEEE Power & Energy Society General Meeting (PESGM), pp. 01–05. IEEE (2022)
22. Paterakis, N.G.: Hybrid quantum-classical multi-cut Benders approach with a power system application. Comput. Chem. Eng. **172**, 108161 (2023)
23. Preskill, J.: Quantum computing in the NISQ era and beyond. Quantum **2**, 79 (2018)
24. Quinton, F.A., Myhr, P.A.S., Barani, M., del Granado, P.C., Zhang, H.: Quantum annealing versus classical solvers: Applications, challenges and limitations for optimisation problems. arXiv preprint arXiv:2409.05542 (2024)
25. Rahmaniani, R., Crainic, T.G., Gendreau, M., Rei, W.: The Benders decomposition algorithm: a literature review. Eur. J. Oper. Res. **259**(3), 801–817 (2017)
26. Rietsche, R., Dremel, C., Bosch, S., Steinacker, L., Meckel, M., Leimeister, J.M.: Quantum computing. Electron. Mark. **32**(4), 2525–2536 (2022)
27. Salgado, B., Sequeira, A., Santos, L.P.: A hybrid classical-quantum approach to highly constrained unit commitment problems. arXiv preprint arXiv:2412.11312 (2024)
28. Schuld, M., Sinayskiy, I., Petruccione, F.: The quest for a quantum neural network. Quantum Inf. Process. **13**, 2567–2586 (2014)

29. Stein, J., et al.: Exponential quantum speedup for simulation-based optimization applications. arXiv preprint arXiv:2305.08482 (2023)
30. Yang, M., Gao, F., Wu, G., Gong, B., Cao, H., Shuang, F.: A quantum-aided algorithm for unit commitment problems. In: 2023 42nd Chinese Control Conference (CCC), pp. 1733–1736. IEEE (2023)
31. Zhang, P., Nikmehr, N., Nikmehr, N., Zhang, P., Bragin, M.: Quantum-enabled distributed unit commitment. Technical Report, Brookhaven National Lab.(BNL), Upton, NY (United States) (2022)

Toward Quantum Utility in Finance: A Robust Data-Driven Algorithm for Asset Clustering

Shivam Sharma[1], Supreeth Mysore Venkatesh[2,3]([✉]), and Pushkin Kachroo[1]

[1] University of Nevada, Las Vegas, Las Vegas, NV, USA
sharms15@unlv.nevada.edu, pushkin@unlv.edu
[2] University of Kaiserslautern-Landau (RPTU), Kaiserslautern, Germany
supreeth.mysore@rptu.de
[3] German Research Center for Artificial Intelligence (DFKI), Saarbruecken, Germany
supreeth.mysore@dfki.de

Abstract. Clustering financial assets based on return correlations is a fundamental task in portfolio optimization and statistical arbitrage. However, classical clustering methods often fall short when dealing with signed correlation structures, typically requiring lossy transformations and heuristic assumptions such as a fixed number of clusters. In this work, we apply the Graph-based Coalition Structure Generation algorithm (GCS-Q) to directly cluster signed, weighted graphs without relying on such transformations. GCS-Q formulates each partitioning step as a QUBO problem, enabling it to leverage quantum annealing for efficient exploration of exponentially large solution spaces. We validate our approach on both synthetic and real-world financial data, benchmarking against state-of-the-art classical algorithms such as SPONGE and k-Medoids. Our experiments demonstrate that GCS-Q consistently achieves higher clustering quality, as measured by Adjusted Rand Index and structural balance penalties, while dynamically determining the number of clusters. These results highlight the practical utility of near-term quantum computing for graph-based unsupervised learning in financial applications.

Keywords: Quantum Utility · Signed Graphs · Graph Clustering · Asset Clustering · Quantum Annealing

1 Introduction and Background

Information about relationships among financial assets has long been central to strategies in portfolio optimization and algorithmic trading. Since companies and their stock prices are interdependent, exploiting these correlations, especially in markets that are not perfectly efficient, can enhance decision-making in portfolio construction. The foundational work of Markowitz [18] introduced the

F. Barbaresco and F. Gerin (Eds.): QUEST-IS 2025, CCIS 2744, pp. 14–22, 2026.
https://doi.org/10.1007/978-3-032-13855-2_2

mean-variance optimization framework, establishing diversification as a means to maximize return while minimizing risk [23].

More recent strategies, such as statistical arbitrage [2] and pair trading [5], use correlation or cointegration metrics but have become less effective due to increased market adoption. Consequently, graph-theoretic methods have emerged as robust alternatives [13], where assets are represented as nodes and correlation matrices are treated as signed, weighted graphs. Clustering these graphs aims to group assets with high intra-cluster and low inter-cluster correlations [14,28], recent works focusing on identifying the best-performing clustering algorithms [4].

Traditional clustering approaches such as k-Means, k-Medoids [10,15], or hierarchical clustering [19] face limitations when applied to signed graphs. These algorithms require positive-definite distance metrics and often rely on transforming signed correlation matrices $\rho \in [-1, 1]$ into non-negative distances using monotonic mappings like [7,17,29]:

$$d_{ij}^{(\alpha)} = \sqrt{\alpha(1 - \rho_{ij})}, \quad \alpha > 0 \tag{1}$$

Although this transformation preserves ranking, it loses semantic fidelity, for example, $\rho_{ij} = 0$ is mapped to $\sqrt{\alpha}$, incorrectly implying a fixed non-zero dissimilarity. Thus, even optimally solving the transformed problem does not recover the optimal clustering under the original signed graph. Centroid-based methods further deviate from the true objective, which is to maximize intra-cluster correlations and minimize inter-cluster correlations, not to minimize distances to centroids. Moreover, these classical methods require manual tuning of hyperparameters such as the number of clusters k, initialization strategies, or threshold values for merging/dividing. These choices are not generalizable across datasets and must be revalidated for unseen data. Spectral clustering methods provide a more natural fit for graph-structured data by using eigenvalue decomposition of the similarity matrix [24]. SPONGE [6], a spectral method for signed graphs, has shown strong performance but is effective mainly when k is large or the graph is very sparse [4,11], conditions not typically met in financial data.

Quantum methods have also been explored, notably in portfolio optimization using gate-based variational circuits [3], though existing hardware limit scalability. One prior attempt at quantum-assisted clustering formulates the problem as a maximum clique and translates it into a large QUBO instance [9], introducing additional binary variables and exceeding the capabilities of current annealers.

In this work, we propose applying the Graph-based Coalition Structure Generation algorithm (GCS-Q) [31] to the task of asset clustering using return correlation data. GCS-Q operates directly on the signed graph, avoiding lossy transformations. It starts with all assets grouped together and iteratively partitions the graph via minimum cuts, maximizing intra-cluster weights at each step and dynamically determining k. This approach is well-suited to the signed graph setting where classical solvers offer only heuristic workarounds. The minimum-cut subproblem is formulated as a QUBO and solved on D-Wave's quantum annealer, leveraging its ability to explore exponentially large solution spaces efficiently. We validate GCS-Q on both synthetic and real-world financial datasets, demonstrating its superior clustering quality compared to classical state-of-the-art methods.

2   Methodology

We consider a set of financial assets $\mathcal{A} = \{a_1, a_2, \ldots, a_n\}$, whose historical returns over a rolling window are denoted by $r_i(t)$ for asset a_i at time t. The pairwise Pearson correlation coefficients [22] are computed as:

$$\rho_{ij} = \frac{\mathrm{Cov}(r_i, r_j)}{\sigma_{r_i}\sigma_{r_j}}, \quad \rho_{ij} \in [-1, 1]. \tag{2}$$

These correlations define a signed, weighted, undirected graph $G = (V, E, w)$, where each vertex $v_i \in V$ uniquely corresponds to asset a_i, and each edge $(v_i, v_j) \in E$ is weighted by $w_{ij} = \rho_{ij}$. In contrast to traditional clustering methods that transform correlations into non-negative distances [7], we retain the signed correlations, thereby preserving both positive (co-movement) and negative (anti-correlation) relationships.

We apply the GCS-Q algorithm, previously developed for coalition structure generation in graphs [31], to this asset clustering setting. The core contribution lies in adapting GCS-Q to leverage its ability to operate directly on the signed graphs without resorting to heuristic transformations.

Formally, GCS-Q aims to find a coalition structure $\Pi = \{C_1, C_2, \ldots, C_k\}$, a partition of V into disjoint subsets, that maximizes the intra-cluster edge weights:

$$\max_{\Pi} \sum_{C \in \Pi} \sum_{\substack{i,j \in C \\ i<j}} w_{ij}. \tag{3}$$

This objective promotes grouping of assets with strong positive correlations while implicitly disincentivizing inclusion of negatively correlated pairs within the same cluster.

The GCS-Q algorithm proceeds iteratively, starting with the full graph $G_0 = G$. At each iteration t, it computes a minimum cut $(S_t, \bar{S}_t)$ on the current subgraph G_{t-1}, using the actual edge weights w_{ij} which is strictly NP-Hard. The resulting partition optimizes the objective in Eq. (3). The process is then recursively applied to the subgraphs induced by S_t and $\bar{S}_t$, eliminating the cut edges at each step, and terminates dividing a subgraph further when there is no cut whose value is lower than the sum of all the edges in that subgraph. This stopping criterion inherently infers the number of clusters k as part of the optimization, obviating the need for manual tuning or initialization required by classical approaches.

Each minimum cut in GCS-Q is formulated as a Quadratic Unconstrained Binary Optimization (QUBO) problem [8], well-suited for near-term quantum devices. These QUBOs can be efficiently solved using quantum annealers (e.g., D-Wave) or variational algorithms like the Quantum Approximate Optimization Algorithm (QAOA) [32], enabling hybrid quantum-classical workflows.

Theoretical work such as [1] has shown that optimally clustering signed graphs with edge weights in $[-1, +1]$ is NP-Hard. Our practical contribution lies in demonstrating that GCS-Q can robustly address this challenge in financial asset clustering, highlighting its potential for real-world deployment and showcasing practical applicability of near-term quantum technologies, particularly quantum annealing.

3 Experiments

We conduct two sets of experiments. First, we benchmark the clustering quality of GCS-Q against classical baselines on synthetically generated signed graphs designed to mimic financial correlation structures. Second, we evaluate the methodology on real-world asset return data to demonstrate its practical utility.

All implementations are in `Python 3.12`. The QUBO subproblems in GCS-Q are solved using the D-Wave `Advantage_system5.4` annealer, which comprises 5614 physical qubits and $40,050$ couplers arranged in the pegasus topology. The device is accessed remotely via the `dwave-ocean-sdk` APIs. As noted in [30], most runtime overhead arises from internet latency and queue wait times due to the lack of onsite quantum hardware. As such, the experiments emphasize solution quality to assess the practical relevance of the methodology using exisitng quantum annealers.

All remaining computations are executed on a standard 64-bit workstation equipped with a 12th Gen Intel Core i7-12800H CPU and 64 GB RAM.

3.1 Synthetic Data

We generate structurally balanced signed graphs, sampling intra-cluster edges from $[0.1, 1.0]$ and inter-cluster edges from $[-1.0, -0.1]$ to emulate realistic correlation patterns. Unlike sparse graphs in [6] that use discrete weights $\in \{-1, 0, +1\}$ and equal cluster sizes, our graphs span a continuous range and incorporate non-uniform cluster sizes via Dirichlet-based stochastic allocation. This introduces heterogeneity while ensuring all clusters remain non-empty. No noise is added, isolating algorithmic behavior under ideal structural conditions. We extrapolate the setup later in Sect. 3.2 to evaluate robustness under noise.

We benchmark GCS-Q against classical baselines commonly used for clustering assets via return correlation data. SPONGE and its symmetric variant SPONGE_{sym} [6] are among the strongest graph-based techniques [4]. We also include k-Medoids (via the PAM algorithm) as a robust baseline well-suited to financial data [7].

Unlike GCS-Q, which determines the number of clusters k dynamically, SPONGE and PAM require k as input. This requirement often demands heuristics like silhouette analysis [25] or the elbow method [27], which are expensive and lack generalizability. Since SPONGE parallels spectral clustering, we estimate k using the eigengap heuristic [12], which selects k based on the largest gap between successive Laplacian eigenvalues.

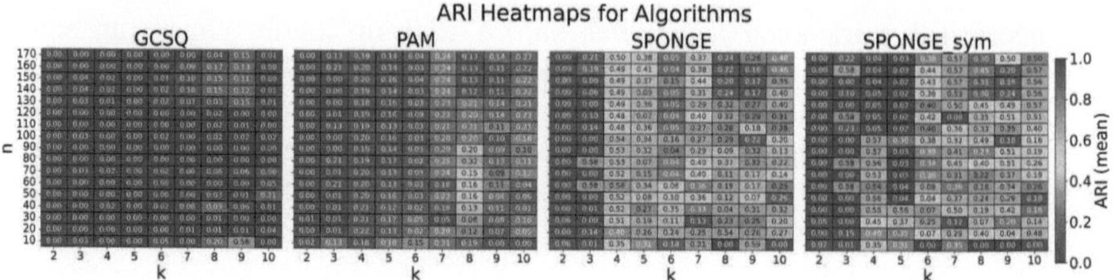

Fig. 1. ARI scores for synthetic data of varying number of nodes n and number of clusters k. Color represents mean and the (tiny) numerical annotations indicate variances across seeds.

We conduct experiments varying total nodes $\in \{10, 20, \ldots, 100\}$, with k ranging from 2 to 10 across 33 random seeds. Since ground-truth clusters are known, we assess accuracy using Adjusted Rand Index (ARI) [26].

Although GCS-Q is limited to roughly 175 nodes due to qubit constraints and hardware topology, Fig. 1 shows it consistently outperforms classical solvers. PAM performs worst, largely due to its reliance on transforming correlations into non-negative distances. While SPONGE and SPONGE$_{sym}$ perform well on graphs with weights in $\{-1, 0, 1\}$ and balanced cluster sizes, their performance degrades on continuous weights and stochastic sizes.

Moreover, SPONGE is designed for sparse graphs [4], which is misaligned with financial data where correlations close to zero are rare. The superior performance of GCS-Q is attributed to the expressive solution space explored by the quantum annealer at each step.

3.2 Yahoo Finance Data

We collect hourly closing prices of 50 assets from diverse sectors from `yfinance`, compute rolling log returns, and derive the Pearson correlation matrix, which is interpreted as a signed adjacency matrix. The same clustering solvers from Sect. 3.1 are applied. In the absence of ground truth, we assess clustering quality using the Penalty metric, which quantifies deviation from structural balance by penalizing negative intra-cluster correlations and positive inter-cluster correlations:

$$\text{Penalty}(\Pi) = \sum_{C \in \Pi} \sum_{\substack{i,j \in C \\ i < j \\ w_{ij} < 0}} |w_{ij}| + \sum_{C_a \neq C_b} \sum_{\substack{i \in C_a, j \in C_b \\ i < j \\ w_{ij} > 0}} w_{ij}$$

A lower penalty indicates better clustering, with zero representing a perfectly balanced structure. Such structurally sound clusters enable more effective asset diversification, aligning with the goal of minimizing portfolio variance in the Markowitz framework [18].

As Fig. 2 illustrates, GCS-Q consistently achieves the lowest penalty, indicating its superiority in uncovering market structures conducive to portfolio optimization and statistical arbitrage. Low penalty values indicate clusters that

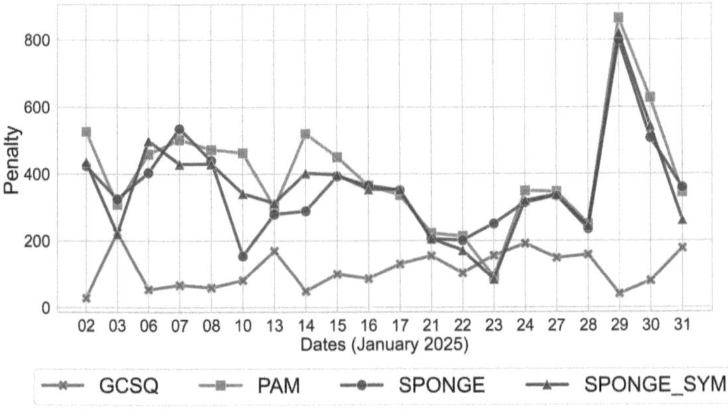

Fig. 2. Penalty comparison across clustering algorithms over business days in January 2025, based on correlations derived from hourly return intervals.

are internally cohesive, which supports mean-reversion strategies [4], and externally distinct, making them well-suited for capturing diverse risk-return profiles essential for effective diversification [23].

4   Discussion and Conclusion

This work is the first, to our knowledge, to demonstrate a quantum advantage in solution quality using real hardware for the practical task of asset clustering, validated on both synthetic and real financial data. GCS-Q's strength lies in its exponential solution space exploration at each iteration. When executed classically, its complexity is $\mathcal{O}(n2^n)$, where $\mathcal{O}(2^n)$ stems from solving the minimum cut and $\mathcal{O}(n)$ from dynamically determining $k \in [1, n]$. While commercial solvers like Gurobi and CPLEX exist for combinatorial optimization, they struggle to scale on dense graphs [33] and varying-sized clusters [16]. Although the adiabatic quantum process theoretically runs in constant time, apart from the hardware noise, current access via shared cloud infrastructure introduces latency and queue delays. In practice, clustering $n = 170$ assets with $k = 10$ took over 10 min on average, despite per-QUBO solve times being capped at 1 s.

GCS-Q is robust to real-world scenarios with heterogeneous cluster sizes and avoids the idealized constraints (e.g., uniform clusters, ternary edge weights) assumed in prior work [4,6]. It requires no manual tuning of k, operating in a fully data-driven manner. Unlike classical methods that convert signed correlations to Euclidean distances [7,17], GCS-Q works directly on the signed graph, preserving relational structure.

Future directions include applying the resulting clusters to downstream tasks like asset allocation and evaluating performance using metrics such as Sharpe Ratio and returns. From a quantum perspective, deploying GCS-Q on D-Wave's hardware with Zephyr topology and better connectivity, exploring custom embeddings, and using gate-model quantum solvers with qubit-efficient strategies [21], as well as benchmarking against classical optimizers like Gurobi, are promising next steps. Moreover, similar iterative approaches are also used in optimizing machine learning models [20]. This work, we believe, marks a concrete advancement toward near-term quantum utility for signed graph clustering with an imperative application in finance.

Code Availability: All code necessary to reproduce the experiments and generate the plots are available at:
https://github.com/supreethmv/Quantum-Asset-Clustering

References

1. Bansal, N., Blum, A., Chawla, S.: Correlation clustering. Mach. Learn. **56**(1), 89–113 (2004). https://doi.org/10.1023/B:MACH.0000033116.57574.95
2. Bertram, W.K.: Analytic solutions for optimal statistical arbitrage trading. XXPhys. A **389**(11), 2234–2243 (2010). https://doi.org/10.1016/j.physa.2010.01.045
3. Buonaiuto, G., Gargiulo, F., De Pietro, G., Esposito, M., Pota, M.: Best practices for portfolio optimization by quantum computing, experimented on real quantum devices. Sci. Rep. **13**(1), 19434 (2023). https://doi.org/10.1038/s41598-023-45392-w
4. Cartea, Á., Cucuringu, M., Jin, Q.: Correlation matrix clustering for statistical arbitrage portfolios. SSRN Electron. J. (2023)
5. Cartea, A., Jaimungal, S.: Algorithmic trading of co-integrated assets. Int. J. Theor. Appl. Fin. **19**(06), 1650038 (2016). https://doi.org/10.1142/S0219024916500382
6. Cucuringu, M., Davies, P., Glielmo, A., Tyagi, H.: Sponge: A generalized eigenproblem for clustering signed networks. In: Chaudhuri, K., Sugiyama, M. (eds.) Proceedings of the Twenty-Second International Conference on Artificial Intelligence and Statistics. Proceedings of Machine Learning Research, vol. 89, pp. 1088–1098. PMLR (2019). https://proceedings.mlr.press/v89/cucuringu19a.html
7. Duarte, F.G., de Castro., L.N.: Asset allocation based on a partitional clustering algorithm. In: Filho, C.J.A.B., Siqueira, H.V., Ferreira, D.D., Bertol, D.W., ao de Oliveira, R.C.L. (eds.) Anais do 15 Congresso Brasileiro de Inteligência Computacional. pp. 1–6. SBIC, Joinville, SC (2021)
8. Glover, F., Kochenberger, G., Du, Yu.: Quantum Bridge Analytics I: a tutorial on formulating and using QUBO models. 4OR **17**(4), 335–371 (2019). https://doi.org/10.1007/s10288-019-00424-y
9. Kalra, A., Qureshi, F., Tisi, M.: Portfolio asset identification using graph algorithms on a quantum annealer. SSRN Electr. J. (2018). https://doi.org/10.2139/ssrn.3333537
10. Kaufman, L., Rousseeuw, P.J.: Partitioning around medoids (program pam). In: Wiley Series in Probability and Statistics, pp. 68–125. John Wiley & Sons, Inc., Hoboken, NJ, USA (1990). https://doi.org/10.1002/9780470316801.ch2
11. Khelifa, N., Allier, J., Cucuringu, M.: Cluster-driven hierarchical representation of large asset universes for optimal portfolio construction. In: Proceedings of the 5th ACM International Conference on AI in Finance. p. 177–185. ICAIF '24, Association for Computing Machinery, New York, NY, USA (2024). https://doi.org/10.1145/3677052.3698676

12. Kong, W., Sun, C., Hu, S., Zhang, J.: Automatic spectral clustering and its application. In: ICICTA 2010. vol. 1, pp. 841–845 (2010). https://doi.org/10.1109/ICICTA.2010.164
13. Korniejczuk, A., Ślepaczuk, R.: Statistical arbitrage in multi-pair trading strategy based on graph clustering algorithms in US equities market. Working Papers 2024-09, Faculty of Economic Sciences, University of Warsaw (2024). https://ideas.repec.org/p/war/wpaper/2024-09.html
14. León, D., Aragón, A., Sandoval, J., Hernández, G., Arévalo, A., Niño, J.: Clustering algorithms for risk-adjusted portfolio construction. In: International Conference on Computational Science, ICCS 2017, 12-14 June 2017, Zurich, Switzerland. Proc. Comput. Sci. **108**, 1334–1343 (2017). https://doi.org/10.1016/j.procs.2017.05.185
15. Lloyd, S.: Least squares quantization in PCM. IEEE Trans. Inf. Theory **28**(2), 129–137 (1982). https://doi.org/10.1109/TIT.1982.1056489
16. Macaluso, A., Venkatesh, S.M., Arenas, D., Klusch, M., Dengel, A.: Quantum-assisted correlation clustering. arXiv preprint arXiv:2509.03561 (2025)
17. Mantegna, R.N.: Hierarchical structure in financial markets. Eur. Phy. J. B - Condensed Matter Compl. Syst. **11**(1), 193–197 (1999). https://doi.org/10.1007/s100510050929
18. Markowitz, H.: Portfolio Selection. J. Fin. **7**(1), 77–91 (1952). https://doi.org/j.1540-6261.1952.tb01525.x
19. Murtagh, F., Contreras, P.: Algorithms for hierarchical clustering: an overview. WIREs Data Min. Knowl. Discovery **2**(1), 86–97 (2012). https://doi.org/10.1002/widm.53
20. Mysore Venkatesh, S., Macaluso, A., Arenas, D., Klusch, M., Dengel, A.: i-qls: Quantum-supported algorithm for least squares optimization in non-linear regression. In: Lees, M.H., et al., (eds.) Computational Science - ICCS 2025, pp. 19–34. Springer Nature Switzerland, Cham (2025)
21. Mysore Venkatesh, S., Macaluso, A., Nuske, M., Klusch, M., Dengel, A.: Qubit-efficient variational quantum algorithms for image segmentation. In: 2024 IEEE International Conference on Quantum Computing and Engineering (QCE). vol. 01, pp. 450–456 (2024). https://doi.org/10.1109/QCE60285.2024.00059
22. Pearson, K.: Vii. note on regression and inheritance in the case of two parents. Proceedings of the Royal Society of London **58**, 240–242 (1895). https://doi.org/10.1098/rspl.1895.0041. publisher: Royal Society
23. Ponsich, A., Jaimes, A.L., Coello, C.A.C.: A survey on multiobjective evolutionary algorithms for the solution of the portfolio optimization problem and other finance and economics applications. IEEE Trans. Evol. Comput. **17**(3), 321–344 (2013). https://doi.org/10.1109/TEVC.2012.2196800
24. Pothen, A., Simon, H.D., Liou, K.P.: Partitioning sparse matrices with eigenvectors of graphs. SIAM J. Matrix Anal. Appl. **11**(3), 430–452 (1990). https://doi.org/10.1137/0611030
25. Rousseeuw, P.J.: Silhouettes: a graphical aid to the interpretation and validation of cluster analysis. J. Comput. Appl. Math. **20**, 53–65 (1987). https://doi.org/10.1016/0377-0427(87)90125-7
26. Steinley, D.: Properties of the Hubert-arable adjusted rand index. Psychol. Methods **9**(3), 386–396 (2004). https://doi.org/10.1037/1082-989X.9.3.386
27. Thorndike, R.L.: Who belongs in the family? Psychometrika **18**(4), 267–276 (1953). https://doi.org/10.1007/BF02289263
28. Tola, V., Lillo, F., Gallegati, M., Mantegna, R.N.: Cluster analysis for portfolio optimization. J. Econ. Dyn. Control **32**(1), 235–258 (2008). https://doi.org/10.1016/j.jedc.2007.01.034

29. Traag, V.A., Bruggeman, J.: Community detection in networks with positive and negative links. Phys. Rev. E **80**, 036115 (2009). https://doi.org/10.1103/PhysRevE.80.036115
30. Venkatesh, S., Macaluso, A., Nuske, M., Klusch, M., Dengel, A.: Q-seg: Quantum annealing-based unsupervised image segmentation. IEEE Comput. Graphics Appl. **44**(5), 27–39 (2024). https://doi.org/10.1109/MCG.2024.3455012
31. Venkatesh, S.M., Macaluso, A., Klusch, M.: Gcs-q: Quantum graph coalition structure generation. In: Computational Science – ICCS 2023, pp. 138–152. Springer Nature Switzerland (2023). https://doi.org/10.1007/978-3-031-36030-5_11
32. Venkatesh, S.M., Macaluso, A., Klusch, M.: Quacs: Variational quantum algorithm for coalition structure generation in induced subgraph games. In: 20th ACM International Conference on Computing Frontiers, p. 197–200. Association for Computing Machinery (2023). https://doi.org/10.1145/3587135.3592192
33. Venkatesh, S.M., Macaluso, A., Nuske, M., Klusch, M., Dengel, A.: Quantum annealing-based algorithm for efficient coalition formation among leo satellites. In: 2024 IEEE International Conference on Quantum Computing and Engineering (QCE) (2024). https://doi.org/10.1109/QCE60285.2024.10279

Hot-Starting Quantum Portfolio Optimization

Sebastian Schlütter[1], Tomislav Maras[2], Alexander Dotterweich[2],
and Nico Piatkowski[3(✉)]

[1] Mainz University of Applied Sciences, 55128 Mainz, Germany
[2] PricewaterhouseCoopers GmbH, 80636 München, Germany
[3] LAMARR Institute, Fraunhofer IAIS, 53757 Sankt Augustin, Germany
`nico.piatkowski@iais.fraunhofer.de`

Abstract. Combinatorial optimization with a smooth and convex objective function arises naturally in applications such as discrete mean-variance portfolio optimization, where assets must be traded in integer quantities. Although optimal solutions to the associated smooth problem can be computed efficiently, existing adiabatic quantum optimization methods cannot leverage this information. Moreover, while various warm-starting strategies have been proposed for gate-based quantum optimization, none of them explicitly integrate insights from the relaxed continuous solution into the QUBO formulation. In this work, a novel approach is introduced that restricts the search space to discrete solutions in the vicinity of the continuous optimum by constructing a compact Hilbert space, thereby reducing the number of required qubits. Experiments on software solvers and a D-Wave Advantage quantum annealer demonstrate that our method outperforms state-of-the-art techniques.

Keywords: Warm-Start · QUBO · Mean-Variance Analysis

1 Introduction

The integration of prior knowledge into mathematical models is essential. Examples include: (I) informed machine learning [28], where knowledge about physical laws allows for useful models even if training data is scarce. And (II) numerical optimization, where the convergence rate of gradient descent can be shown to depend on the squared Euclidean distance between the initial solution $x^{(0)}$ and the optimal solution x^* [20]. In the context of quantum computing, integrating prior knowledge about the solution is known as warm-start [2,6]. An outstanding discussion of the topic can be found in [29].

In case of *Adiabatic Quantum Computation* (AQC) warm-start techniques act on the annealing schedule as the entry point since the initial state is built into the hardware and cannot be influenced—state preparation methods like [6] cannot be applied. Only a handful of practical methods is known, e.g. for training ML models to generate annealing schedules [5,14] as well as techniques that rely

F. Barbaresco and F. Gerin (Eds.): QUEST-IS 2025, CCIS 2744, pp. 23–33, 2026.
https://doi.org/10.1007/978-3-032-13855-2_3

on classical pre-processing [16,29]. They can be seen as annealing equivalents to the techniques focusing on the parameter initialization for variational circuit, likewise pre-computing or generating annealing schedules instead of initial variational parameters.

At the core of our work lies the observation that some hard combinatorial optimization problems correspond to a surprisingly simple problem when allowing for continuous solutions. For example, *Quadratic Unconstrained Binary Optimization* (QUBO) with objective $f(\boldsymbol{x}) = \boldsymbol{x}^\top \boldsymbol{Q}\boldsymbol{x}$ with $\boldsymbol{x} \in \{0,1\}^n$ is NP-hard [25]. However, when $\boldsymbol{x}$ comes from a simplex, the problem becomes a convex *Quadratic Program* (QP) that can be solved to optimality in polynomial time. Here, this observation is utilized by constructing a new QUBO that incorporates the location of the QP's optimal solution $\bar{\boldsymbol{x}}^* \in \mathbb{R}^n$. This approach allows us to restrict the discrete search space to solutions in the direct neighborhood of $\bar{\boldsymbol{x}}^* \in \mathbb{R}^n$, which results in a lower-dimensional Hilbert-space and thus in a reduced number of qubits when compared to the standard QUBO formulation. The situation is depicted in Fig. 1. The search space is endogenously determined to guarantee that it contains the optimum; in general it is not sufficient to round up or down the continuous solution.

Existing quantum approaches to portfolio optimization either model binary inclusion variables per asset [6,30], or assume predefined investment bands of lower and upper bounds per asset [1,23,26]. To improve solution quality at limited hardware, papers have used insights to guide the quantum search, for instance, by identifying high-Sharpe-Ratio candidates in a pre-processing step [11] or using the results of a classical heuristic to seed the quantum annealer via reverse annealing [30]. Warm-starting methods using the solution from the continuous relaxation have been explored, including initial state preparation for simulated annealing [27] and the preparation of quantum states for algorithms on gate-based quantum computers [6]. To the best of our knowledge, our paper is the first one that integrates the solution of the continuous relaxation into the QUBO construction and thereby endogenously determines relevant investment bands.

We conduct experiments based on real-world S&P 500 stock market data on software solvers and a D-Wave Advantage quantum annealer. Our results show that our novel formulation can lead to a drastic reduction of qubits—effectively allowing for larger problem instances on actual quantum hardware.

2   Background

Quantum Computing. Let us quickly introduce the basic notion of what can be considered as quantum computation [22]. Today, practical quantum computing (QC) consists of two dominant paradigms: Adiabatic QC and gate-based QC. In both scenarios, a quantum state $|\psi\rangle$ for a system with n qubits is a 2^n dimensional complex and ℓ_2-normalized vector. The domain of $|\psi\rangle$ is known as Hilbert space. In the gate-based (universal) framework, a quantum computation is defined as a matrix multiplication $|\psi_{\text{out}}\rangle = C|\psi_{\text{in}}\rangle$, where the C is a $(2^n \times 2^n)$-dimensional

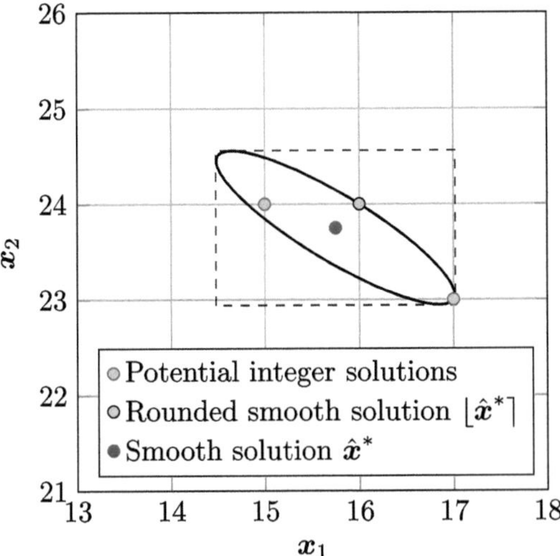

Fig. 1. Exemplary visualization of our proposed method. x_1 and x_2 denote integer quantities of two assets. A naive approach would require roughly 5 qubits per dimension to allow for quantities up to $2^5 = 32$. Our approach constructs an ellipsoid around a smooth solution that can be obtained efficiently via a standard QP solver. Only integer points within the ellipsoid's dashed bounding-box are considered during optimization. Thus, the number of qubits required for each dimension is reduced to 2 and 3, respectively.

unitary matrix (the circuit), given via a series of inner and outer products of low-dimensional unitary matrices (the gates). The circuit is derived manually or via heuristic search algorithms [8]. In adiabatic quantum computation (AQC)—the framework that we consider in the paper at hand—the result of computation is defined to be the eigenvector $|\phi_{\min}\rangle$ that corresponds to the smallest eigenvalue of some $(2^n \times 2^n)$-dimensional Hermitian matrix $\boldsymbol{H}$ (also known as Hamiltonian). In practical AQC, $\boldsymbol{H}$ is further restricted to be a real diagonal matrix whose entries can be written as $\boldsymbol{H}_{i,i} = H(\boldsymbol{Q})_{i,i} = (\boldsymbol{x}^{(i)})^\top \boldsymbol{Q} \boldsymbol{x}^{(i)}$ where $\boldsymbol{x}^{(i)} = \text{binary}(i, n) \in \{0,1\}^n$ is some (arbitrary but fixed) n-bit binary expansion of the unsigned integer i. Here, $\boldsymbol{Q} \in \mathbb{R}^{n \times n}$ is the so-called QUBO matrix. By construction, computing $|\phi_{\min}\rangle$ is equivalent to solving

$$\min_{\boldsymbol{x} \in \{0,1\}^n} \boldsymbol{x}^\top \boldsymbol{Q} \boldsymbol{x} \,. \tag{1}$$

Adiabatic quantum algorithms rely on this connection by encoding combinatorial problems into QUBO matrices.

In both paradigms, the output vector is 2^n-dimensional and can thus not be read-out efficiently for non-small n. Instead, the output of a practical quantum computation is a random integer i between 1 and 2^n, which is drawn from the probability mass function $\mathbb{P}(i) = |\langle i|\boldsymbol{\psi}_{\text{out}}\rangle|^2$. This sampling step is also known as collapsing the quantum state $|\boldsymbol{\psi}_{\text{out}}\rangle$ to a classical binary state $\text{binary}(i, n)$.

AQC has been applied to numerous combinatorial optimization problems [15], ranging over satisfiability [10], routing problems [21] to machine learning [19].

The number of physically available qubits of an actual *Quantum Processing Unit* (QPU) is limited. As a result, a theoretically sound QUBO formulation might not be able to run on a real-world QPU as n exceeds the number of available qubits. Decomposition techniques can be applied in these cases, cutting the original QUBO into smaller pieces. It is, however, not guaranteed that re-combining partial solutions will lead to a good (or even feasible) global solution. Hence, another approach is to come up with an alternative QUBO formulation that requires a smaller number of qubits. One example is the integration of inequality constraints via the unbalanced penalty method [13], which requires less qubits than naive integration of the constraints. In the paper at hand, the QUBO dimension is reduced by reducing the search space to discrete solutions in the vicinity of the optimum of a relaxed convex QP.

Portfolio Theory. In the sense of the classical mean-variance approach introduced by [17], investors select non-negative portfolio weights $\boldsymbol{w}$ on a continuous scale, with each entry $\boldsymbol{w}_i$ defining the fraction of wealth invested in asset i. Let $\boldsymbol{\mu}$ denote the expected returns and $\boldsymbol{\Sigma}$ the covariance matrix of asset returns. The selection can be formalized by an optimization problem identifying the maximal expected portfolio return subject to a specified level of portfolio variance:

$$\max_{\boldsymbol{w}} \ \boldsymbol{\mu}^\top \boldsymbol{w} \quad \text{subject to} \quad \boldsymbol{w}^\top \boldsymbol{\Sigma} \boldsymbol{w} \leq c, \sum_{i=1}^{n} \boldsymbol{w}_i = 1, \boldsymbol{w}_i \geq 0 \text{ for all } i \ . \quad (2)$$

Problem (2) is the classical starting point to identify an optimal portfolio with integer amounts of asset purchases. Let $\boldsymbol{p} \in \mathbb{R}_+^n$ collect the price per unit of each asset, and let $B > 0$ denote the budget to be invested. Similar to [4], to identify the optimal units of purchased assets $\boldsymbol{x} \in \mathbb{Z}_+^n$, one considers the problem

$$\max_{\boldsymbol{x}} \ \tilde{\boldsymbol{\mu}}^\top \boldsymbol{x} \quad \text{subject to} \quad \boldsymbol{x}^\top \tilde{\boldsymbol{\Sigma}} \boldsymbol{x} \leq c \cdot B^2, \sum_{i=1}^{n} \boldsymbol{p}_i \cdot \boldsymbol{x}_i = B, \boldsymbol{x}_i \in \mathbb{Z}_+ \text{ for all } i \ . \quad (3)$$

Notice how (2) gives rise to (3) by adding the integrality constraint $\boldsymbol{x}_i \in \mathbb{Z}_+$. Their solutions can be translated back and forth via the following equations[1]:

$$\boldsymbol{w} = \frac{\boldsymbol{x} \circ \boldsymbol{p}}{B}, \quad \tilde{\boldsymbol{\Sigma}} = \mathrm{diag}(\boldsymbol{p}) \cdot \boldsymbol{\Sigma} \cdot \mathrm{diag}(\boldsymbol{p}), \quad \tilde{\boldsymbol{\mu}} = \boldsymbol{\mu} \circ \boldsymbol{p}, \quad (4)$$

In order to construct a convex problem over the simplex, the expected portfolio return and the portfolio variance are combined into a quadratic objective function via the risk aversion parameter $\gamma > 0$. Consequently, we arrive at

$$\max_{\boldsymbol{w}} \ \boldsymbol{\mu}^\top \boldsymbol{w} - \frac{\gamma}{2} \cdot \boldsymbol{w}^\top \boldsymbol{\Sigma} \boldsymbol{w} \quad \text{subject to} \quad \sum_{i=1}^{n} \boldsymbol{w}_i = 1, \boldsymbol{w}_i \geq 0 \text{ for all } i \ . \quad (5)$$

---

[1] The operator $\circ$ denotes element-wise multiplication.

In (3), it would be restrictive to require that the budget is invested exactly, $\sum_{i=1}^{n} p_i \cdot x_i = B$. Instead, we assume that the difference between the invested amount and the budget B is invested at the risk-free interest rate r_f. In this case, the w_i represent the weights of each risky asset, while $(1 - \sum_{i=1}^{n} w_i)$ represents the weight to be invested at the risk-free rate. Consequently, the constraint $\sum_{i=1}^{n} w_i = 1$ can be removed and the expected return becomes

$$\boldsymbol{\mu}^\top \boldsymbol{w} + \left(1 - \mathbf{1}^\top \boldsymbol{w}\right) \cdot r_f = \left(\boldsymbol{\mu} - r_f \cdot \mathbf{1}\right)^\top \boldsymbol{w} + r_f \tag{6}$$

with $\mathbf{1}$ denoting an n-dimensional vector of ones. Moreover, we omit the no-shortselling constraint $\boldsymbol{w}_i \geq 0$, as in practice transaction costs may imply that the investor slowly trades from an existing portfolio to the preferred one. Therefore, the optimal solution typically remains long-only, so the constraint is empirically non-binding (as seen in Sect. 4). The new problem can be formulated as

$$\max_{\boldsymbol{w}} \ \underbrace{\left(\boldsymbol{\mu} - r_f \cdot \mathbf{1}\right)^\top}_{\boldsymbol{\mu}_f} \boldsymbol{w} - \frac{\gamma}{2} \cdot \boldsymbol{w}^\top \boldsymbol{\Sigma} \boldsymbol{w} \ . \tag{7}$$

Setting $\tilde{\gamma} = \gamma/B$ and invoking Eq. 4 for $\boldsymbol{\mu}_f$, we arrive at

$$\max_{\boldsymbol{x}} \ \tilde{\boldsymbol{\mu}}_f^\top \boldsymbol{x} - \frac{\tilde{\gamma}}{2} \cdot \boldsymbol{x}^\top \tilde{\boldsymbol{\Sigma}} \boldsymbol{x} \quad \text{subject to} \quad \boldsymbol{x}_i \in \mathbb{Z} \text{ for all } i \ . \tag{8}$$

The latter represents the NP-hard portfolio *Quadratic Integer Program* (QIP). However, since $\boldsymbol{\Sigma}$ is a covariance matrix which is typically positive definite, dropping the integrality constraint of (8) leads to a smooth convex quadratic program for which solutions can be readily obtained via standard solvers. For empirical situations with sample covariance matrices being ill-conditioned or singular, regularization methods are widely used to ensure positive definiteness; see, e.g., [12]. Furthermore, the structural form of the mean-variance framework in (8) can be derived, at least approximately, for homogeneous risk measures such as Value-at-Risk or Expected Shortfall, as well as for deterministic regulatory capital standards (see [24]). Even when (8) is extended to include additional linear or convex quadratic terms, such as transaction costs, a smooth convex quadratic program can be established (see Sect. 4).

In order to construct an equivalent QUBO, $\boldsymbol{x}$ must be binarized, e.g., by reserving a fixed amount of k bits for each asset and interpreting the bits as unsigned binary expansion of some integer. Clearly, the choice of k has an immediate impact on both, the accuracy to which we can represent the asset weights and the dimension of the resulting QUBO. We will re-visit this topic in our experiments.

3   Methodology

By adopting ideas from [3], we derive a novel method to reduce the number of qubits required for the portfolio optimization QUBO. The proposed method relies on an informed re-formulation of the portfolio QIP.

Theorem 1. (Informed Ellipsoid Formulation). *Let x^* be an optimal solution to (8) with $\tilde{\Sigma}$ positive definite, $\hat{x}^*$ the smooth maximizer of $f(x) = \tilde{\mu}_f^\top x - \frac{\tilde{\gamma}}{2} \cdot x^\top \tilde{\Sigma} x$, and $\hat{f}^* := f(\hat{x}^*)$ the corresponding optimal function value. Assume that at least one entry of $\hat{x}^*$ is not integer-valued. Let further*

$$E := \left\{ x \in \mathbb{R}^n : (x - \hat{x}^*)^\top \left(\frac{\tilde{\gamma}}{2C} \cdot \tilde{\Sigma} \right) (x - \hat{x}^*) \le 1 \right\} \tag{9}$$

with $C = \hat{f}^* - f(\lfloor \hat{x}^* \rceil) > 0$. Then, $x^* \in E$. Thus, optimizing over the ellipsoid E is sufficient to compute x^*. Here, $\lfloor \cdot \rceil$ denotes component-wise rounding to the nearest integer.

Proof. First, notice that

$$(x - \hat{x}^*)^\top \left(\frac{\tilde{\gamma}}{2} \tilde{\Sigma} \right) (x - \hat{x}^*) \le C \Leftrightarrow (x - \hat{x}^*)^\top \left(\frac{\tilde{\gamma}}{2} \tilde{\Sigma} \right) (x - \hat{x}^*) - \hat{f}^* \le -f(\lfloor \hat{x}^* \rceil)$$

$$\Leftrightarrow (\hat{x}^*)^\top \left(\tilde{\gamma} \tilde{\Sigma} \right) x - \frac{\tilde{\gamma}}{2} x^\top \tilde{\Sigma} x \ge f(\lfloor \hat{x}^* \rceil) .$$

Moreover, since $\hat{x}^*$ satisfies the first order condition of f, it follows that

$$\tilde{\mu}_f - \tilde{\gamma} \tilde{\Sigma} \hat{x}^* = 0 \Leftrightarrow \hat{x}^* = \frac{1}{\tilde{\gamma}} \tilde{\Sigma}^{-1} \tilde{\mu}_f .$$

Plugging this into the last inequality yields

$$\tilde{\mu}_f x - \frac{\tilde{\gamma}}{2} x^\top \tilde{\Sigma} x = f(x) \ge f(\lfloor \hat{x}^* \rceil) .$$

We see that all integer points contained in E (if any) achieve a better objective function value than the rounded smooth solution $\lfloor \hat{x}^* \rceil$. Since all points outside of E are strictly worse than $\lfloor \hat{x}^* \rceil$, the best integer solution must be included in E as well. ∎

Based on these insights, we compute the extent of E for each asset i in terms of upper and lower bounds. The lower and upper boundaries of the ellipsoid in all dimensions can then be computed via

$$L = \hat{x}^* - \sqrt{\mathrm{diag}(M^{-1})} \quad \text{and} \quad U = \hat{x}^* + \sqrt{\mathrm{diag}(M^{-1})}$$

with $M = \frac{\tilde{\gamma}}{2C} \cdot \tilde{\Sigma}$. Finally, instead of optimizing over $x \in \mathbb{Z}^n$, we exploit the limits of the ellipsoid and optimize over

$$x_i \in [L_i, U_i] \cap \mathbb{Z} \text{ for all } i \tag{10}$$

This corresponds to optimization over a bounding hypercube that tightly encloses the ellipsoid[2] (see Fig. 1). Since $[L_i, U_i]$ can be different for each i,

---

[2] While this introduces sub-optimal solutions into the search space, it does not harm the soundness of our result. Nevertheless, the effectiveness of our approach can be enhanced through several methods; for example, by successively refining the search space based on the best integer solution found, and/or by including penalties in the objective function to more tightly bound the search space.

the effective search space is adapted to the individual extent of each dimension. Thus, this procedure has the potential to reduce the search space, and hence the number of required qubits, by a large margin. We call this procedure *hot-start*.

4 Experiments

Two experiments are conducted to demonstrate the viability of our approach. First, the number of qubits required by a baseline k-bits per asset construction is compared to the number of qubits required for the proposed approach for asset universes of increasing size. Second, we compare the solution quality obtained on a D-Wave Advantage Quantum Annealer and a randomized search heuristic, both utilizing the baseline and our novel construction.

We calibrate $\tilde{\mu}$ and $\tilde{\Sigma}$ using monthly S&P 500 equity returns from August 2003 to July 2023, sourced from Refinitiv. The vector μ and the matrix Σ are estimated as the sample mean and sample covariance matrix of the stock returns over this period. The vector p contains stock prices as of July 2023; the risk-free interest rate is set to $r_f = 5.32\%$ based on the Federal Funds Effective Rate, as reported by the Federal Reserve Bank of St. Louis (FRED).

Our investment universe consists of the n stocks with the highest market capitalization as of July 2023. The investor has a budget of $B = 250{,}000\,\text{USD}$ and initially holds a portfolio with (approximately) equal weighting across all n assets in the investment universe. The initial portfolio is integer-valued, i.e., rounded as

$$x_0 = \left\lfloor \frac{B}{n} \cdot p^{-1} \right\rfloor \tag{11}$$

where p^{-1} denotes the vector of element-wise reciprocals of the price vector p and $\lfloor \cdot \rfloor$ denotes component-wise flooring. We consider problem (8), supplemented by transaction costs. As in [9], we assume that transaction costs are determined by the covariance matrix Σ and thus achieve

$$\max_{x} \ \tilde{\mu}_f^\top x - \frac{\tilde{\gamma}}{2} \cdot x^\top \tilde{\Sigma} x - \tilde{\kappa} \cdot (x - x_0)^\top \tilde{\Sigma} (x - x_0) \quad \text{subject to} \quad x_i \in \mathbb{Z} \text{ for all } i \,. \tag{12}$$

We set $\gamma = 3$ and $\tilde{\kappa} = 50 \cdot \gamma/B$.[3]

For the $n = 4$ stocks with highest market capitalization, Table 1 shows how Hot-Start constructs a QUBO with six qubits. Starting from the initial portfolio x_0, the smooth solution $\hat{x}^*$ of (12) suggests reducing positions in Microsoft, NVIDIA, and Amazon, while increasing the position in Apple. Enclosing hypercube (10) around ellipsoid E from (9) implies that the optimal integer-valued solution of (12) must be either 194 or 195 for Microsoft, either 372 or 373 for

---

[3] $\gamma = 3$ lies within the range of economically realistic and empirically estimated risk-aversion coefficients; for a meta-analysis see [7]. Compared to empirical estimates, our calibration of transaction costs is moderate; for example, [9, p. 2328] estimate κ for several commodities and find that, in the median, κ is a factor of 500 larger than γ.

Apple, and so on. Thus, (12) can be solved as a QUBO requiring $1+1+3+1 = 6$ qubits, where the number of qubits is determined by the binary encoding rule $\#\text{qubits} = \lceil \log_2(\#\text{integers}) \rceil$ for each stock.

Increasing n successively to 100, the maximum number of qubits required by Hot-Start for any single asset increases from 3 (for $n = 4$) to 10 (for $n = 100$). To provide a conservative baseline for comparison, we calibrate a fixed k-bits per asset construction with $k = 10$. This baseline is as efficient as possible while still ensuring that the optimal solution is not overlooked.

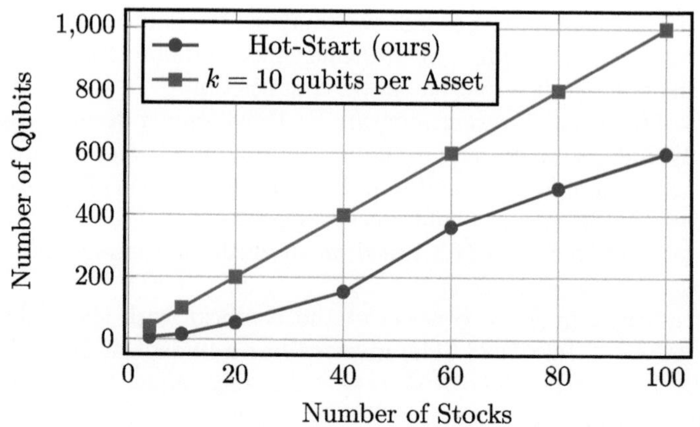

Fig. 2. Number of qubits required to encode the portfolio QUBO, as a function of the number of stocks. Comparison between our approach (red boxes) and a baseline (blue circles) where each stock quantity is encoded by 10 qubits. (Color figure online)

Figure 2 shows that in the first experiment, Hot-Start requires a substantially smaller number of qubits than the baseline. For the second experiment, we consider the D-Wave Advantage2 Prototype 2.6 QPU with 128 readouts (shots) per QUBO. Before sending the QUBO to the QPU, it must be mapped onto the Zephyr hardware topology. In case of the baseline, this step failed for all problems with more than 40 stocks. In case of Hot-Start, all problems with up to 80 stocks could be mapped successfully to the QPU. The energies of the found solutions coincide with those found by the D-Wave simulated annealing software solver. We refer to [18] for a comprehensive discussion on how to reduce integrated control errors and thus hardware noise on D-Wave annealers.

5   Concluding Remarks

We presented Hot-Start, the first warm-start technique that lowers the qubit demand for quantum portfolio optimization. Experiments show favorable scaling: the conventional encoding lacks prior knowledge of the qubit count needed to represent a solution and therefore over-allocates, whereas Hot-Start infers

Table 1. Derivation of the number of qubits for problem (12) with $n = 4$ stocks.

Stock	x_0	$\hat{x}^*$	Integers in box		Decision variables	
			Smallest	Largest	# integers	# qubits
Microsoft	197	194.6	194	195	2	1
Apple	365	372.6	372	373	2	1
NVIDIA	1436	1434.2	1432	1436	5	3
Amazon	491	489.7	489	490	2	1
Total						6

that number automatically and uses qubits more efficiently. With Hot-Start we successfully embedded QUBOs for portfolios of up to 80 stocks on current QPUs, while the baseline approach already failed beyond 40 stocks. Because the approach relies only on the existence of a convex QP representation, we expect Hot-Start to transfer to a broad class of optimization problems well beyond finance, providing an effective pathway for harnessing near-term and future QPUs.

Acknowledgment. This research has been funded by the Federal Ministry of Education and Research of Germany and the state of North-Rhine Westphalia as part of the LAMARR Institute for Machine Learning and Artificial Intelligence.

References

1. Aguilera, E., de Jong, J., Phillipson, F., Taamallah, S., Vos, M.: Multi-objective portfolio optimization using a quantum annealer. Mathematics **12**(9), 1291 (2024)
2. Beaulieu, D., Pham, A.: Max-cut clustering utilizing warm-start QAOA and IBM runtime (2021). https://arxiv.org/abs/2108.13464
3. Buchheim, C., Caprara, A., Lodi, A.: An effective branch-and-bound algorithm for convex quadratic integer programming. Math. Program. **135**(1–2), 369–395 (2012). https://doi.org/10.1007/S10107-011-0475-X
4. Castro, F., Gago, J., Hartillo, I., Puerto, J., Ucha, J.: An algebraic approach to integer portfolio problems. European J. Oper. Res. **210**(3), 647–659 (2011). https://doi.org/10.1016/j.ejor.2010.11.007
5. Chen, Y.Q., Chen, Y., Lee, C.K., Zhang, S., Hsieh, C.Y.: Optimizing quantum annealing schedules with monte carlo tree search enhanced with neural networks (2022). https://arxiv.org/abs/2004.02836
6. Egger, D.J., Mareček, J., Woerner, S.: Warm-starting quantum optimization. Quantum **5**, 479 (2021). https://doi.org/10.22331/q-2021-06-17-479
7. Elminejad, A., Havranek, T., Irsova, Z.: Relative risk aversion: a meta-analysis. J. Econom. Surv. (2022)
8. Franken, L., et al.: Quantum circuit evolution on NISQ devices. In: IEEE CEC, pp. 1–8 (2022). https://doi.org/10.1109/CEC55065.2022.9870269
9. Gârleanu, N., Pedersen, L.H.: Dynamic trading with predictable returns and transaction costs. J. Finance **68**(6), 2309–2340 (2013)

10. Kochenberger, G., Glover, F., Alidaee, B., Lewis, K.: Using the unconstrained quadratic program to model and solve Max 2-SAT problems. Int. J. Oper. Res. **1**(1–2), 89–100 (2005)
11. Lang, J., Zielinski, S., Feld, S.: Strategic portfolio optimization using simulated, digital, and quantum annealing. Appl. Sci. **12**(23), 12288 (2022)
12. Ledoit, O., Wolf, M.: Robust performance hypothesis testing with the sharpe ratio. J. Empirical Finance **15**(5), 850–859 (2008)
13. Lee, L.K., et al.: Formalizing, normalizing, and splitting the energy network re-dispatch for quantum annealing (2024). https://arxiv.org/abs/2409.09857
14. Lin, J., Zhang, Z., Zhang, J., Li, X.: Hard-instance learning for quantum adiabatic prime factorization. Phys. Rev. A **105**(6) (2022). https://doi.org/10.1103/physreva.105.062455
15. Lucas, A.: Ising formulations of many NP problems. Front. Phys. **2** (2014). https://doi.org/10.3389/fphy.2014.00005
16. de Luis, A.P., Garcia-Saez, A., Estarellas, M.P.: Steered quantum annealing: improving time efficiency with partial information (2022). https://arxiv.org/abs/2206.07646
17. Markowitz, H.: Portfolio selection. J. Finance **7**, 77–91 (1952)
18. Mücke, S., Gerlach, T., Piatkowski, N.: Optimum-preserving QUBO parameter compression. Quantum Mach. Intell. **7**(1), 1 (2025). https://doi.org/10.1007/S42484-024-00219-3
19. Mücke, S., Piatkowski, N., Morik, K.: Hardware acceleration of machine learning beyond linear algebra. In: Workshops of the ECML PKDD. vol. 1167, pp. 342–347. Springer (2019). https://doi.org/10.1007/978-3-030-43823-4_29
20. Nesterov, Y.: Introductory lectures on convex optimization: a basic course. applied optimization, Springer New York, NY, 1 edn. (2004). https://doi.org/10.1007/978-1-4419-8853-9
21. Neukart, F., Compostella, G., Seidel, C., von Dollen, D., Yarkoni, S., Parney, B.: Traffic flow optimization using a quantum annealer. Front. ICT **4** (2017)
22. Nielsen, M.A., Chuang, I.L.: Quantum computation and quantum information. Cambridge University Press (2016)
23. Palmer, S., Sahin, S., Hernandez, R., Mugel, S., Orus, R.: Quantum portfolio optimization with investment bands and target volatility. arXiv preprint arXiv:2106.06735 (2021)
24. Paulusch, J., Schlütter, S.: Sensitivity-implied tail-correlation matrices. J. Bank. Finance **134**, 106333 (2022)
25. Punnen, A.P.: The Quadratic Unconstrained Binary Optimization Problem: Theory, Algorithms, and Applications. Springer Cham, 1 edn. (2022). https://doi.org/10.1007/978-3-031-04520-2
26. Rosenberg, G., Haghnegahdar, P., Goddard, P., Carr, P., Wu, K., De Prado, M.L.: Solving the optimal trading trajectory problem using a quantum annealer. In: Proceedings of the 8th Workshop on High Performance Computational Finance, pp. 1–7 (2016)
27. Rubio-García, Á., García-Ripoll, J.J., Porras, D.: Portfolio optimization with discrete simulated annealing. arXiv preprint arXiv:2210.00807 (2022)
28. von Rüden, L., et al.: Informed machine learning - a taxonomy and survey of integrating prior knowledge into learning systems. IEEE Trans. Knowl. Data Eng. **35**(1), 614–633 (2023). https://doi.org/10.1109/TKDE.2021.3079836
29. Truger, F., Barzen, J., Bechtold, M., Beisel, M., Leymann, F., Mandl, A., Yussupov, V.: Warm-starting and quantum computing: a systematic mapping study. ACM Comput. Surv. **56**(9) (2024). https://doi.org/10.1145/3652510

30. Venturelli, D., Kondratyev, A.: Reverse quantum annealing approach to portfolio optimization problems. Quantum Mach. Intell., 17–30 (2019). https://doi.org/10.1007/s42484-019-00001-w

Session: 5 Quantum Algorithms, Computing; Simulation – Scheduling, Logistics and Complex Systems

Constrained Quantum Optimization for Scheduling Aircraft Arrivals

Gaspard Cassassolles[1,2], Hian Lee Kwa[2(✉)], Teck Yoong Chai[2],
and Yung Sze Gan[2]

[1] CentraleSupélec, Paris, France
`gaspard.cassassolles@student-cs.fr`
[2] CortAIx Labs Singapore, Thales Solutions Asia, Singapore, Singapore
`{hianlee.kwa,teckyoong.chai}@thalesgroup.com,`
`yungsze.gan@asia.thalesgroup.com`

Abstract. Air traffic scheduling near airports is a complex task, particularly during landings where aircraft must maintain strict temporal separations at waypoints. Disruptions such as in-flight delays can lead to trajectory conflicts, which are typically mitigated by adjusting arrival times, which effectively issues speed advisories to pilots. This paper introduces a quantum optimization approach for this problem by encoding it as a Quadratic Unconstrained Binary Optimization (QUBO) instance. We further investigate how two variational algorithms, Quantum Approximate Optimization Algorithm (QAOA) and Variational Quantum Eigensolver (VQE), perform on this highly constrained problem compared to an unconstrained benchmark (MaxCut), highlighting the additional complexity introduced by constraints in quantum optimization.

Keywords: Air Traffic Management · Aircraft Scheduling · Combinatorial Optimization · Quantum Approximate Optimization Algorithm · Quantum Optimization · Variational Quantum Eigensolver

1 Introduction

When arriving at airports, flights typically follow predetermined routes, known as Standard Terminal Arrival Routes (STARs), which are defined by a series of waypoints. However, disruptions such as weather delays can cause deviations in the timing and spacing between aircraft. This may lead to a loss of separation standards, particularly in high-traffic airspace near airports. A common approach to maintain separation standards is to adjust an aircraft's estimated time of arrival at the STAR entry point by issuing a speed advisory, i.e., an instruction for the pilot to adjust their speed to meet a revised target time. This ensures that separation standards are adhered to without any deviations in the planned route, thereby reducing the amount of fuel consumed. This task is normally carried out by an Arrival Manager (AMAN) that considers multiple factors such as aircraft weight, weather, and runway availability.

F. Barbaresco and F. Gerin (Eds.): QUEST-IS 2025, CCIS 2744, pp. 37–47, 2026.
https://doi.org/10.1007/978-3-032-13855-2_4

The AMAN schedulers currently rely on classical optimization algorithms to find the optimal maneuvers for each aircraft in a given situation. Recently, there has also been multiple attempts of including the use of Reinforcement Learning models to enhance the performance of these tools [10]. However, these aforementioned classical computing tools may struggle to keep pace with the ever increasing volume of air traffic.

Quantum computing offers a promising approach for tackling the growing computational complexity of air traffic management, as it can consider large variable spaces and numerous possible solutions in parallel. This capability could improve the aircraft speed advisory process by reducing delays, minimizing conflict resolution maneuvers, and enhancing fuel efficiency. While Stollenwerk et al. [11] investigated quantum annealing for optimizing individual aircraft trajectories, our focus is on high-density traffic scenarios where arrivals follow fixed routes such as STARs.

In this paper, we explore the use of quantum computing in air traffic management, specifically the aircraft speed advisory problem, and formulate it as a combinatorial optimization task. Our main contribution is twofold: (1) we express the speed advisory problem using a QUBO formulation, making it suitable for quantum algorithms and (2) we provide a comparison of QAOA and VQE on this constrained problem against an unconstrained baseline (MaxCut) to illustrate the challenges of handling constraints in quantum optimization. We implement these formulations on hybrid quantum-classical solvers and evaluate their performance across different problem sizes and circuit depths. In doing so, we examine the trade-offs between solution quality and feasibility.

2   Problem Formulation

Aircraft are required to maintain minimum temporal and spatial separations throughout their flight to ensure safety. Conflicts occur when these separation standards are violated, which is more likely during approach due to the high air traffic density and limited runway resources. In this study, we focus on maintaining separation at the fixed waypoints defined by the STAR, under the assumption that doing so ensures conflict-free arrival sequencing. To achieve this, we apply controlled time shifts to each aircraft's arrival at the waypoint. In practice, this is implemented through speed advisories, which pilots execute to satisfy Time To Gain (TTG) or Time To Lose (TTL) requirements [7]. In our formulation, a speed advisory corresponds to selecting a discrete time shift for each aircraft, which becomes the decision variable in our combinatorial optimization model.

2.1   Combinatorial Optimization Formulation

While in practice, there may be multiple entry points along a STAR that results in the merging of routes at specific waypoints, for simplicity we only consider a single waypoint in this study. Let n be the number of aircraft, each with an estimated time of arrival (ETA) at the waypoint. We define a discrete set of

m possible time shifts for each aircraft, allowing both advances and delays. We define the binary decision variables as:

$$X = \{x_{ij} \mid i = 1, \ldots, n; \; j = 1, \ldots, m; \; x_{ij} \in \{0, 1\}\}. \tag{1}$$

Here, $x_{ij} = 1$ indicates that aircraft i is assigned time shift j; otherwise, $x_{ij} = 0$. To ensure that each aircraft is assigned exactly one time shift, we impose the following one-hot encoding constraint:

$$\sum_{j=1}^{m} x_{ij} = 1 \quad \forall i \in \{1, \ldots, n\}. \tag{2}$$

We then define the actual time shift T_i applied to aircraft i as:

$$\Delta T_i = \Delta_d \sum_{j=1}^{m} \left(j - \frac{m+1}{2} \right) x_{ij}, \tag{3}$$

where Δ_d denotes the discretization step in minutes. This expression captures both advances and delays relative to the original ETA. Here, the maximum value possible for a time shift is the same for a delay or an advance, and is equal to $\Delta_d(\frac{m-1}{2})$ minutes.

A conflict between aircraft i and i' occurs if their temporal separation falls below the required minimum, T_{min}. This condition can be expressed as:

$$\Delta T_i - \Delta T_{i'} \in [-T_{\min} - (T_i - T_{i'}), \; T_{\min} - (T_i - T_{i'})], \tag{4}$$

where T_i, T_j denote the ETA of aircraft i and j respectively and $T_{min} = 4$ minutes. We then seek to minimize the total time variation compared to the initial ETAs. This is expressed as:

$$f_{\text{total variation}} = \sum_{i=1}^{n} |\Delta T_i|. \tag{5}$$

2.2   Quantum Optimization

When solving QUBO problems, two main approaches emerge corresponding to the two primary types of quantum hardware. In their work finding optimal aircraft takeoff delays Stollenwerk et al. employed a quantum annealer developed by D-Wave [11]. Such quantum annealers encode the cost function into the energy of a Hamiltonian to be minimized. This method relies on the adiabatic principle, which states that a quantum system in an eigenstate undergoing slow enough changes will remain in that eigenstate [1]. By initializing the system in the ground state of a simple Hamiltonian and gradually transforming it into a *Cost Hamiltonian* that encodes the objective function, the system ideally transitions into the ground state of the cost Hamiltonian, thus identifying the optimal solution.

Another prominent method of solving quantum optimization problems is the Variational Quantum Eigensolver (VQE) that is run on gate-based quantum computers. This is a hybrid quantum-classical algorithm that iteratively searches for the ground state of a Hamiltonian, i.e., the solution that yields the optimum value for a given cost function, using a parameterized quantum circuit. Using this framework, a parameterized circuit $U(\theta)$, also known as an *ansatz*, is used to generate trial wavefunctions $|\psi(\theta)\rangle = U(\theta)|0\rangle^{\otimes n}$, guided by a classical optimization algorithm that aims to solve the minimization problem:

$$\min_{\theta}\langle\psi(\theta)|H_C|\psi(\theta)\rangle = \min_{\theta} C(\theta), \tag{6}$$

where the left hand side expression is the expected value of the Cost Hamiltonian, representing the energy of the system. The solution to the QUBO problem is given by the wavefunction $|\psi(\theta^*)\rangle$, where $\theta^* = \arg min_{\theta} C(\theta)$ are the optimized parameters for the quantum circuit that yield the minimum cost. Here, $C(\theta)$ is the cost of the ansatz with the parameters θ. In practice, this cost function is estimated by running the quantum circuit multiple times, sampling the resulting bitstrings, and computing the average cost function value over the number of samples obtained.

QAOA is a specific case of VQE meant to replicate the principles of quantum annealing on gate-based quantum computers through creating an appropriate Ansatz from the cost Hamiltonian [3,5]. It can be shown that QAOA determines quantum state with guaranteed approximation ratio with respect to the ground state [3,4]. Furthermore, QAOA is the equivalent of adiabatic computation as the circuit depth, p, approaches ∞. This guarantees that QAOA will find the optimal solution of the combinatorial optimization problem on the condition that the ansatz is sufficiently deep and the optimal circuit parameters can be found. This ansatz is given by:

$$|\beta,\gamma\rangle = U_p(\beta,\gamma)|+\rangle^{\otimes N} = \left(\prod_{k=1}^{p} e^{-i\beta_k H_C} e^{-i\gamma_k H_M}\right)|+\rangle^{\otimes N}, \tag{7}$$

where β and γ are the variational parameters found by a COBYLA optimizer, p is the number of layers, or depth of the circuit, and H_M is the Mixer Hamiltonian, which is used to encourage exploration of the solution space by driving transitions between basis states.

2.3   QUBO Formulation

Applying VQEs to the formulated combinatorial optimization problem requires first recasting it as a QUBO. In this form, the objective is to find a binary vector $\mathbf{z} \in \{0,1\}^n$, that minimizes a quadratic cost function. This is given as:

$$\min_{\mathbf{z}\in\{0,1\}^n} \mathbf{z}^\top \mathbf{Q}\mathbf{z}, \tag{8}$$

where n is the number of variables and Q is an $n \times n$ real matrix that encodes the problem's objective functions and constraints. Since QUBO problems are inherently unconstrained, we incorporate one-hot encoding and enforce no-conflict rules by appending weighted penalties to the cost function, which discourage any constraint violations. In this work, two penalty functions are used:

$$f_{\text{encoding}} = \lambda_{\text{encoding}} \sum_{i=1}^{n} \left(\sum_{j=1}^{m} x_{ij} - 1 \right)^2 , \tag{9}$$

which enforces one-hot encoding, and

$$f_{\text{conflict}} = \lambda_{\text{conflict}} \sum_{\substack{i,j,\,i',j' \\ \Delta T_i - \Delta T_{i'} \in [-T_{\min} - (T_i - T_{i'}),\ T_{\min} - (T_i - T_{i'})]}} x_{ij} x_{i'j'}, \tag{10}$$

which penalizes aircraft conflicts. In the two penalty functions $\lambda_{\text{encoding}}$ and $\lambda_{\text{conflict}}$ are sufficiently large penalty weights.

The overall cost function to be minimized can be written as

$$f = f_{\text{total variation}} + f_{\text{encoding}} + f_{\text{conflict}}. \tag{11}$$

2.4   Choice of Penalty Weights

Implementing constraints as penalty terms induces a trade-off between solution validity and objective optimality. While increasing penalty weights reduces invalid solutions, the use of excessive weights can dominate the final cost function and the cost distribution, thereby producing an ill-conditioned optimization landscape that impedes parameter tuning. Moreover in hardware implementations, large penalty coefficients map to stronger coupling terms in the Hamiltonian, which may be challenging to realize. As such, we aim to find the smallest pair of lambdas such that the ground-state bitstring satisfies all constraints. For this, we test different values of a 2D-grid, using an exact solver. In Fig. 1, the box-like boundary between the valid and invalid zones suggests that the validity of the two penalties is independent. For all four scheduling problems, we set $\lambda_{\text{encoding}} = 9$ and $\lambda_{\text{conflict}} = 2$, as these are the lowest integer values for these penalty weights that ensure a valid solution with an exact solver for all scheduling problems.

3   Experiment Setup

To solve this QUBO problem, we use IBM's Qiskit simulator [8] to emulate noise-free qubits. Each execution of a quantum circuit uses 10,000 shots. For parameter optimization, COBYLA is employed with a convergence tolerance of 1e−5. We define four problems of various sizes, as seen in Table 1, to be solved using both QAOA and VQE.

We also define four MaxCut problems of the same sizes, which are naturally unconstrained, to compare the solving of our constrained scheduling problem to

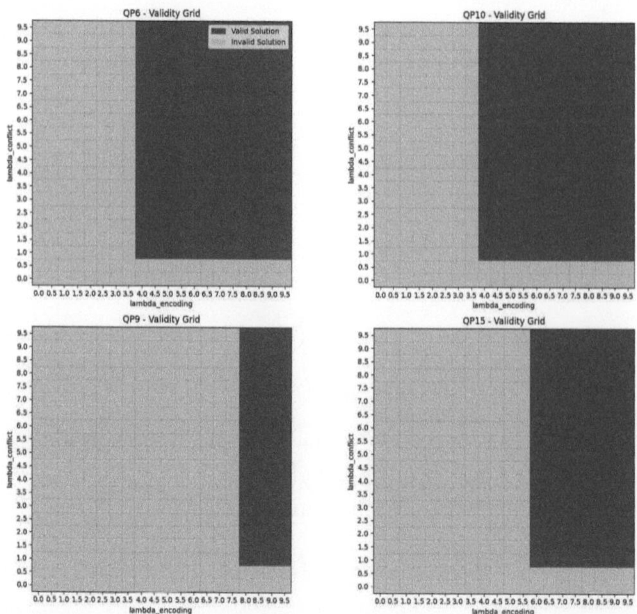

Fig. 1. Effect of penalty weights ($\lambda_{\text{encoding}}, \lambda_{\text{conflict}}$) on solution validity for four scheduling problems.

Table 1. Four aircraft speed advisory problem instances of different sizes.

Name	n (No. of Planes)	m (Possible Time Shifts)	Number of variables
QP6	2	3	6
QP9	3	3	9
QP10	2	5	10
QP15	3	5	15

a well-known unconstrained baseline. Including this comparison to the MaxCut problem serves to demonstrate that unconstrained problems are inherently easier for variational quantum algorithms to solve compared to highly constrained ones like our scheduling problem, highlighting the additional complexity introduced by constraint handling in quantum optimization. Readers unfamiliar with the MaxCut problem are directed to [3] for an introduction.

To assess the performance of a given quantum solver, two metrics are used: (1) the probability of sampling the optimal solution, and (2) the normalized approximation ratio, defined as:

$$r = \frac{C(\theta) - C_{max}}{C(\theta^*) - C_{max}}. \tag{12}$$

As previously mentioned, we set the penalty coefficients as $\lambda_{\text{encoding}} = 9$ and $\lambda_{\text{conflict}} = 2$.

4   Results

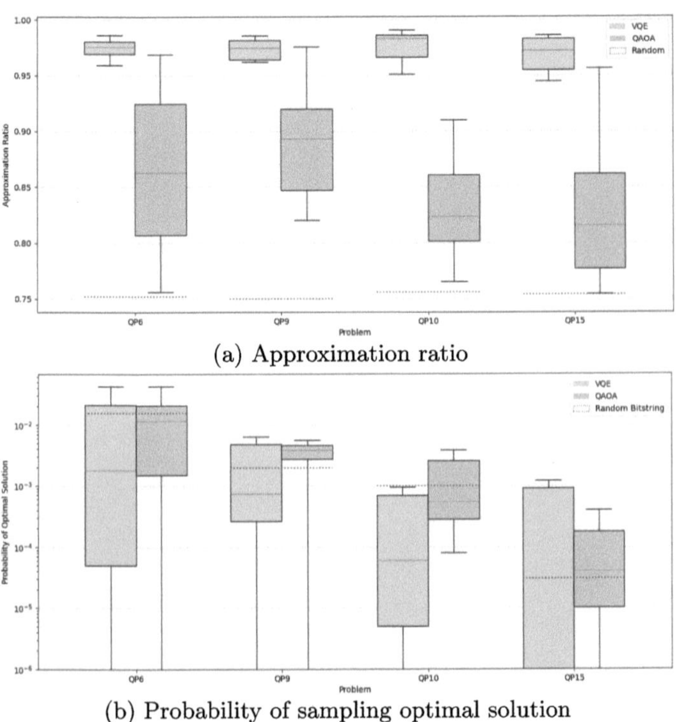

(a) Approximation ratio

(b) Probability of sampling optimal solution

Fig. 2. Performance of using QAOA (green) and VQE (blue) when applied to the speed advisory problem. (Color figure online)

When comparing the performance of both QAOA and a VQE (both with $p = 1$), Fig. 2a illustrates that both quantum algorithms give a better approximation ratio than a random choice; the expected cost of the result generated by these heuristics is better than that of a random bitstring. In addition, it can be observed that the VQE solver consistently outputs a better approximation ratio than the QAOA solver, with less variation in the performance. This is due to the higher expressivity of the VQE circuit compared to the QAOA circuit, which stems from the higher number of rotational angles used in the VQE.

However, the results shown in Fig. 2b indicate that the probability of selecting the optimal bitstring is essentially equal to the probability obtained by randomly choosing a bitstring. This shows that, although the optimization process tends to produce solutions with a lower expected cost than randomly chosen bitstrings, it does not significantly increase the likelihood of sampling the true optimum. This is because the classical optimizer minimizes the expected cost of the circuit, i.e., the average cost over all bitstrings sampled from the circuit, instead of directly maximizing the probability of the optimal solution.

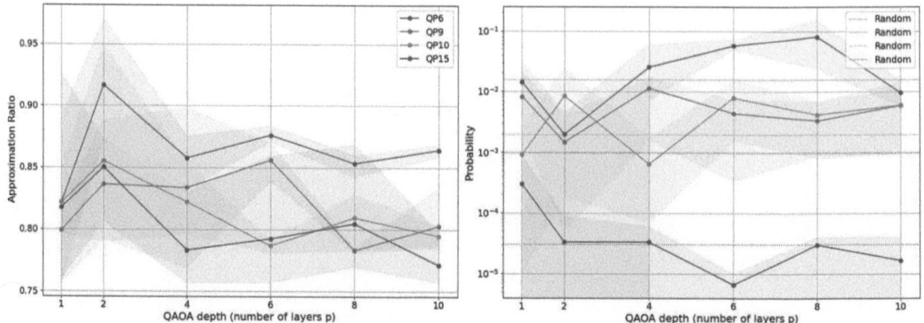

Fig. 3. Comparison of the approximation ratio and probability of sampling the optimum bitstring with increasing circuit depth.

Although deeper circuits are typically expected to yield higher-quality solutions [3], Fig. 3 shows no significant improvement in QAOA performance with increasing p. This lack of improvement may be due to the sparse and irregular energy landscape. Indeed, Fig. 4 shows a highly irregular and non-convex energy landscape where small shifts in the value of β and γ causes a drastic impact on the expected cost of the circuit. Increasing the circuit depth introduces more variational parameters, which can make it increasingly difficult for classical optimizers to find the global minimum and reliably reach the optimal bitstring.

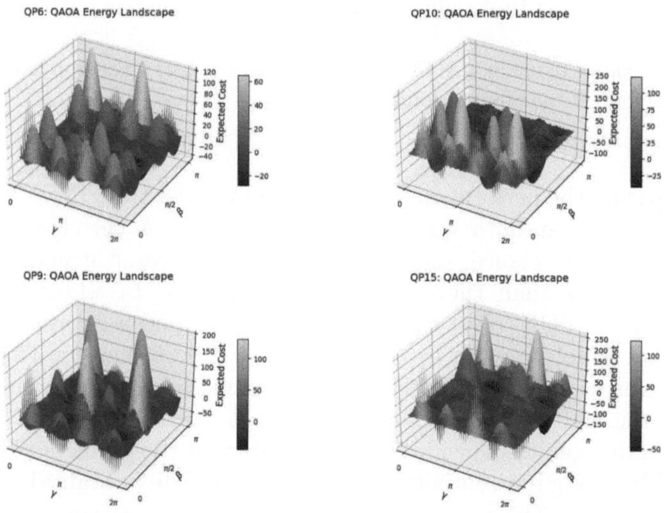

Fig. 4. Energy landscape among the parameter space for QAOA with p=1.

Part of the challenge in sampling the optimum bitstring stems from the highly constrained problem formulation. As shown in Fig. 5, there are more possible values for the cost of an individual bitstring for the highly constrained speed advisory problem compared to the MaxCut problem, which is naturally unconstrained. This contrast validates our motivation for including MaxCut as a baseline. Unconstrained problems such as MaxCut present a more regular solution space with no invalid solutions, making them significantly easier for variational quantum algorithms to handle. In contrast, the strict constraints in the speed advisory problem create a sparse and irregular landscape. This reduces the proportion of valid solutions, thereby impeding the optimizer's ability to find low-cost configurations. This observation reinforces the inherent difficulty of constraint handling in quantum optimization and explains the performance gap between the two problem classes.

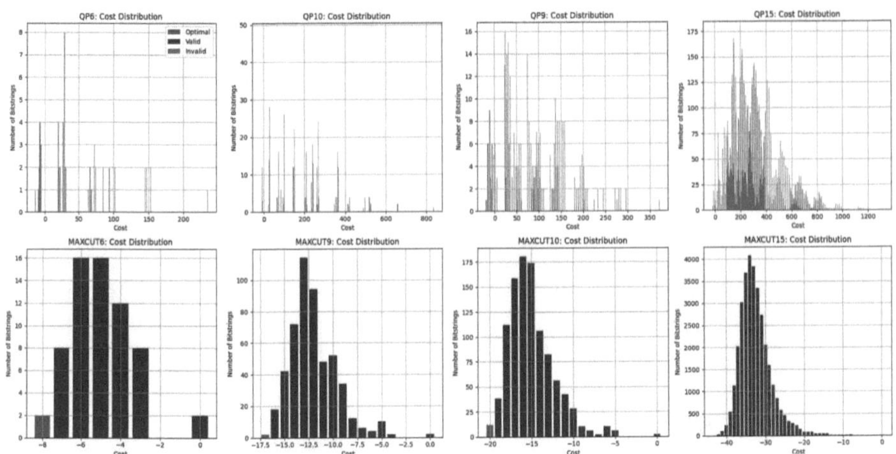

Fig. 5. Cost distribution for all possible solution bitstrings for all 8 problems.

5   Conclusion and Discussion

In this paper, we presented a framework that translates the aircraft arrival scheduling problem into a combinatorial optimization problem, formulates it as a QUBO, and evaluates solutions obtained from both QAOA and VQE ansätze on a noise-free quantum emulator. Our findings show that while these algorithms can achieve good approximation ratios, the probability of sampling the exact optimal solution remains low due to the highly irregular cost landscape, which makes classical parameter optimization challenging.

Consistent with prior studies, VQE outperformed QAOA for small problem instances, owing to its higher circuit expressivity. While our results do not indicate a quantum advantage at this scale, they highlight the additional complexity introduced by constraints. These additional constraints makes the speed advisory problem significantly harder to solve with quantum optimization algorithms compared to unconstrained benchmarks such as MaxCut.

The present study is limited to single-waypoint scenarios and noiseless simulations. Future work will extend the model to multiple waypoints and route merging for greater realism and evaluate implementations on quantum hardware to capture noise and connectivity constraints. We also plan to develop the presented quantum methods in an integrated arrival and departure management (AMAN/DMAN) framework, where the optimizer coordinates both arrival and departure flights while considering runway capacity limits [9]. Even though classical optimization heuristics remain superior for small-scale cases, our long-term goal is to assess whether quantum approaches can offer scalability benefits as problem sizes and system complexity grow.

Future work could also explore advanced techniques to improve solution quality. For example, the Conditional Value at Risk (CVaR) cost function focuses optimization on the best-performing subset of sampled bitstrings, rather than the full expectation, which can increase the probability of sampling low-cost solutions [2]. Similarly, replacing the standard X-mixer in QAOA with an XY-mixer Hamiltonian, which is done using the Quantum Alternating Operator Ansatz, preserves Hamming weights of the qubits during evolution [6]. This ensures that the one-hot encoding constraints are respected and may reduce the number of invalid states sampled.

Acknowledgements. The authors would like to thank Prof. Dimitris Angelakis, Mr. Gordon Ma, and Mr. Ioannis Leonidas from AngelQ for their valuable insights on VQEs, as well as Mr. Francois Balland and Mr. Andrei Purica from Thales Land and Air Systems for the discussions about ATM operations.

References

1. Albash, T., Lidar, D.A.: Adiabatic quantum computation. Rev. Mod. Phys. **90**(1), 015002 (2018)
2. Barkoutsos, P.K., Nannicini, G., Robert, A., Tavernelli, I., Woerner, S.: Improving variational quantum optimization using CVaR. Quantum **4**, 256 (2020)
3. Farhi, E., Goldstone, J., Gutmann, S.: A quantum approximate optimization algorithm. arXiv preprint arXiv:1411.4028 (2014)
4. Farhi, E., Goldstone, J., Gutmann, S.: A quantum approximate optimization algorithm applied to a bounded occurrence constraint problem. arXiv preprint arXiv:1412.6062 (2014)
5. Farhi, E., Goldstone, J., Gutmann, S., Lapan, J., Lundgren, A., Preda, D.: A quantum adiabatic evolution algorithm applied to random instances of an np-complete problem. Science **292**(5516), 472–475 (2001)
6. Hadfield, S., Wang, Z., O'gorman, B., Rieffel, E.G., Venturelli, D., Biswas, R.: From the quantum approximate optimization algorithm to a quantum alternating operator ansatz. Algorithms **12**(2), 34 (2019)

7. Huo, Y., Delahaye, D., Sbihi, M.: A dynamic control method for extended arrival management using enroute speed adjustment and route change strategy. Trans. Res. Part C: Emerg. Technol. **149**, 104064 (2023)
8. Javadi-Abhari, A., et al.: Quantum computing with qiskit (2024)
9. Ma, J., Delahaye, D., Sbihi, M., Scala, P.: Integrated optimization of arrival, departure, and surface operations. In: International Conference for Research in Air Transportation 2018. ICRAT (2018)
10. Machrouh, J., Purica, A., Tevissen, J., Barbaresco, F.: Qualification/validation of AI-augmented ATM solutions for sustainable aviation. In: Towards Sustainable Aviation Summit 2025. 3AF (2025)
11. Stollenwerk, T., et al.: Quantum annealing applied to de-conflicting optimal trajectories for air traffic management. IEEE Trans. Intell. Transp. Syst. **21**(1), 285–297 (2020)

QAOA in Quantum Datacenters:
Parallelization, Simulation,
and Orchestration

Amana Liaqat[1], Ahmed Darwish[2], Adrian Roman[1], and Stephen DiAdamo[2(✉)]

[1] Qoro Quantum, London, England
[2] Qoro Quantum, Munich, Germany
stephen@qoroquantum.de

Abstract. Scaling quantum computing requires networked systems, leveraging HPC for distributed simulation now and quantum networks in the future. Quantum datacenters will be the primary access point for users, but current approaches demand extensive manual decisions and hardware expertise. Tasks like algorithm partitioning, job batching, and resource allocation divert focus from quantum program development. We present a massively parallelized, automated QAOA workflow that integrates problem decomposition, batch job generation, and high-performance simulation. Our framework automates simulator selection, optimizes execution across distributed, heterogeneous resources, and provides a cloud-based infrastructure, enhancing usability and accelerating quantum program development. We find that QAOA partitioning does not significantly degrade optimization performance and often outperforms classical solvers. We introduce our software components—Divi, Maestro, and our cloud platform—demonstrating ease of use and superior performance over existing methods.

Keywords: Quantum computing · hybrid computing · QAOA · distributed quantum computing · parallel quantum computing · cloud quantum computing · quantum simulation · quantum datacenter

1 Introduction

To scale quantum computing infrastructure to practical levels, both in the near and long term, networked systems are essential. In the near term, quantum computers can be integrated with classical High Performance Computing (HPC) infrastructure, forming quantum datacenters [1,8]. These datacenters offer users unified access to classical and quantum resources through a single interface. In the long term, quantum systems are expected to evolve into fully networked clusters, where quantum computers share entanglement over quantum networks to enable distributed quantum computing.

In the current near-term scenario, quantum computers in datacenters are often not connected via quantum networks. Instead, they function as distributed

F. Barbaresco and F. Gerin (Eds.): QUEST-IS 2025, CCIS 2744, pp. 48–55, 2026.
https://doi.org/10.1007/978-3-032-13855-2_5

systems in the traditional sense, relying on classical connections. While this setup offers performance gains [2,3,7], it also imposes limitations. Without quantum entanglement between devices, qubit resources cannot be effectively shared, limiting scalability. Nevertheless, even classical networking can provide runtime improvements, though the specific benefits depend heavily on the application and how computational tasks are managed [5,9].

Our proposed solution, aligned with efforts like [8,11], integrates high-performance simulation and hardware execution under a unified interface. This approach allows users to focus on solving problems without needing to manage infrastructure complexities. While some solutions already exist, there remains a gap in the ecosystem when it comes to optimizing specific applications for massively parallel and distributed systems. Developers still face fundamental questions: How should quantum algorithms be parallelized for specific hardware? How can batch jobs be structured to avoid overwhelming resources? Which simulators are best suited for validating solutions? And, when transitioning from simulation to real hardware, how much of the workflow needs modification?

Our workflow incorporates the three essential components in the following way: (1) Problem decomposition, which partitions a constraint graph into manageable subgraphs; (2) Efficient subproblem generation; and (3) Automated batch job simulation, enabling the execution of diverse circuit sets at scale. We present a software library that automates these processes, reducing the user's workload and facilitating the efficient generation and execution of QAOA batch jobs through a unified interface.

On the simulation side, we develop a common interface that integrates multiple simulation engines. A key feature of this interface is its ability to act as a single point of entry for diverse simulators and simulation types. It automates the selection of the most suitable simulator for each circuit within a batch job, optimizing execution based on circuit characteristics such as complexity, size, and entanglement. Managing interoperability between different simulators is challenging due to variations in input formats, capabilities, and memory requirements. Our approach abstracts these differences, allowing users to run circuits on different backends without manual intervention, and maintaining performance of the simulator.

To ensure accessibility, we deploy these services through a cloud-based infrastructure. This system allows users to execute QAOA workflows without needing access to local high-performance computing resources. The cloud platform manages batch job submissions, allocates computational resources dynamically, and provides seamless scaling based on demand. This setup enables users to run large-scale QAOA experiments without worrying about the underlying infrastructure, ensuring efficient resource utilization and optimized performance.

The remainder of this paper is organized as follows: Sect. 2 details our automated parallelization workflow for QAOA and introduces Divi, our software library for managing this process. Section 3 discusses Maestro, our interface for integrating multiple simulators, and presents benchmark results. Finally, Sect. 4 concludes the paper and discusses future research directions.

2 Efficient Execution Workflow for QAOA

To simplify the execution of large-scale QAOA problems on distributed systems, we develop a workflow and software package that automates the parallelization process. A key question that arises is whether problem partitioning affects solution quality. We first evaluate this on the MaxCut problem, detailing our approach to partitioning. Beyond partitioning, we introduce parallelization at the optimization step, integrating these components into our software library, Divi.

2.1 Constraint-Graph Partitioning for QAOA

An integration of graph partitioning techniques with a quantum resource-aware approach is essential. To address the challenges posed by large-scale optimization problems within the constraints of near-term quantum computing, a flexible methodology that balances both the problem's specific characteristics and the hardware's limitations is required. A balanced and adaptive strategy is crucial for optimizing performance and achieving practical advancements in the current quantum computing era.

In our methodology, we employ spectral partitioning to decompose a large graph into smaller subgraphs, enabling the application of QAOA to solve the MaxCut problem on quantum processors with practical limitations. This carries over to simulators that also have strict limitations in qubit counts. Spectral partitioning methods are advantageous for partitioning graphs as they balance subgraphs' sizes effectively while preserving the overall structure of the principal graph. In our implementation, rather than employing spectral bisection, we use k-means clustering to partition the graph using the Python library *scikit-learn*, specifically its SpectralClustering algorithm [10]. This enables us to partition based on the desired number of clusters, which is constrained by the qubit size that the subgraph must adhere to. This approach helps to mitigate the information loss associated with recursive bisection methods.

Once the graph is partitioned, each graph is optimized independently. The results of each sub-problem are then aggregated, resulting in one final result. The approach leverages graph partitioning to break down a complex optimization problem into smaller, independent subproblems, which are solved separately before recombining the results.

2.2 Partitioned Optimization Evaluation

To evaluate the performance of partitioned QAOA, we use the MaxCut problem on various graph structures of varying size. The MaxCut problem entails partitioning a graph into two sets to maximize the number of edges between them. The MaxCut problem has applications in logistics, network design, and machine learning [4]. Additionally, the MaxCut problem serves as a standard benchmark for evaluating optimization algorithms due to its NP-hard complexity and will be used as the primary performance metric in our results.

When solving MaxCut with QAOA, each graph node corresponds to a qubit, totaling n qubits for n nodes. Edges are represented by controlled-Z (CZ) gates, entangling the qubits. This setup encodes the problem into a quantum circuit, with each qubit's state ($|0\rangle$ or $|1\rangle$) representing the node's set assignment.

While assessing the performance of solvers, the cut-size serves as the primary metric for solution quality, where the higher the cut size, the better the result. Figure 1 illustrates the comparative outcomes between a hybrid approach (graph partitioning followed by QAOA execution under qubit-count constraints) and the standalone Goemans-Williamson (GW) algorithm on circulant graphs. GW is an approximate solver that provides a guaranteed performance ratio of at least 0.878 for the MaxCut problem through semidefinite programming relaxation [6]. We use this solver as our classical benchmark due to its computational efficiency and well-established theoretical guarantees, making it a standard reference point for evaluating quantum approaches to combinatorial optimization problems.

The data demonstrates QAOA achieves parity with GW for small-scale graphs of circulant type ($n \lesssim 500$), but diverges progressively as problem size increases, revealing an optimality gap. Crucially, QAOA implementations with larger subproblem dimensions (e.g., with qubit counts $n = 14$ vs. $n = 6$) exhibit enhanced approximation ratios, narrowing the GW performance gap. This empirical correlation underscores an inverse relationship between partition-induced subproblem reduction and solution fidelity—a trade-off inherent to divide-and-conquer solution strategies.

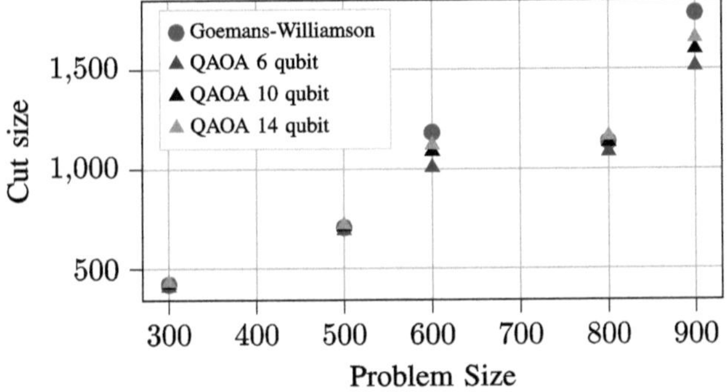

Fig. 1. Approximation quality comparison between partitioned QAOA and the Goemans-Williamson solver across different graph sizes. Larger subproblems yield higher fidelity.

2.3 Divi: Automated Parallelization of Quantum Algorithms

As quantum computing hardware advances, the complexity of quantum algorithms and their execution requirements increase accordingly. Many quantum algorithms, particularly hybrid quantum-classical algorithms such as QAOA,

VQAs, or quantum machine learning, are well structured to be parallelized and distributed across hybrid computing infrastructure. To leverage this, it requires an understanding of the underlying structure of the algorithms and a software infrastructure to perform the various tasks involved to generate and aggregate parallelized quantum programs.

To address these challenges, we have developed Divi, a Python library designed to streamline the execution of large-scale quantum algorithms by automating the parallelization and batching of quantum programs. Divi is designed to integrate seamlessly with cloud infrastructure, automating away the complexity of parallelizing, batching, and aggregating quantum programs. Divi brings the following key features:

(1) Automatic Parallelization of Quantum Programs. Divi automates the parallelization of hybrid quantum programs by decomposing complex problems into smaller, independently executable tasks, enabling efficient optimization while reducing user burden. Using a hierarchical structure, it supports both parallel and serial optimization methods, including Monte Carlo, genetic algorithms, and Nelder-Mead, and allows users to define multiple decision paths to generate additional jobs when beneficial. Divi manages iterative updates, computes expectation values across measurement bases, handles fallback cases, and aggregates results into a final output, automating job tracking and synchronization through a cloud interface to ensure completeness and reliability.

(2) Efficient Batch Job Generation. To serialize and execute quantum subprograms, Divi uses Pennylane to generate circuit abstractions due to its flexible modeling and QASM transpilation capabilities. However, Pennylane's QASM export (v0.40) mishandles expectation value computations by producing only one QASM file per Hamiltonian, rather than one per term with distinct measurement bases. Divi resolves this by implementing a custom export routine that correctly generates separate QASM programs per term. Additionally, Pennylane's repeated transpilation of fixed-structure circuits incurs heavy computational overhead, which is inefficient for variational algorithms. Divi avoids this by creating a symbolic QASM scaffold once per subprogram and binding parameters at runtime, dramatically reducing redundant compilation. This approach improves performance significantly, e.g., reducing generation time for 1,500 circuits by a factor of 13—highlighting a broader inefficiency in frameworks lacking symbolic caching (Fig. 2).

3   Simulating Batches of Circuits

Quantum circuit simulation uses various methods, each offering trade-offs between accuracy, scalability, and computational efficiency. State-vector simulation is the most direct approach, representing the quantum state as a complex vector of dimension 2^n, where n is the number of qubits. This method allows exact simulation and full access to state amplitudes, making it ideal for debugging and analyzing quantum algorithms. However, its exponential memory requirements limit scalability—simulating more than 30 qubits can require

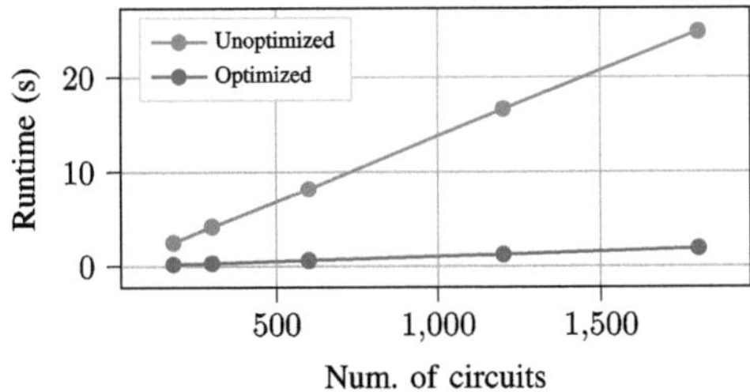

Fig. 2. Time it takes to generate a batch of circuits for QAOA for 5 nodes and varying the number of samples with MC optimization.

hundreds of gigabytes of memory—making it impractical for large systems but effective for small- to medium-scale circuits.

Choosing the right simulation method involves balancing precision, scalability, and computational resources. State-vector simulators are ideal for exact results in small circuits, while MPS and tensor networks extend capabilities to larger or structured systems. GPU-accelerated methods can enhance performance for large, parallelizable workloads but may struggle with smaller or memory-intensive tasks. Additionally, factors like hardware configuration—cores, memory, and architecture—impact performance, complicating simulator selection, especially in batch processing scenarios with diverse circuits.

Maestro, our platform, addresses these challenges by integrating multiple simulation paradigms into a unified interface. It allows users to leverage the most appropriate simulator for their specific needs while maintaining interoperability across quantum hardware and software ecosystems.

Maestro provides an automated simulator selection mechanism to optimize performance. When a circuit is to be run, at runtime, Maestro has two ways to decide which simulator to use. Firstly it can execute a single shot across multiple supported simulators in parallel, records execution times, and selects the right backend for the remaining shots, with an understanding of the multi-shot optimization performance for each. This method is simple and flexible in that it works regardless of software updates to the simulators and the underlying hardware. The downside is, running one shot on many simulators occupies computing resources and can delay the computation in some case.

The second method we use, as an alternative, uses simulation algorithm complexity with circuit information in addition to hardware features (determined at compile time) to make the decision. At compile time, the constants for the algorithm complexity are determined based on the hardware, and then used to select the right simulator for the input circuit. Regression methods are used to extrapolate runtime. This method is very efficient, as it is essentially a lookup task, but it can be challenging to determine the complexity of the simulation algorithm for each simulator, especially when they use parallelization and multi-threading, and requires care. This is an ongoing research direction which we aim to explore more deeply (Fig. 3).

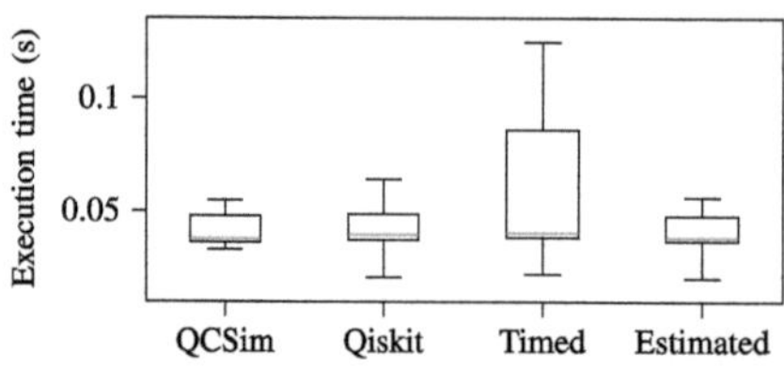
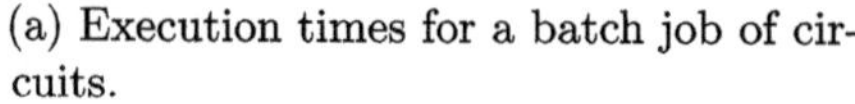

(a) Execution times for a batch job of circuits.

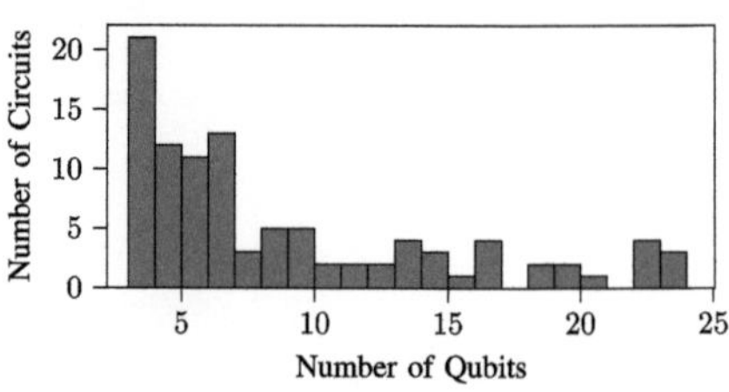

(b) Distribution of circuits executed.

Fig. 3. Execution time distribution for a batch job of circuits. All circuits run for 5,000 shots.

4   Outlook and Conclusion

In this work, we present a comprehensive framework that transforms the execution of large-scale QAOA workflows by addressing key scalability challenges in quantum computing. By automating complex decision-making, problem partitioning, and batch job execution, we reduce the need for user intervention and streamline quantum workflows for greater scalability and usability. Our integrated system includes Divi, which parallelizes and batches QAOA through intelligent decomposition; Maestro, which dynamically selects optimal simulators via a unified interface; and our Cloud Platform, which orchestrates distributed execution across heterogeneous resources, including both simulators and real devices.

This framework bridges the gap between theoretical algorithm design and practical implementation, offering a critical step toward realizing useful quantum computing. We are actively extending its capabilities with realistic noise modeling and GPU-accelerated simulators to further improve performance. Looking ahead, we see distributed quantum computing—enabled by multi-vendor, multimodal networks—as a promising path to scale quantum circuits beyond classical limits. Achieving this requires robust software infrastructure that abstracts complexity while coordinating diverse resources. Our work contributes to this goal by promoting standardization and automation, paving the way for quantum computing to deliver real-world impact.

References

1. Alexeev, Y., et al.: Quantum-centric supercomputing for materials science: a perspective on challenges and future directions. Futur. Gener. Comput. Syst. **160**, 666–710 (2024)
2. Carrera Vazquez, A., Tornow, C., Ristè, D., Woerner, S., Takita, M., Egger, D.J.: Combining quantum processors with real-time classical communication. Nature, 1–5 (2024)
3. Chen, K.C., et al.: Noise-aware distributed quantum approximate optimization algorithm on near-term quantum hardware. arXiv preprint arXiv:2407.17325 (2024)

4. Dunning, I., Gupta, S., Silberholz, J.: What works best when? A systematic evaluation of heuristics for max-cut and QUBO. INFORMS J. Comput. **30**(3), 608–624 (2018)
5. Giortamis, E., Romão, F., Tornow, N., Lugovoy, D., Bhatotia, P.: Orchestrating quantum cloud environments with qonductor. arXiv preprint arXiv:2408.04312 (2024)
6. Goemans, M.X., Williamson, D.P.: Improved approximation algorithms for maximum cut and satisfiability problems using semidefinite programming. J. ACM (JACM) **42**(6), 1115–1145 (1995)
7. Mineh, L., Montanaro, A.: Accelerating the variational quantum Eigensolver using parallelism. Quan. Sci. Technol. **8**(3), 035012 (2023)
8. Mohseni, M., et al.: How to build a quantum supercomputer: Scaling challenges and opportunities. arXiv preprint arXiv:2411.10406 (2024)
9. Moretti, R., Giachero, A., Radescu, V., Grossi, M.: Enhanced feature encoding and classification on distributed quantum hardware. arXiv preprint arXiv:2412.01664 (2024)
10. Pedregosa, F., et al.: Scikit-learn: machine learning in Python. J. Mach. Learn. Res. **12**, 2825–2830 (2011)
11. Schulz, M., Ruefenacht, M., Kranzlmüller, D., Schulz, L.B.: Accelerating HPC with quantum computing: it is a software challenge too. Comput. Sci. Eng. **24**(4), 60–64 (2022)

Towards Large-Scale Satellite Acquisition Scheduling with Hybrid Quantum-Classical Optimization

Amer Delilbasic[1,2]($\boxtimes$) iD, Morris Riedel[1,2,3] iD, Kristel Michielsen[3,4] iD, and Gabriele Cavallaro[1,2,3] iD

[1] Forschungszentrum Jülich, Jülich 52428, Germany
a.delilbasic@fz-juelich.de
[2] University of Iceland, Reykjavík 102, Iceland
[3] AIDAS, Jülich 52425, Germany
[4] RWTH Aachen University, Aachen 52056, Germany

Abstract. The need for on-demand satellite acquisitions for Earth observation is rapidly increasing, along with the availability of public and commercial satellite constellations capable of fulfilling such requests. Scheduling acquisitions optimally remains a computationally challenging task, especially in complex scenarios with multiple satellites, constraints, and high number of requests. Today, heuristic algorithms are commonly used to find sub-optimal solutions that balance computational time and acquisition value, for example in terms of economic return. The potential role of quantum optimization in this domain is still an open question. While recent demonstrations have explored quantum approaches, they are limited to small problem instances due to current hardware constraints and the difficulty of simulating large quantum systems with classical hardware.

In this work, we present a hybrid quantum-classical optimization approach to address acquisition scheduling problems of any size. At each iteration, a classical heuristic generates a global candidate solution, from which a smaller local sub-problem is extracted and solved using a quantum optimization algorithm. We validate the method on large-scale instances, leveraging D-Wave Advantage to solve the sub-problems with quantum annealing. Initial results show the feasibility of the approach and highlight the capability of quantum annealing to guide the hybrid solver towards better overall acquisition schedules than simulated annealing in a comparable amount of time, in particular as the complexity increases.

Keywords: quantum optimization · hybrid quantum-classical computation · quantum annealing · scheduling

1 Introduction

The growing demand for timely and high-resolution Earth observation data has led to a rapid increase in both public and commercial imaging satellite constella-

F. Barbaresco and F. Gerin (Eds.): QUEST-IS 2025, CCIS 2744, pp. 56–64, 2026.
https://doi.org/10.1007/978-3-032-13855-2_6

tions [4,7]. These systems are expected to support diverse applications, ranging from environmental monitoring and disaster response to defense and agriculture [3]. However, fully exploiting these satellite resources requires intelligent and efficient scheduling of imaging tasks [19]. Acquisition scheduling involves assigning requests to specific satellites in a way that maximizes acquisition value while satisfying a wide set of operational constraints. The total value of an acquisition schedule can be related to economic return, for the case of commercial imaging systems, or to urgency in performing the acquisition, for the case of rapidly evolving extreme events such as fires or eruptions. The underlying constraints can involve satellite orbits, sensor agility, on-board resources, weather conditions, and ground station availability.

Due to the problemâĂŹs combinatorial complexity, exact solutions are often infeasible for large scenarios. Heuristic and metaheuristic algorithms, such as greedy methods, genetic algorithms, or Simulated Annealing (SA), are commonly used to find good-enough solutions within reasonable time frames [2]. However, as the problems become larger in practical cases, the performance of existing methods degrades. In recent years, quantum optimization has been explored. Early studies have shown promising results on small satellite acquisition scheduling problems, e.g., using Quantum Annealing (QA) [11,16] and Quantum Approximate Optimization Algorithm (QAOA) [12,14]. However, scalability remains limited, due to algorithmic limitations, current quantum hardware constraints, and the difficulty of mapping large problems to existing quantum systems. For example, QAOA has only been tested in simulation, already showing degrading performance with more than 10 requests, and QA can only realistically solve problems with less than 100 requests.

In the optimization literature, hybrid quantum-classical methods have emerged as a practical strategy for tackling large, NPâĂŠhard problems that exceed current quantum hardware limits. A common pattern is to use a classical algorithm, such as metaheuristics, mathematical programming, or constraintâĂŠbased search, to generate and maintain a feasible global solution, then identify smaller, wellâĂŠdefined subâĂŠproblems that can be mapped to a quantum solver. This approach has been applied in domains such as scheduling [18], routing [8], and portfolio selection [15]. Makarov et al. [12] also successfully solved satellite acquisition scheduling problems with up to 209 requests with the D-Wave `LeapBQMHybrid` solver. However, it cannot be determined whether and to which extent this specific solver benefits from quantum optimization, leaving the door open to research in the direction of hybrid optimization.

In this paper, we present a hybrid quantum-classical method designed for large-scale satellite acquisition scheduling problems. The method relies on a classical heuristic to generate a global candidate solution, from which local, constraint-aware sub-problems are extracted and solved using a quantum annealer. This hybridization allows us to exploit quantum hardware without being limited by its size. We validate the approach on large problem instances with up to 20000 requests using the Advantage quantum annealer from D-Wave Systems [5]. The rest of the paper is structured as follows: Sect. 2 provides a problem formulation for satellite acquisition scheduling. Section 3 introduces the proposed approach. Section 4 provides results from experimental validation. Section 5 concludes the paper and summarizes possible future research directions.

2   Problem Formulation for Satellite Acquisition Scheduling

Satellite acquisition scheduling varies significantly across systems and applications. It fundamentally involves assigning acquisition requests to satellites in a way that maximizes the overall value of collected data while minimizing acquisition time. These problems are subject to a range of operational constraints, which may involve satellite configuration, sensor characteristics, on-board memory capacity, cloud coverage, illumination conditions, or ground station communication windows.

In this work, we focus on a simplified acquisition scheduling formulation, first introduced in the work of Quetschlich et al. [14]. The setup involves a single satellite tasked with fulfilling up to R acquisition requests $\{r_i : i = 1, ..., R\}$ during one orbital pass. Each acquisition i is associated with a binary decision variable $x_i \in 0, 1$, where $x_i = 1$ indicates that the acquisition is selected. Each request also has an associated value v_i, representing, for instance, economic importance or urgency, as highlighted in Sect. 1.

The primary constraint arises from the time required to rotate the satellite's sensor between targets located at different off-nadir angles. Let $T_{rotation}(i, j)$ denote the sensor rotation time between acquisitions i and j, and $T_{transition}(i, j)$ the available orbital transition time between the two. The optimization problem is:

$$\max_{x \in 0,1^R} \sum_{i=1}^{R} x_i v_i \quad \text{subject to: } T_{rotation}(i, j) \leq T_{transition}(i, j) \quad \text{if } x_i = x_j = 1 \tag{1}$$

A visual depiction of this acquisition setup is provided in Fig. 1.

The problem is reformulated as a Quadratic Unconstrained Binary Optimization (QUBO) problem, the constraint being integrated into the energy (objective) function as a penalty term:

$$\max_{x \in \{0,1\}^R} \sum_{i=1}^{R} x_i v_i - p \cdot \sum_{i=1}^{R-1} \sum_{j=i}^{R} x_i x_j \cdot c(i, j) \tag{2}$$

Here, p is a penalty coefficient, and $c(i, j) = 1$ if $T_{rotation}(i, j) > T_{transition}(i, j)$, and 0 otherwise. This ensures compatibility with quantum optimization methods such as QA [10], as this formulation is fundamentally equivalent to the Ising model. Note that we effectively convert Eq. 2 to a minimization problem by inverting the sign of the terms, as this is a stardard for optimization. Despite being a simplified formulation with a linear cost function, one satellite

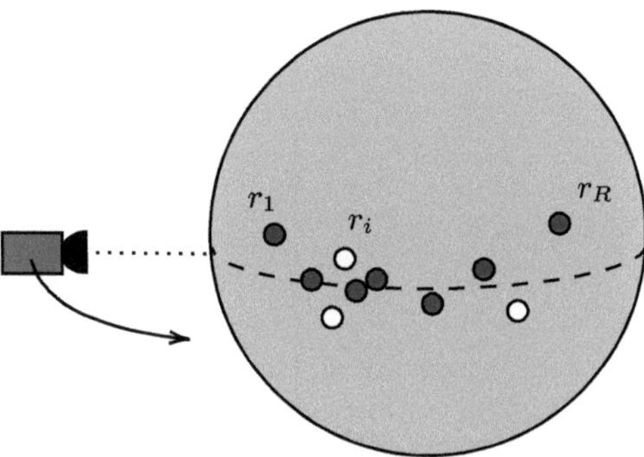

Fig. 1. Considered setting for satellite acquisition scheduling, with acquisition requests r_i represented as circles along the satellite trajectory projection on the Earth's surface. The objective consists in finding a subset of acquisition requests (here, a possible subset is indicated in red) that maximizes the acquired value and satisfies the rotation constraint. The image was first introduced in [6].

and one set of constraints, computational limitations arise when the number of requests R increases, as sensor rotation contraints increase at an even faster rate.

3 Hybrid Quantum-Classical Optimization for Satellite Acquisition Scheduling

As mentioned in the previous sections, finding optimal acquisition schedules for a high number of acquisition requests R is prohibitive. Heuristic methods based on quantum optimization cannot be directly applied, due to the aforementioned limitations in computational capabilities. Here, we introduce an iterative approach to integrate quantum optimization within an acquisition scheduling algorithm.

Full Problem Solution Sampling. A conventional (classical) heuristic method can start tackling the large problem and sampling candidate solutions. The solution associated with the lowest energy, corresponding to a sub-optimal acquisition schedule, can be the initial solution for a subsequent local refinement.

Sub-problem Generation. In order to locally improve on the initial solution, a suitable sub-problem is generated. In Fig. 2, the proposed sub-problem generation method is shown. The method selects a local subset of acquisition requests within a window, for which rotation constraints (see Eq. 1) are in place with respect to the acquisitions delimiting the window. First, two acquisitions r_i and r_j belonging to the initial solution are chosen. This ensures that the updated schedules for the selected acquisitions present no constraint violations with the initial acquisition schedule outside the window. The window is iteratively extended until a S-dimensional sub-problem is reached. An independent sub-problem is thus generated.

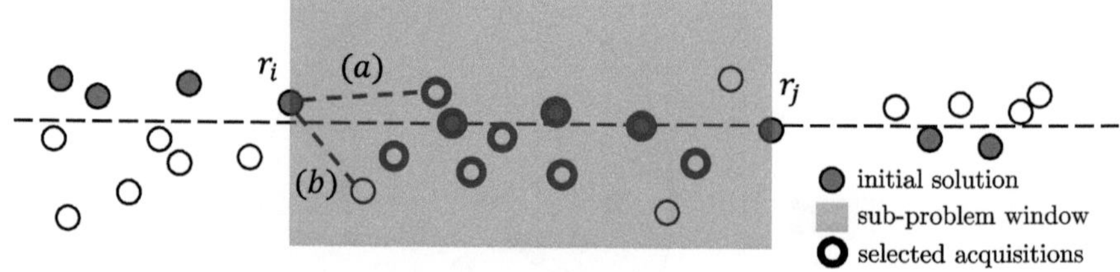

Fig. 2. Example of sub-problem generation. Circles represent acquisition requests distributed along the satellite orbit projection. Two acquisitions r_i and r_j belonging to the initial solution to the acquisition scheduling (in red) define a window (in blue). The selected acquisitions for the sub-problem generation (circled in blue) are all acquisitions within the window for which the rotation constraint is satisfied, i.e. (Color figure online), $T_{rotation}(r_{i,j}, r) \leq T_{transition}(r_{i,j}, r)$ (a), excluding the incompatible ones (b).

Sub-problem Solution Sampling. The size S of the sub-problem can be adaptively selected so that a quantum optimization algorithm, such as QA, can sample solutions for it. The initial solution can also be exploited, e.g., using advanced schedules such as reverse quantum annealing [6,13] or for warm-starting QAOA [14]. The best solution is used to update the schedule, which then becomes the initial solution of the next iteration of full and sub-problem solution sampling.

4   Experimental Validation

In this section, we show initial results for hybrid optimization on large acquisition scheduling problems, following the approach of Sect. 3. We generated two problems, with requests $R = 10000$ and $R = 20000$, each request having value $v_i = \{1,2\}$, with overall penalty coefficient $p = \max(x_i + x_j) + 1 = 5$. In order to validate the iterative approach and understand the impact of hybrid optimization, we worked with two experimental setups:

1. *Simulated annealing - simulated annealing (SA-SA)*: SA is used for both full problem and sub-problem solution sampling. It is executed on 2 Intel Xeon Cascade Lake Platinum 8260M CPUs from the DEEP HPC system located at the JÃijlich Supercomputing Centre (JSC) [17].
2. *Simulated annealing - quantum annealing (SA-QA)*: in this case, forward QA is used for sub-problem solution sampling. QA is executed using the `Advantage_system_6.4` solver, accessed through D-Wave Leap [1].

We generate subproblems of a fixed size of 60 requests, to ensure a feasible problem embedding into the Advantage system. We perform a fixed number of runs for each solver (100 runs at 1000 sweeps for SA, 1000 runs at $100\mu s$ annealing time for QA) and 10 iterations. The high number of runs ensures statistical significance of the obtained results, from which we collect suitable metrics for practical use. We compute a discrete cumulative distribution function

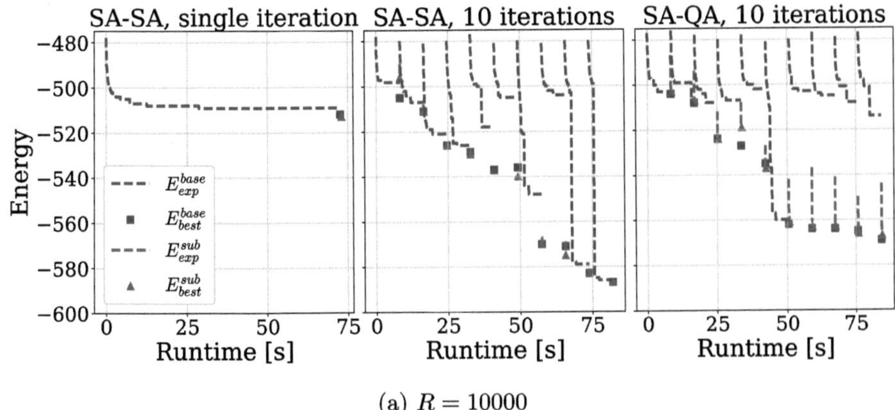

(a) $R = 10000$

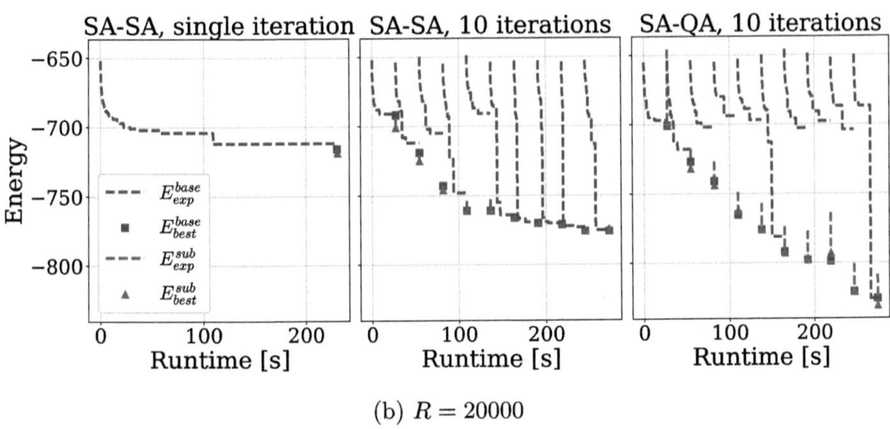

(b) $R = 20000$

Fig. 3. Hybrid optimization results for two different acquisition scheduling problems, i.e., number of requests $R = 10000$ (a) and $R = 20000$ (b), and two algorithm combinations (SA-SA, SA-QA). Results are compared to one SA-SA iteration, executed for a comparable amount of time. Blue plots show the expected (E_{exp}^{base}) and best (E_{best}^{base}) energies recorded on the full problem of size 60. Red plots show the expected (E_{exp}^{sub}) and best (E_{best}^{sub}) energies recorded on the generated sub-problem of size 60.

(CDF) for the statistical distribution of the sample energies collected by all the runs. We can derive the expected energy for each iteration I as the energy value for which $CDF(\text{expected\_energy}_I) = 1 - 0.5^I$. A continuous inverse CDF, needed for finding the expected energy, is obtained by interpolating the discrete CDF using PCHIP 1-D monotonic cubic interpolation [9] and inverting it. The results for the two setups are compared to a single SA-SA iteration with 1000 runs. The code for the method and the experimental setup is made available for reproducibility[1].

A plot for the expected and best solution energies vs. runtime is shown in Fig. 3. The best solution found in our runs for each subroutine was used as initial solution of the subsequent subroutine. This initial experiment confirms the feasibility of the iterative sub-problem generation approach, as the solution search generally improves at each iteration. QA is also able to run and sample feasible solutions for the generated sub-problems. The question is whether it is worth running QA over SA. For the $R = 10000$ problem instance, this is not the case, except for a better energy found by the SA-QA setup at iteration 6, after which no significant improvement is reached. In the $R = 20000$ problem instance, solutions updated by QA actually lead to better final acquisition schedules compared to the SA-SA setup. This could be related to the fact that a higher acquisition density leads to higher problem complexity, where QA can have an edge. The higher solution variability of QA, signaled by the varying expected sub-problem energy, also indicated a difference in the exploration of the solution space. Another indicator is the high difference between expected and best energies obtained in the SA-QA case, showing that QA shifts from the conventional statistical distribution of SA solutions.

5   Conclusion

The presented hybrid approach to optimization enables experimental validation of quantum optimization subroutines in acquisition scheduling, with initial promising results for QA. A higher contribution of quantum optimization, finding better solutions and working on larger subproblems, is an expected outcome of future quantum machines. Practical use will also benefit from tight on-premise integration of quantum machines, which will reduce communication delays. Future work will focus on extending the validation to other optimization methods, beyond standard QA, including a deeper analysis of the benefits of quantum optimization. In addition, further scalability will also be studied, working with a higher number of requests and more complex acquisition scenarios.

Acknowledgments. The authors gratefully acknowledge support from the project JUNIQ that has received funding from the German Federal Ministry of Education and Research (BMBF) and the Ministry of Culture and Science of the State of North Rhine-Westphalia. This work is part of the Quantum Computing for Earth Observation (QC4EO) initiative from the ESA Φ-lab. This work is co-financed by the EUROCC2 project funded by the European High-Performance Computing Joint Undertaking (JU) and EU/EEA states under grant agreement No 101101903. This work is partially supported by the Icelandic Centre for Research, project *"Hybrid Quantum-Classical Workflows for EO"*, RANNÍS grant number: 2511078–051 (http://en.rannis.is/).

---

[1] https://gitlab.jsc.fz-juelich.de/sdlrs/qmp.

Disclosure of Interests. The authors have no competing interests to declare that are relevant to the content of this article.

References

1. D-Wave Leap (2025). https://cloud.dwavesys.com/leap/
2. Agnèse, J.C., Bataille, N., Blumstein, D., Bensana, E., Verfaillie, G.: Exact and approximate methods for the daily management of an earth observation satellite. Proc. SpaceOPS (1996)
3. Bovolo, F., et al.: Big data from space for precision agriculture applications. In: IOP Conference Series: Earth and Environmental Science **509**(1), 012004 (2020). https://doi.org/10.1088/1755-1315/509/1/012004
4. Celesti, M., et al.: The copernicus hyperspectral imaging mission for the environment (chime): Status and planning. In: IGARSS 2022 - 2022 IEEE International Geoscience and Remote Sensing Symposium, pp. 5011–5014 (2022). https://doi.org/10.1109/IGARSS46834.2022.9883592
5. D-Wave Systems: https://www.dwavesys.com/
6. Delilbasic, A., Le Saux, B., Riedel, M., Michielsen, K., Cavallaro, G.: Reverse quantum annealing for hybrid quantum-classical satellite mission planning. In: IGARSS 2024 - 2024 IEEE International Geoscience and Remote Sensing Symposium, pp. 432–436 (2024). https://doi.org/10.1109/IGARSS53475.2024.10640974, https://ieeexplore.ieee.org/abstract/document/10640974, iSSN: 2153-7003
7. Denby, B., Lucia, B.: Orbital edge computing: Nanosatellite constellations as a new class of computer system. In: Proceedings of the Twenty-Fifth International Conference on Architectural Support for Programming Languages and Operating Systems, pp. 939–954. ASPLOS '20, Association for Computing Machinery, New York (2020). https://doi.org/10.1145/3373376.3378473
8. Feld, S., et al.: A hybrid solution method for the capacitated vehicle routing problem using a quantum annealer. Front. ICT **6** (2019). https://doi.org/10.3389/fict.2019.00013, https://www.frontiersin.org/journals/ict/articles/10.3389/fict.2019.00013/full
9. Fritsch, F.N., Butland, J.: A method for constructing local monotone piecewise cubic interpolants. SIAM J. Sci. Stat. Comput. **5**(2), 300–304 (1984). https://doi.org/10.1137/0905021
10. Kadowaki, T., Nishimori, H.: Quantum annealing in the transverse ising model. Phys. Rev. E **58**(5), 5355–5363 (1998)
11. Makarov, A., Taddei, M.M., Osaba, E., Franceschetto, G., Villar-Rodríguez, E., Oregi, I.: Optimization of image acquisition for earth observation satellites via quantum computing. In: Quaresma, P., Camacho, D., Yin, H., Gonçalves, T., Julian, V., Tallón-Ballesteros, A.J. (eds.) Intelligent Data Engineering and Automated Learning - IDEAL 2023, pp. 3–14. Springer Nature Switzerland, Cham (2023)
12. Makarov, A., et al.: Quantum optimization methods for satellite mission planning. IEEE Access **12**, 71808–71820 (2024). https://doi.org/10.1109/ACCESS.2024.3402990, https://ieeexplore.ieee.org/abstract/document/10534762
13. Pelofske, E., Hahn, G., Djidjev, H.N.: Advanced anneal paths for improved quantum annealing. In: Proceedings - IEEE International Conference on Quantum Computing and Engineering, QCE 2020, pp. 256–266 (2020). https://doi.org/10.1109/QCE49297.2020.00040

14. Quetschlich, N., Koch, V., Burgholzer, L., Wille, R.: A hybrid classical quantum computing approach to the satellite mission planning problem. arXiv preprint (2023). https://arxiv.org/abs/2308.00029v1
15. Sakuler, W., Oberreuter, J.M., Aiolfi, R., Asproni, L., Roman, B., Schiefer, J.: A real-world test of portfolio optimization with quantum annealing. Quantum Mach. Intell. **7**(1), 43 (2025). https://doi.org/10.1007/s42484-025-00268-2
16. Stollenwerk, T., Michaud, V., Lobe, E., Picard, M., Basermann, A., Botter, T.: Agile Earth observation satellite scheduling with a quantum annealer. IEEE Trans. Aerospace Electron. Syst. **57**, 3520–3528 (10 2021). https://doi.org/10.1109/TAES.2021.3088490
17. Suarez, E., Kreuzer, A., Eicker, N., Lippert, T.: The DEEP-EST project. In: Porting Applications to a Modular Supercomputer-Experiences from the DEEP-EST Project; Schriften des Forschungszentrums Jülich IAS Series, pp. 9–25 (2021). https://juser.fz-juelich.de/record/905812/files/IAS_Series_48-1.pdf
18. Tran, T., et al.: A hybrid quantum-classical approach to solving scheduling problems. In: Proceedings of the International Symposium on Combinatorial Search. vol. 7, pp. 98–106 (2016)
19. Zhang, B., et al.: Progress and challenges in intelligent remote sensing satellite systems. IEEE J. Selected Top. Appl. Earth Observ. Remote Sens. **15**, 1814–1822 (2022). https://doi.org/10.1109/JSTARS.2022.3148139

Quantum Model for CVRPTW

Imran Meghazi[1,2(✉)] and Éric Bourreau[2]

[1] La Poste, Paris, France
[2] LIRMM, Paris, France
{imran.meghazi,eric.bourreau}@lirmm.fr

Abstract. This paper proposes a quantum algorithm for the capacitated vehicle routing problem with time windows (CVRPTW) based on Grover Search framework. This problem is often faced by Postal services in the context of package delivery or other time-sensitive operations. We provide an implementation on gate based quantum computer of a model inspired by classical route first, cluster second technique. The quantum paradigm allows to overcome suboptimality inherent property of this decomposition. In the current NISQ (Noisy Intermediate-Scale Quantum) era, the most important limitation is the number of available qubits which makes time windows and capacity constraints hard to tackle. We introduce a qubit-efficient split-inspired modeling which adds only a linear number of decision qubits to standard quantum formulations for Traveling Salesman Problem (TSP).

Keywords: operations research · quantum computing · optimization

1 Introduction

Operations research aims to tackle a wide range of combinatorial problems, among which the Vehicle Routing Problem (VRP) is a prominent example. The VRP generalizes the well known Travelling Salesman Problem (TSP). The TSP seeks the shortest cycle that visits every node in a graph exactly once. In a logistics context, it can be interpreted as finding the most efficient route for a single vehicle to deliver one item to each customer. The VRP extends this formulation by considering multiple vehicles operating simultaneously, each subject to capacity constraints, in order to serve all customers while minimizing the total travel cost. This problem arises daily in logistics companies and organizations, such as the French postal service *La Poste*. In practice, delivery services face additional constraints. In this paper, we focus on two of them : a capacity constraint, which reflects the limited space of each vehicle, and a time window constraint that allows customers to specify a preferred delivery period. Solving the VRP with such constraints, referred to as the Capacitated Vehicle Routing Problem with Time Windows (CVRPTW), is computationally difficult and is usually adressed using heuristics [7] and bio-inspired algorithms [1].

Quantum Computing is a promising paradigm that has already produced algorithms outperforming their classical counterpart. In some cases, such as

F. Barbaresco and F. Gerin (Eds.): QUEST-IS 2025, CCIS 2744, pp. 65–76, 2026.
https://doi.org/10.1007/978-3-032-13855-2_7

Shor's algorithm [11] it offers exponential speedup, in others, like Grover's algorithm [6], a quadratic improvement. Grover's algorithm provides an oracle-based general framework for addressing combinatorial problems. While quantum oracle-based models for the Traveling Salesman Problem and QUBO formulations for the CVRPTW have been proposed, to the best of our knowledge, no quantum oracle-based formulation has yet been introduced for the CVRPTW. In this work, we propose such a model using Grover Adaptive Search [5] on gate-based quantum computers.

2   The Capacitated VRP with Time Windows

2.1   Problem Definition

A *Capacitated Vehicle Routing Problem with Time Windows* can be defined as follows: a fleet of vehicles must deliver packages to a set of n customers $V = \{v_1, v_2, \ldots, v_n\}$, where all packages are initially stored at a central depot d. Each customer v_i requests a specific quantity of goods denoted by q_i. In addition to capacity constraints, customers specify a preferred delivery time window during the day. In this work, we consider the case where each customer is assigned a single time window, meaning that customer v_i must be served within the interval $[a_i, b_i]$.

2.2   Methods

In classical operations research, the Vehicle Routing Problem is typically addressed using two main approaches. The most well-known is the *cluster-first, route-second* heuristic, which involves grouping customers into clusters and then solving a separate routing problem, such as a TSP, within each cluster. For instance, customers may be clustered based on their geographical proximity. It is then assumed that solving a TSP for each cluster yields a VRP solution that is reasonably close to optimal. However, this assumption breaks down in the presence of time window constraints, which can invalidate geographically efficient clusterings. Several heuristics based on this strategy can be found in the literature, such as in [7].

An alternative strategy, albeit less intuitive, is the *route-first, cluster-second* approach, originally introduced by Beasley [2]. In this method, a so-called *giant tour* is constructed that visits all customers, and this tour is subsequently divided into feasible *sub-tours* that satisfy the problem constraints. We specifically focus on the *split* procedure presented in [10], where the partitioning of the giant tour is typically performed according to vehicle capacity limits.

2.3   Split Procedure

In this section, we detail the *split* procedure, as our quantum approach is heavily inspired.

The *split* algorithm takes as input an auxiliary graph $H = (X, A, W)$ constructed from the solution obtained in the first routing phase, namely the *giant tour*. The set X contains the nodes, ordered from node 0 to n according to this tour. Let P be this tour. The set of arcs A contains an arc (i, j), with $i < j$, if a trip visiting every customers between P_{i+1} and P_j doesn't violate any constraints. The corresponding weight w_{ij} represents the total cost of that sub-tour, computed as $D_{d,i+1} + \sum_{k=i+1}^{j-1} D_{P_k, P_{k+1}} + D_{j,d}$ with D the distance matrix. Finding a minimum cost path from node 0 to n in H yields an optimal splitting of the giant tour, where every arc corresponds to a feasible sub-tour that respects the original problem's constraints.

As an illustrative example, let us consider a typical instance of the Capacitated Vehicle Routing Problem.

Distance matrix

	d	1	2	3	4	5	6
d	0	23	30	23	14	20	26
1	23	0	17	27	27	38	36
2	30	17	0	21	40	46	37
3	23	27	21	0	35	40	12
4	14	27	40	35	0	16	31
5	20	38	46	40	16	0	33
6	26	36	32	12	31	33	0

Nodes

v_i	q_i
d	0
1	2
2	3
3	1
4	3
5	2
6	3

$C^{max} = 5$

The only constraints is that, for every subtour, the total must not exceed the vehicle's maximum capacity C^{max}. Stated differently, the cumulative load at k^{th} node, denoted c_{P_k}, must not exceed C^{max}.

$$\sum_{k=i+1}^{j} c_{P_k} \leq C^{max}, \ \forall (i, j) \in X, i < j \tag{1}$$

We first compute a giant tour, let P be the ordering of this tour, P_i is the ith node of the tour. Figure 1 ashow this giant tour and Fig. 1 bshows the auxiliary graph.

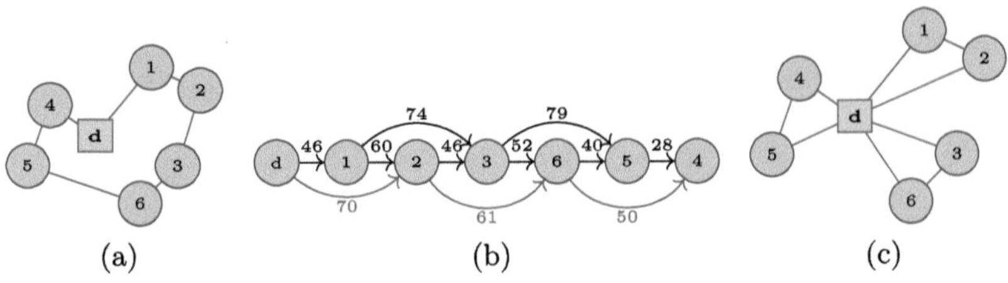

Fig. 1. The steps to compute a solution. 1 athe giant tour. 1 bthe auxiliary graph H. 1 cthe solution.

Finding an optimal split for this giant tour can be effectively done by computing the shortest path over Fig. 1 b

3   Quantum Model

In classical computing, a split-based formulation is inherently suboptimal, as the optimal VRP solution is not necessarily the optimal split of the optimal *giant tour*. Consequently, it requires evaluating every possible *giant tour* and all its splits, an exponential task. As a result, it is typically employed only as a highly efficient heuristic. In contrast, quantum superposition enables the parallel evaluation of all these possibilities, allowing the identification of the optimal *giant tour* for every possible splits.

3.1   Split Inspired Model

Objective Function. Let $G = (V, E)$ be the graph instance. Let D denotes the distance matrix of this graph. Following the approach introduced in the previous section, we define separately the decision variables related to the ordering of customers and those related to the clustering (or splitting) of the route. Let P represent the *giant tour*, and let P_i denote its ith node.

Then we define the following variables :

$$y_i = \begin{cases} 1 & \text{if } P_i \text{ is the end of a sub-tour} \\ 0 & \text{otherwise} \end{cases}$$

The variables y_i encode the splitting decisions. Specifically, if $y_i = 1$, it indicates that the vehicle must return to the facility immediately after visiting node P_i. Consequently, we enforce $y_0 = 1$ and $y_n = 1$ to ensure that the tour starts and ends at the depot.

Thus the value to minimize is defined as follows :

$$\sum_{i=1}^{n-1} \left[D_{P_i, P_{i+1}} (1 - P_i) + \left(D_{P_i, f} + D_{f, P_{i+1}} \right) y_i \right] + D_{f, P_1} + D_{P_n, f} \tag{2}$$

As discussed in the previous section, the clustering process (or splitting, in this context) must be performed in accordance with the problem's constraints.

Uniqueness Constraint. Each customer must be visited exactly once, meaning that no customer may appear more than once in the tour.

$$P_i \neq P_j, \ \forall (i, j) \in V \tag{3}$$

Capacity Constraint. This constraint ensures that the total load of each sub-tour does not exceed the vehicle's maximum capacity. Given a fixed customer ordering, it can be formalized as shown in Eq. 1. In practical terms, this means that at each node i, the cumulative load c_i must be less than or equal to the vehicle's capacity limit C^{max}.

$$c_i \leq C^{max}, \ \forall i \in V \tag{4}$$

In the quantum computing context, all operations must be reversible. As a result, it is necessary to store the load at each preceding node in order to compute the current load. The load at node i is thus calculated as the sum of the loads of the previous nodes, conditioned by the y_i variables, plus the load associated with node i itself. As previously defined, $y_{i-1} = 1$ indicates that node $i - 1$ marks the end of a sub-tour, implying that the next node and—thus next subtour— is served by a new vehicle.

$$c_i = q_{P_i} + (c_{i-1} \times (1 - y_{i-1})) \tag{5}$$

Time Window Constraint. This constraint behaves similarly to the capacity constraint. The visiting time at node i, denoted by t_i, must not exceed the latest acceptable arrival time, denoted by b_i.

$$t_i \leq b_{P_i}, \ \forall i \in V \tag{6}$$

As with the capacity constraint, it is necessary to store the visiting times of all previous nodes in order to compute the arrival time at node i. In this case, when $y_i = 1$, the current time is reset, as each sub-tour is assumed to be performed by a separate vehicle. Therefore, the computation of the time window constraint follows a similar logic to that of the capacity constraint. The key difference lies in the use of a maximum: the arrival time at node i is defined as the maximum between the earliest acceptable time a_i and the cumulative travel time.

$$t_i = \max\left(a_{P_i}, T_{f,P_i} \times y_{i-1} + \left((t_{i-1} + T_{P_{i-1},P_i}) \times (1 - y_{i-1})\right)\right) \tag{7}$$

This gives us the following model :

$$\min_{P,y} \sum_{i=1}^{n-1} \left[D_{P_i,P_{i+1}} (1 - y_i) + \left(D_{P_i,depot} + D_{depot,P_{i+1}}\right) y_i \right] + D_{depot,P_1} + D_{P_n,depot}$$

$$\text{s.t.} \quad P_i \neq P_j, \quad \forall(i,j) \tag{8a}$$

$$c_i \leq C^{max}, \quad \forall i \tag{8b}$$

$$t_i \leq b_{P_i}, \quad \forall i \tag{8c}$$

$$c_i = q_{P_i} + (c_{i-1} \times (1 - y_i)) \tag{8d}$$

$$t_i = \max\left(a_{P_i}, T_{f,P_i} \times y_i + \left((t_{i-1} + T_{P_{i-1},P_i}) \times (1 - y_i)\right)\right) \tag{8e}$$

$$P_i \in [\![1,n]\!], \forall i [\![1,n]\!] \tag{8f}$$

3.2 Quantum Search

The quantum computing paradigm has led to the development of several noteworthy algorithms. Among them, Grover's search algorithm [6] is particularly relevant in the context of operations research, as it enables the identification of solutions within an unstructured search space. The only requirement is to construct an oracle function that returns 1 for all states satisfying the problem's constraints, and 0 otherwise. The algorithm amplifies the amplitudes—and consequently the probabilities—of the desirable states, starting from a uniform superposition over all possible configurations. Theoretically, Grover's algorithm achieves a success probability approaching 1 after $O(\sqrt{N/M})$ [3] iterations, where N is the size of the search space and M is the number of valid solutions. As a result, the algorithm's complexity is directly influenced by the number of decision variables used in the problem formulation.

To perform function minimization, we adopt the quantum search-based optimization algorithm introduced by Dürr and Høyer [5]. The principle of the algorithm is as follows:

1. Set an initial threshold value k and add a constraint to the Grover search that only considers solutions with a cost less than k.
2. If a valid solution is found, update k to this solution's cost.
3. Repeat the process until no better solution can be found.

We define the oracle function O as follows :

$$O(|P\rangle, |y\rangle, k) = \begin{cases} 1 & \text{if } (l(|P\rangle, |y\rangle)) < k) \\ & \land (P_i \neq P_j, \ \forall(i,j)) \\ & \land (c_i \leq C^{max}, \ \forall i \in V) \\ & \land (t_i \leq b_{P_i}, \ \forall i \in V) \\ 0 & \text{otherwise} \end{cases}$$

where l computes the objective function 2.

Let $\hat{O}_k$ denote the oracle operator for a given threshold k, and let $\hat{S}_{|\Psi\rangle}$ be the amplitude amplification operator introduced by Grover. Then, the Grover operator is defined as $\hat{G} = \hat{O}_k \hat{S}_{|\Psi\rangle}$. Applying $\hat{G}$ iteratively amplifies the amplitudes of all states that satisfy the constraints given in Eqs. (3), (4), and (6).

Algorithm 1 outlines the procedure used to implement and solve the model. We define $|AD\rangle$ as the register encoding the *All Different* constraint, $|\delta\rangle$ as the qubit for the distance constraint, and $|\kappa\rangle$ and $|\tau\rangle$ as the vectors encoding the capacity and time constraints, respectively. For instance, $|\kappa_i\rangle = 1$ if the accumulated load of the vehicle since the start of the subtour is valid: it equals

0 if P_i is the first node of a subtour, or $|c_{i-1}\rangle$ plus the demand at the ith node, provided this sum does not exceed the vehicle's capacity.

Algorithm 1: Grover Search CVRPTW Algorithm

Input: q the quantity vector, T the travelling time matrix, D the distance matrix, k the cost threshold

Output: P and y quantum register

1   $|P\rangle \leftarrow H^{\otimes n \log n}$
2   $|y\rangle \leftarrow H^{\otimes n}$
3   $|q\rangle \leftarrow |-\rangle$
4   $m \leftarrow 1$
5   **while** $m \leq \sqrt{2^{n \log n + n}}$ **do**
6   **for** $i \leftarrow 1$ **to** n **do**
7   **for** $j \leftarrow i+1$ **to** n **do**
8        $|AD\rangle \leftarrow P_i \neq P_j$
9      $|c_i\rangle \leftarrow q_i$
10   **if** $\neg |y_{i-1}\rangle$ **then**
11       $|w\rangle \leftarrow |w\rangle + D_{P_{i-1},P_i}$
12       $|c_i\rangle \leftarrow |c_{i-1}\rangle + |c_i\rangle$
13       $|t_i\rangle \leftarrow \max(a_{P_i}, |t_{i-1}\rangle + T_{P_{i-1},P_i})$
14   **else**
15       $|w\rangle \leftarrow |w\rangle + D_{P_{i-1},P_{depot}} + D_{P_{depot},P_i}$
16       $|t_i\rangle \leftarrow \max(a_{P_i}, T_{depot,P_i})$
17      $|\kappa_i\rangle \leftarrow (|c_i\rangle \leq C^{max})$
18      $|\tau_i\rangle \leftarrow (|t_i\rangle \leq b_{P_i})$
19    $|\delta\rangle \leftarrow (|w\rangle \leq k)$
20    $|q\rangle \leftarrow |AD\rangle \wedge \bigwedge_{i=0}^{n} |\kappa_i\rangle \wedge \bigwedge_{i=0}^{n} |\tau_i\rangle \wedge |\delta\rangle$
21   Apply Grover operator to $|P\rangle \otimes |y\rangle$
22    $m \leftarrow m+1$

---

4   Implementation

In this section, we will discuss how to implement such algorithm on a gate based quantum computer. We assume that *multi-controlled* X gates also known as MCX gates are available.

P register There are various ways to represent a giant tour, which ultimately corresponds to a solution of the TSP. The implementation of constraints depends heavily on the chosen representation. For example, [12] introduces a successor-based representation that enables efficient cost evaluation. However, as noted in [14], it requires additional transformations to ensure that only valid solutions are marked.

In this work, we adopt a basic position-based representation, where $V \subset \mathbb{N}$ and $P_i \in V \; \forall i$. The uniqueness constraint can be easily enforced by ensuring that each register holds a distinct value.

4.1 Capacity and Time Constraints

Comparisons. As described in the previous section, both the capacity and time window constraints ultimately reduce to comparisons, which can be efficiently implemented using a single ancilla qubit and components from the ripple-carry adder circuit introduced in [4].

Capacity Constraint. As discussed in Sect. 3.1, each value c_i representing the cumulative load at node P_i must be stored, which requires n dedicated quantum registers. To compute each c_i as defined in Eq. 5, we must first encode the demand associated with node P_i. For this, we use a *conditional encoder*, a type of circuit that conditionally superposes multiple values onto a quantum register based on the value of one or more index registers. An implementation of such an encoder is described in paragraph 3 of the METHODS section in [14], it is denoted as F gate.

Depending on the value of the register $|P_i\rangle$, we use the encoder to assign the corresponding demand value to a load register. If $y_i = 0$, indicating that P_i is not the start of a new sub-tour, the previous accumulated load is added to this demand value. Conversely, if $y_i = 1$, the vehicle is assumed to have returned to the depot and restarted its route, so the cumulative load is reset. Figure 2 illustrates this implementation, which directly corresponds to lines 9, 12, and 17 of Algorithm 1.

[row sep=0.6cm,between origins, column sep=0.15cm] $|P\rangle$ 3

$|y\rangle$ X   2   X

$|c_{i-1}\rangle$ [2]Adder

$|c_i\rangle$ F $\leq C^{max}$1

$|\kappa\rangle$

Fig. 2. Quantum circuit for capacity constraint.

Time Window Constraint. The time window constraint is structurally similar to the capacity constraint, with the key difference being the need to compute a max() function. This is required to determine whether the vehicle arrives within the allowed time window. The arrival time at node P_i is computed by comparing two values: the earliest allowed delivery time a_{P_i} and the current travel time. The maximum of these two values is selected as the effective arrival time. This comparison and assignment can be implemented using a basic comparator. Figure 3 depicts this implementation, which directly maps to lines 13, 16, and 18 of Algorithm 1. The $Enc()$ gate denotes the encoding of a value into a register.

[row sep=0.6cm,between origins, column sep=0.1cm] $|P\rangle$ 3   3
$|y\rangle$ X 2 1 X 2
$|t_{i-1}\rangle$ [2]Adder
$|t_i\rangle$ F   F $\leq a_{P_i} 1 Enc(a_{P_i}) \leq b_{P_i} 2$
$|\theta_i\rangle$ 1 -1 X 1 X
$|\tau\rangle$

Fig. 3. Quantum circuit for time constraint.

4.2 Complexity

Space Complexity. In this paper, we adopt a position-based encoding of the solution, where customer P_i is interpreted as the ith customer visited in the giant tour. This representation requires $n \log n$ qubits to encode the tour sequence. Additionally, we introduce n qubits to store the binary variables y_i, which represent the splitting decisions.

A significant number of qubits is also required to enforce the constraints. Specifically, $\mathcal{O}(n^2)$ qubits are needed to guarantee tour validity, since there are $n(n-1)/2$ pairwise inequality constraints to ensure that each customer appears exactly once.

For the capacity constraints, the cumulative load must be computed and stored at each step along the tour. This requires $n(\log d_{\max} + 1)$ qubits, where $d_{\max}$ denotes the maximum vehicle load capacity. In addition, n extra qubits are needed to store boolean results indicating whether the load at each step violates the capacity constraint. The total qubit requirement for capacity is therefore: $n(\log d_{\max} + 1) + n = \mathcal{O}(2n + n \log d_{\max})$.

Regarding the time window constraints, a similar number of registers is required. The arrival time at each node must be stored using n registers of size $\log t_{\max} + 1$, where $t_{\max}$ is the maximum allowed time. Since the effective arrival time is computed as the maximum between the accumulated travel time and the customer's earliest delivery time, n additional boolean qubits are used to indicate which value is selected. Moreover, n further boolean qubits are needed to store whether the time window is respected. The total number of qubits for time-related constraints is thus: $n(\log t_{\max} + 1) + 2n = \mathcal{O}(3n + n \log t_{\max})$.

Finally, the cost evaluation requires $\log w_{max}$ where $w_{\max}$ represents an upper bound on the solution's cost.

Putting all components together, the total number of qubits required by the algorithm is: $\mathcal{O}(n^2 + n \log n + 6n + n \log d_{\max} + n \log t_{\max} + \log w_{max})$. However, the number of decision variables remains relatively small, namely $n \log n + n$, as only n qubits are added compared to the most compact exact quantum formulation for the TSP [12]. For instance, the example presented in 2.3 would require at least 147 qubits to solve.

To the best of our knowledge, no oracle-based formulations of the CVRPTW exist for comparison. A few QUBO-based formulations have been proposed: route-based formulations, where the number of routes grows exponentially and must therefore be generated using heuristics [8], and time-expanded formulations

[13], where the decision variables represent every possible arc for each vehicle between nodes at different times. In the former case, the number of decision variables depends on the number of routes considered, while in the latter it is significantly larger than in our formulations, even before accounting for potential slack variables. In classical computing, the state of the art relies on route-based formulations, where routes generation is handled via column generation [9].

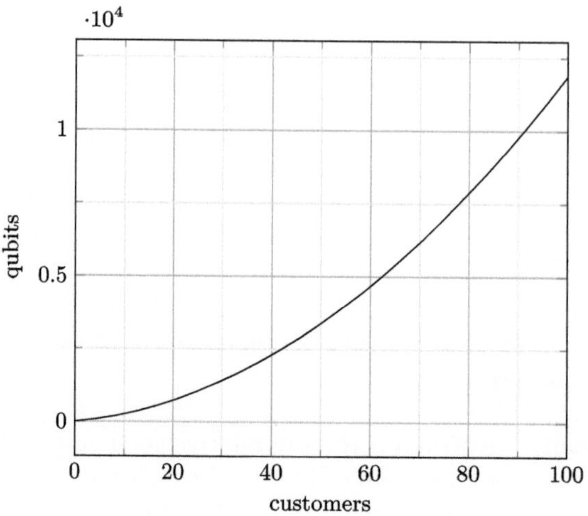

Fig. 4. Number of qubits relative to the number of customers for the oracle with the maximum capacity and number of time windows both fixed to 8.

Time Complexity. We use the number of MCX gates as a unit for time complexity analysis. For each gate set, we do not account for the inverse (uncomputation) circuits required to *free* ancilla qubits, as they mirror the forward operations and are already covered by the $\mathcal{O}$ notation. There are three categories of constraints contributing to the overall circuit depth.

First, the *all-different* constraint, ensuring uniqueness of the tour, can be implemented using $\mathcal{O}(n^2)$ MCX gates.

Second, the *capacity constraint* involves several components. The conditional encoder requires $\mathcal{O}(n \log d_{\max})$ MCX gates. We use Cuccaro et al.'s adder [4], which is efficient for this representation and contributes $\mathcal{O}(6 \log d_{\max})$ MCX gates. The comparison circuit can also be implemented using a modified version of the same adder. Since these operations are applied at each node, the total cost for capacity-related operations becomes: $\mathcal{O}\left(n^2 \log d_{\max} + 12n \log d_{\max}\right)$.

Third, the *time window constraint* has a similar structure to the capacity constraint, with the difference that both the encoder and the comparison are applied *twice* per node. This results in a total cost of: $\mathcal{O}\left(2n^2 \log d_{\max} + 24n \log d_{\max}\right)$.

Finally, for cost evaluation, we require n *conditional matrix encoders* and adders to compute the tour cost. These operations contribute a total of:

$\mathcal{O}(n^3 \log n + 6n \log n)$. This component is relatively expensive and could potentially be optimized using a phase-based cost encoding via the Quantum Fourier Transform (QFT). However, such an approach would require translating our representation into a form compatible with QFT-based evaluation, which would in turn increase both qubit requirements and gate complexity [14].

In order to achieve a success probability close to 1, the Grover operator $\hat{G}$ must be applied $\mathcal{O}(\sqrt{N})$ times, where N is the size of the search space. In our case, $N = 2^{n \log n + n}$, which dominates the overall time complexity.

5 Conclusion

The proposed model allows one to tackle a complex routing problem, namely the CVRPTW. By adopting a split-inspired modeling approach on quantum computers, the model introduces only a linear additive overhead in the number of decision qubits compared to standard quantum TSP encodings while being optimal. Consequently, the time complexity is $\mathcal{O}(\sqrt{2^{n \log n + n}})$. Although large-scale implementations remain limited by current quantum hardware due to the gate complexity, this research lays the groundwork for future advancements in quantum logistics optimization.

References

1. (PDF) Applying the ANT system to the vehicle routing problem. In: ResearchGate. https://doi.org/10.1007/978-1-4615-5775-3_20
2. Beasley, J.E.: Route first–cluster second methods for vehicle routing. Omega **11**(4), 403–408 (1983). https://doi.org/10.1016/0305-0483(83)90033-6
3. Boyer, M., Brassard, G., Høyer, P., Tapp, A.: Tight bounds on quantum searching. Fortschr. Phys. **46**(4–5), 493–505 (1998). https://doi.org/10.1002/(SICI)1521-3978(199806)46:4/5⟨493::AID-PROP493⟩3.0.CO;2-P
4. Cuccaro, S.A., Draper, T.G., Kutin, S.A., Moulton, D.P.: A new quantum ripple-carry addition circuit (2004). https://doi.org/10.48550/arXiv.quant-ph/0410184
5. Durr, C., Hoyer, P.: A quantum algorithm for finding the minimum (1999). https://doi.org/10.48550/arXiv.quant-ph/9607014
6. Grover, L.K.: A fast quantum mechanical algorithm for database search. In: Proceedings of the Twenty-Eighth Annual ACM Symposium on Theory of Computing, pp. 212–219. STOC '96, Association for Computing Machinery, New York (1996). https://doi.org/10.1145/237814.237866
7. Koskosidis, Y.A., Powell, W.B., Solomon, M.M.: An optimization-based heuristic for vehicle routing and scheduling with soft time window constraints. Transp. Sci. **26**(2), 69–85 (1992)
8. Lucas, A.: Ising formulations of many NP problems. Front. Phys. **2** (2014). https://doi.org/10.3389/fphy.2014.00005
9. Pessoa, A., Sadykov, R., Uchoa, E., Vanderbeck, F.: A generic exact solver for vehicle routing and related problems. Math. Program., 483–523 (2020). https://doi.org/10.1007/s10107-020-01523-z
10. Prins, C.: A simple and effective evolutionary algorithm for the vehicle routing problem. Comput. Oper. Res. **31**(12), 1985–2002 (2004). https://doi.org/10.1016/S0305-0548(03)00158-8

11. Shor, P.W.: Polynomial-time algorithms for prime factorization and discrete logarithms on a quantum computer. SIAM J. Comput. **26**(5), 1484–1509 (1997). https://doi.org/10.1137/S0097539795293172
12. Srinivasan, K., Satyajit, S., Behera, B.K., Panigrahi, P.K.: Efficient quantum algorithm for solving travelling salesman problem: an IBM quantum experience (2018). https://doi.org/10.48550/arXiv.1805.10928
13. Vargas, A., Shukla, P., Allmendinger, R., Jaeger, A.: On solving the capacitated vehicle routing problem with time windows using quantum annealing. In: Proceedings of the Genetic and Evolutionary Computation Conference Companion, pp. 1979–1983. GECCO '24 Companion, Association for Computing Machinery, New York (2024). https://doi.org/10.1145/3638530.3664139
14. Zhu, J., Gao, Y., Wang, H., Li, T., Wu, H.: A Realizable GAS-based quantum algorithm for traveling salesman problem (2022). https://doi.org/10.48550/arXiv.2212.02735

Session: 6 Quantum Algorithms, Computing; Simulation – Quantum Machine Learning A

IQNN-CS: Interpretable Quantum Neural Network for Credit Scoring

Abdul Samad Khan[1], Nouhaila Innan[2,3]($\boxtimes$), Aeysha Khalique[1], and Muhammad Shafique[2,3]

[1] Lahore University of Management Sciences, Lahore, Pakistan
`{24120006,aeysha.khalique}@lums.edu.pk`
[2] eBRAIN Lab, Division of Engineering, New York University Abu Dhabi (NYUAD), Abu Dhabi, UAE
`{nouhaila.innan,muhammad.shafique}@nyu.edu`
[3] Center for Quantum and Topological Systems (CQTS), NYUAD Research Institute, NYUAD, Abu Dhabi, UAE

Abstract. Credit scoring is a high-stakes task in financial services, where model decisions directly impact individuals' access to credit and are subject to strict regulatory scrutiny. While Quantum Machine Learning (QML) offers new computational capabilities, its black-box nature poses challenges for adoption in domains that demand transparency and trust. In this work, we present IQNN-CS, an interpretable quantum neural network framework designed for multiclass credit risk classification. The architecture combines a variational QNN with a suite of post-hoc explanation techniques tailored for structured data. To address the lack of structured interpretability in QML, we introduce Inter-Class Attribution Alignment (ICAA), a novel metric that quantifies attribution divergence across predicted classes, revealing how the model distinguishes between credit risk categories. Evaluated on two real-world credit datasets, IQNN-CS demonstrates stable training dynamics, competitive predictive performance, and enhanced interpretability. Our results highlight a practical path toward transparent and accountable QML models for financial decision-making.

Keywords: Quantum Machine Learning · Credit Scoring · Interpretability · Quantum Finance

1 Introduction

Credit scoring is a foundational task in financial services, directly influencing decisions related to loan approvals, credit limits, and interest rates [1]. Its impact is not only financial but also societal, affecting individuals' access to economic opportunities. As such, credit scoring systems must meet two core requirements: high predictive performance and strict compliance with fairness

© The Author(s), under exclusive license to Springer Nature Switzerland AG 2026
F. Barbaresco and F. Gerin (Eds.): QUEST-IS 2025, CCIS 2744, pp. 79–95, 2026.
https://doi.org/10.1007/978-3-032-13855-2_8

and transparency regulations, such as the European Union's General Data Protection Regulation (GDPR) and the Basel III banking regulations. In this context, interpretability becomes essential, not as an optional enhancement, but as a mandatory feature for deployment in regulated environments.

Quantum Neural Networks (QNNs) have recently emerged as promising models for structured learning tasks, owing to their potential to encode and process high-dimensional data using fewer resources [2,3]. Their expressive power enables them to represent complex decision boundaries for domains like finance, where patterns in structured tabular data can be subtle and interdependent.

However, existing QNN-based models often prioritize performance over interpretability, which is crucial for sensitive decision-making contexts such as credit scoring. Moreover, the development of interpretable QNNs remains in its infancy. Unlike classical models, where attribution methods and explanation techniques are well-studied, the quantum learning community lacks standardized tools to reason about QNN decisions, especially in multiclass classification settings. In this work, we ask a central question:

How can QNNs be designed for real-world credit scoring applications, where achieving high accuracy is not sufficient, and interpretability is a core requirement?

To address this, we propose **IQNN-CS**, an interpretable architecture of QNNs specifically tailored for multiclass credit scoring. Our goal is not to demonstrate quantum supremacy or to outperform classical machine learning models, but rather to explore how QNNs can be made suitable for high-stakes financial applications through interpretability-aware design.

The main contributions of this paper are as follows:

- We propose a hybrid classical-quantum pipeline where a variational QNN performs classification on structured financial data, with interpretability achieved using adapted classical and quantum techniques.
- We introduce Inter-Class Attribution Alignment (ICAA), a new metric that quantifies attribution divergence across predicted classes, enabling structured analysis of model reasoning in multiclass tasks.
- We evaluate IQNN-CS on two real-world credit datasets, focusing on robustness, attribution behavior, and interpretability, while emphasizing deployment feasibility in regulated financial settings.

2   Background and Related Work

2.1   QML for Credit Scoring

Quantum Machine Learning (QML) has been increasingly applied in the financial sector [4], covering applications such as loan eligibility prediction [5], credit default classification [6], market prediction [7,8], fraud detection [9–12], and other financial applications [13,14]. Some of these tasks are naturally formulated as classification problems, while others are formulated as combinatorial optimization problems. Within this broad spectrum, credit scoring has received limited but growing attention in QML research.

When it comes to credit scoring, only a few works have directly addressed the problem using quantum models. The systemic quantum score model introduced a quantum kernel-based approach aimed at improving generalization in low-data and imbalanced settings, demonstrating performance benefits in a production-grade financial scenario [15]. A separate study explored a hybrid quantum-classical model for credit scoring among small- and medium-sized enterprises, highlighting reductions in training time by embedding a quantum layer into a classical neural network [16]. Both approaches treated credit scoring as a binary classification task and focused on performance rather than transparency or explainability.

Other research efforts have framed credit scoring as an optimization problem. One study utilized a QUBO formulation to optimize scoring card thresholds and combinations, employing both quantum annealing and classical heuristics to enhance bank profitability under credit risk constraints [17]. A related work investigated cost-effective feature selection for mobile credit scoring using a quantum-inspired evolutionary algorithm based on quantum gates and representations, achieving reduced feature costs and competitive accuracy [18].

While these studies highlight the potential of quantum and quantum-inspired models for credit scoring, they either treat the task as binary classification or focus on optimization formulations. Crucially, they overlook the interpretability requirements critical for real-world deployment in regulated financial environments. In contrast, we address credit scoring as a multiclass classification problem and focus on the interpretability of quantum models, an aspect that has not been covered in prior work.

2.2   Interpretability in QML

Interpretability in QML is increasingly important for deploying models in sensitive domains such as finance. However, most classical explainability methods do not directly extend to quantum models due to their probabilistic nature and circuit-based representations.

Initial efforts, such as Q-LIME, adapt local explanation techniques to quantum settings by approximating the influence of features on predictions [19]. Other works combine quantum autoencoders with classifiers and apply LIME and SHAP for interpretability in quantum representation learning [20].

To address global explainability, frameworks like QuXAI use Q-MEDLEY to trace feature attributions through quantum feature maps [21]. Visual tools such as QGrad-CAM further enable fine-grained class-specific explanations using variational quantum circuits [22]. Despite these advances, interpretability for structured, multiclass QML tasks, such as credit scoring, remains largely unexplored. Our work addresses this gap through a domain-specific, interpretable QNN framework.

3   Methodology

The IQNN-CS framework is structured as a five-stage pipeline designed to deliver interpretable and accurate credit scoring using a hybrid quantum-classical architecture (see Fig. 1 and Appendix Algorithm 1).

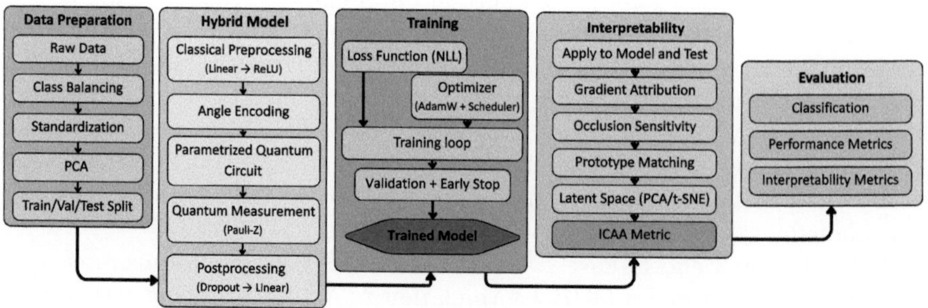

Fig. 1. Overview of the IQNN-CS pipeline.

3.1   Data Preparation

The process begins with preprocessing two benchmark credit datasets containing numerical and categorical financial attributes. To address class imbalance, we apply undersampling to Dataset 1 and SMOTE to Dataset 2, depending on the severity of imbalance. All numerical features are standardized by transforming each feature x_j to $x'_j = (x_j - \mu_j)/\sigma_j$, where μ_j and σ_j denote the empirical mean and standard deviation. To reduce dimensionality and ensure compatibility with quantum hardware constraints, Principal Component Analysis (PCA) is applied, resulting in quantum-compatible input vectors. PCA projects the data $\mathbf{x}$ onto a lower-dimensional subspace using orthogonal components $\mathbf{w}_k$ as $z_k = \mathbf{w}_k^\top \mathbf{x}$, for $k = 1, \ldots, d$. The resulting representation aligns with the available number of qubits. Stratified sampling is used to divide the dataset into training, validation, and test sets, preserving class distribution.

3.2   Hybrid Quantum-Classical Model

The model architecture consists of three sequential components. First, a classical preprocessing block transforms the PCA-compressed input $\mathbf{z} \in \mathbb{R}^d$ into an encoded latent vector $\mathbf{h}_{cl1} \in \mathbb{R}^{d'}$ through fully connected layers with ReLU activations. This output $\mathbf{h}_{cl1}$ is then passed to the quantum layer, where the features are encoded onto N_Q qubits, typically initialized to the $|0\rangle^{\otimes N_Q}$ state, using angle encoding, where each input feature ϕ_j (from $\mathbf{h}_{cl1}$) rotates a qubit via an operation like $R_P(\phi_j) = e^{-i\phi_j P/2}$, with P being a Pauli operator (e.g., $\sigma_x, \sigma_y, \sigma_z$). This is followed by a multi-layer entangling variational quantum circuit (VQC),

$U(\boldsymbol{\theta})$, parameterized by trainable angles $\boldsymbol{\theta}$. Expectation values of Pauli-Z observables, $\langle \sigma_z^{(k)} \rangle = \langle \psi_0 | U^\dagger(\boldsymbol{\theta}) \sigma_z^{(k)} U(\boldsymbol{\theta}) | \psi_0 \rangle$ for each qubit k, are measured to extract quantum features. These features are concatenated and forwarded to the classical postprocessing head. This final block, which includes dropout regularization and linear projections, produces logits $\mathbf{o} \in \mathbb{R}^C$ over the C credit risk classes. The full system is end-to-end differentiable and implemented via PennyLane's `TorchLayer`, enabling seamless integration with PyTorch's automatic differentiation capabilities.

3.3 Training Procedure

The model is trained using the negative log-likelihood (NLL) loss function, given for a batch of M samples as $L_{NLL} = -\frac{1}{M} \sum_{i=1}^{M} \sum_{c=1}^{C} y_{i,c} \log(\hat{y}_{i,c})$, where $y_{i,c}$ is the ground-truth indicator for class c and $\hat{y}_{i,c}$ is the predicted probability from the softmax output. Optimization is performed with AdamW, which introduces decoupled weight decay. Learning rate scheduling, such as cosine annealing, is employed to stabilize convergence. Quantum gradients with respect to variational parameters $\boldsymbol{\theta}$ are computed using the parameter-shift rule: $\frac{\partial \langle \hat{O} \rangle}{\partial \theta_k} = \frac{1}{2s} \left(\langle \hat{O} \rangle_{\theta_k + s} - \langle \hat{O} \rangle_{\theta_k - s} \right)$, with $s = \pi/2$ for single-qubit gates. Early stopping is applied based on validation loss to prevent overfitting, and the best-performing model checkpoint is retained.

3.4 Post-Hoc Interpretability

Interpretability is assessed using four complementary techniques (see Appendix Algorithm 2). First, gradient-based attribution methods, including saliency maps, gradient$\times$input, integrated gradients, and SmoothGrad, estimate the importance of each input feature x_j through $\partial L / \partial x_j$ or related formulations. Second, prototype matching is performed in the quantum feature space by extracting activation vectors $\mathbf{a}_{QNN}(x)$ and computing cosine similarity with training instances: $S_C(\mathbf{u}, \mathbf{v}) = \frac{\mathbf{u} \cdot \mathbf{v}}{\|\mathbf{u}\| \|\mathbf{v}\|}$. This allows retrieval of similar historical examples for a given test input. Third, we employ latent space visualization by projecting the high-dimensional quantum embeddings into lower-dimensional manifolds using t-SNE. This allows us to inspect the quantum representation geometry and assess class separability within the feature space. Fourth, we introduce the ICAA metric. For each class c, let $\mathbf{A}_c(x_{inst})$ be its attribution vector. The ICAA matrix is defined by: $ICAA_{ij}(x_{inst}) = S_C(\mathbf{A}_i(x_{inst}), \mathbf{A}_j(x_{inst}))$, capturing the pairwise similarity in feature attributions across classes. Additionally, we assess the region of indecision by perturbing inputs and measuring the variance in attribution vectors.

3.5 Evaluation Metrics

The final model performance is evaluated using accuracy and macro F1-score. Accuracy is computed as $\text{Acc} = \frac{TP+TN}{TP+TN+FP+FN}$, where TP is the number of true positives, TN is the number of true negatives, FP is the number of false positives, and FN is the number of false negatives. While the macro F1-score aggregates class-wise F1-scores: $F1_c = \frac{2P_c R_c}{P_c + R_c}$,   Macro F1 $= \frac{1}{C} \sum_{c=1}^{C} F1_c$, where P_c and R_c are precision and recall for class c. Confusion matrices offer class-level diagnostic insights. Finally, we visualize interpretability outputs, gradient maps, and matched prototypes, along with latent space projections of QNN activations using t-SNE. These projections are color-coded by true labels to assess the separability of quantum features.

4   Results

4.1   Experimental Settings

Experiments were conducted on two benchmark credit scoring datasets: Dataset 1 [23] and Dataset 2 [24]. All models were implemented using PyTorch and PennyLane [25], with quantum circuits simulated via `default.qubit`. Training was executed on a CPU (Apple M3, 16 GB RAM) without GPU acceleration. Each dataset was split into training (70%), validation (15%), and test (15%) sets using stratified sampling. The hybrid model combined six qubits and four `StronglyEntanglingLayers` with classical dense layers and dropout. Optimization was performed over 50 epochs using the AdamW optimizer (learning rate 0.01) and class-weighted cross-entropy loss. Full configuration details are provided in Appendix Table 2.

4.2   Training and Convergence

Figure 2 shows the training, validation, and test curves for both datasets. For Dataset 1, accuracy curves in (a) exhibit rapid improvement within the first few

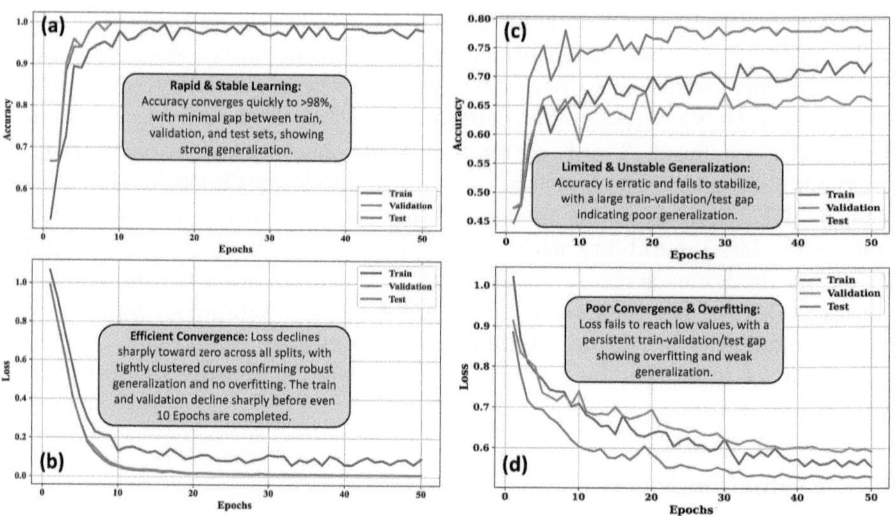

Fig. 2. Accuracy and loss curves for IQNN-CS. (a) accuracy and (b) loss for Dataset 1; (c) accuracy and (d) loss for Dataset 2.

epochs, stabilizing above 98% across all splits. The corresponding loss curves in (b) show a sharp decrease followed by a smooth plateau, with minimal discrepancy between training and validation, indicating efficient convergence and strong generalization. In contrast, Dataset 2 exhibits more unstable training behavior. Accuracy trends in (c) plateau early, with validation and test performance remaining below 80%. The loss curves in (d) reveal a slower and noisier decline, particularly in the validation set. The wider gap between training and validation suggests either overfitting or greater dataset complexity due to feature diversity and class imbalance.

4.3   Classification Performance

Table 1 summarizes the classification results for both datasets. On Dataset 1, the model achieved perfect performance, with 100% accuracy and F1-score. This confirms the model's ability to separate risk levels cleanly in well-structured data. For Dataset 2, overall accuracy was 77.3%, with variable class-wise metrics. The model showed high recall for the Low class (0.97) but low precision (0.64), suggesting misclassifications. The High class had strong precision (0.95) but lower recall (0.67), reflecting under-identification of high-risk cases. These results highlight Dataset 2's complexity and imbalance.

Table 1. Classification performance on both Datasets.

Class	Dataset 1			Dataset 2		
	Precision	Recall	F1-score	Precision	Recall	F1-score
Low	1.00	1.00	1.00	0.64	0.97	0.77
Average	1.00	1.00	1.00	0.73	0.84	0.78
High	1.00	1.00	1.00	0.95	0.67	0.79
Accuracy (%)	**100**			**77.3**		

4.4   Interpretability Analysis

To assess the transparency and trustworthiness of the IQNN-CS model, we conduct a comprehensive interpretability analysis. This includes examining latent quantum geometry, attribution consistency, example influence, class attribution alignment, and softmax entropy behavior. We compare findings across both datasets to understand how dataset quality impacts interpretability.

Feature Attribution Analysis: As illustrated in Fig. 3, saliency maps for a representative test instance show focused and interpretable patterns in Dataset 1. In contrast, Dataset 2 exhibits diffuse, noisy attributions that vary across runs, suggesting that the model relies less consistently on meaningful input features.

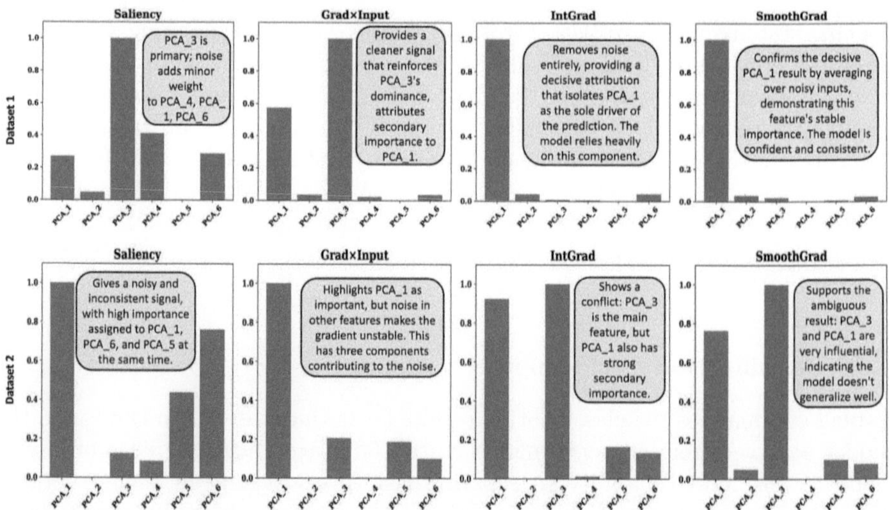

Fig. 3. Saliency maps for a selected test sample. Dataset 1 shows concentrated attributions; Dataset 2 appears more diffuse.

Quantum Representation Geometry: Latent embeddings extracted from the quantum layer are visualized using t-SNE as shown in Fig. 4. Dataset 1 shows well-separated manifolds, indicating robust internal representations. In contrast, Dataset 2 presents entangled clusters, particularly between the Average and High classes, reflecting the model's confusion in classification.

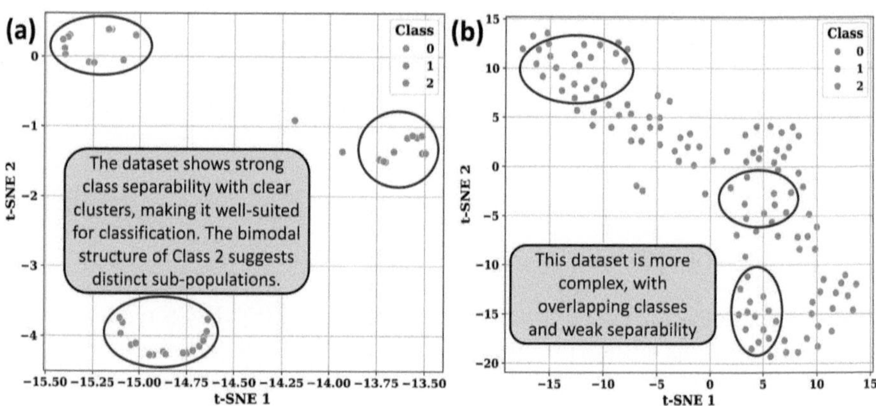

Fig. 4. t-SNE projection of quantum embeddings. (a) Dataset 1 yields a clear separation; (b) Dataset 2 shows an overlap.

Attribution Sensitivity: To assess how strongly predictions depend on top-ranked features, we occlude inputs and track confidence degradation. As shown in

Fig. 5, Dataset 1 experiences sharp confidence drops, indicating reliance on a few informative features. In Dataset 2, degradation is gradual and less structured.

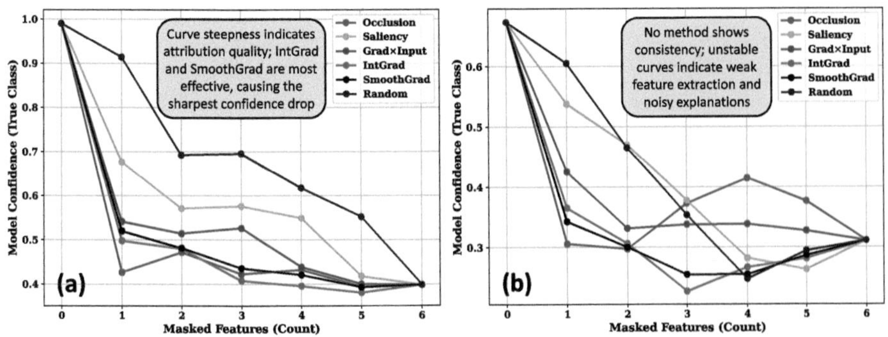

Fig. 5. Prediction confidence vs. occluded features. Left: Dataset 1 shows sharp drops; Right: Dataset 2 decays smoothly.

ICAA: We assess whether the model produces disentangled explanations for each class using ICAA. As shown in Fig. 6, Dataset 1 yields low inter-class attribution similarity, while Dataset 2 exhibits significant overlap, revealing a less distinct decision rationale.

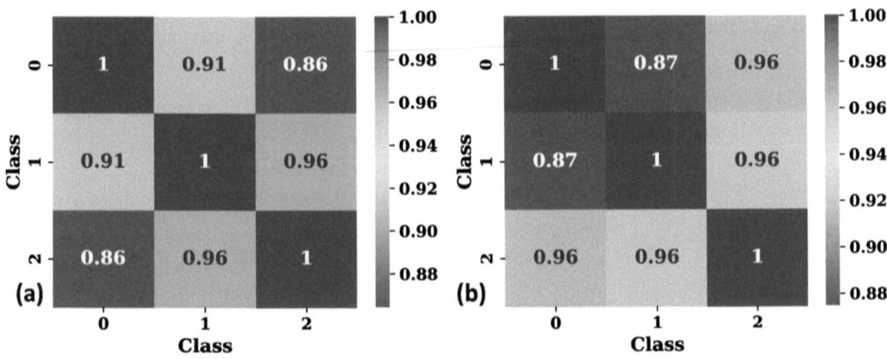

Fig. 6. ICAA matrices. Dataset 1 (a) shows clean class-wise separation; Dataset 2 (b) reveals overlapping attribution logic.

4.5 Discussion

Our experiments highlight a stark contrast in how the IQNN-CS model performs across two credit scoring datasets. While Dataset 2 proved more challenging, despite using the same model, its lower accuracy and less structured latent space suggest that the success of QNNs strongly depends on how well the data aligns with the model's inductive bias. Dataset 2 presented complex and overlapping feature spaces that the model struggled to disentangle. Interpretability tools helped uncover these differences. Occlusion and our proposed ICAA metric revealed that, in Dataset 2, attribution patterns overlapped, indicating that the model relied on similar features across classes, a potential source of misclassification. Gradient-based explanations were inconsistent, especially in noisy predictions, suggesting they should be used cautiously in quantum settings. Additionally, example-based attribution showed that Dataset 2 often aligned with incorrect classes, despite high-confidence predictions. This shows that a model's internal logic can break down even when output probabilities appear reliable.

In summary, these findings underscore that interpretability is not simply about explanation; it is a crucial diagnostic tool. Metrics like ICAA provide deeper insight into how well a model separates reasoning across classes, which is particularly important in high-stakes domains like finance.

5   Conclusion

We introduced IQNN-CS, a hybrid quantum-classical model for credit scoring that combines strong predictive performance with interpretable outputs. Across two real-world datasets, the model demonstrated high accuracy when the data structure aligned with its quantum encoding and revealed clear decision patterns through interpretability analysis. A key contribution of this work is the ICAA metric, which quantifies how distinctly the model reasons about different classes. It proved useful for identifying when the model's internal logic breaks down, even when standard metrics suggest good performance. Together with attribution sensitivity and example-based influence tracing, ICAA strengthens the tools available for understanding quantum models. IQNN-CS shows that accurate quantum models can also be transparent and trustworthy. As quantum computing advances, such interpretability-first frameworks will be essential for deploying reliable AI in sensitive fields like finance and healthcare.

Acknowledgments. This work was supported in part by the NYUAD Center for Quantum and Topological Systems (CQTS), funded by Tamkeen under the NYUAD Research Institute grant CG008, and the Center for Cyber Security (CCS), funded by Tamkeen under the NYUAD Research Institute Award G1104.

A   Appendix

A.1   IQNN-CS Training Procedure

The IQNN-CS architecture is trained using a hybrid quantum-classical pipeline as detailed in Algorithm 1.

Algorithm 1. IQNN-CS Training Procedure

Require: Raw Datasets (D_{raw1}, D_{raw2}), NumEpochs (N_{epochs}), BatchSize (B), LearningRate (η),
PCA Dimensions (d_{PCA1}, d_{PCA2}), Qubit Counts (N_{Q1}, N_{Q2})
Ensure: Trained IQNN-CS model $(M_{IQNN-CS})$
1: Initialize the IQNN-CS model structure $(M_{IQNN-CS})$:
2:    Define classical pre-processing network $M_{classical\_pre}$ using d_{PCA} input dimensions.
3:    Define Quantum Neural Network M_{QNN} with N_Q qubits, specified layers, and encoding method.
4:    Define classical post-processing network $M_{classical\_post}$ for the specified number of output classes.
5: Assemble $M_{IQNN-CS}$ by sequentially connecting $M_{classical\_pre} \rightarrow M_{QNN} \rightarrow M_{classical\_post}$.
6: **for all** dataset D_{raw} in $\{D_{raw1}, D_{raw2}\}$ **do**
7:                              $\triangleright$ Configure dataset-specific parameters: d_{PCA}, N_Q
8:    Perform preprocessing on D_{raw} to obtain D_{proc}:
9:       Apply class balancing technique (e.g., Undersampling, SMOTE).
10:       Apply feature standardization (e.g., StandardScaler).
11:       Apply PCA to reduce dimensionality to d_{PCA}.
12:    Split D_{proc} into training D_{train}, validation D_{val}, and test D_{test} sets.
13:    Initialize optimizer (e.g., AdamW) with $M_{IQNN-CS}$ parameters and learning rate η.
14:    Initialize learning rate scheduler (e.g., StepLR or CosineAnnealingLR).
15:    **for** $epoch = 1 \rightarrow N_{epochs}$ **do**
16:       Set $M_{IQNN-CS}$ to training mode.
17:       **for all** batch (X_b, y_b) in D_{train} **do**
18:          Clear gradients in the optimizer.
19:          Obtain predictions $\hat{y}_b$ via a forward pass of X_b through $M_{IQNN-CS}$.
20:          Calculate loss (e.g., Negative Log-Likelihood) between $\hat{y}_b$ and y_b.
21:          Compute gradients via hybrid backpropagation through $M_{IQNN-CS}$.
22:          Update model parameters using the optimizer.
23:       **end for**
24:       Adjust learning rate using the LR scheduler.
25:       Set $M_{IQNN-CS}$ to evaluation mode.
26:       Calculate validation loss $loss_{val}$ on D_{val} using $M_{IQNN-CS}$.
27:       **if** validation loss $loss_{val}$ meets early stopping criteria **then**
28:          **break** $\triangleright$ Cease training if validation performance degrades or stagnates
29:       **end if**
30:    **end for**
31: **end for**
32: **return** Trained $M_{IQNN-CS}$

A.2   Interpretability Pipeline

The interpretability module, summarized in Algorithm 2, combines gradient-based saliency maps, occlusion analysis, example-based attribution, ICAA, and latent space visualization. These methods are executed on selected test instances, allowing insight into feature influence and consistency across representations.

A.3   Experimental Settings

The experimental configuration used across datasets is outlined in Table 2. This setup ensures consistency for reproducibility and fair interpretability comparisons.

A.4   Extended Interpretability Analysis

Attribution Stability and Regions of Indecision: To assess robustness, we introduced Gaussian noise and computed the standard deviation of saliency maps. Most samples showed stable attribution under perturbation. However, Sample 12 in Dataset 2 showed significantly higher variance (Table 3), indicating unreliable class evidence.

Algorithm 2. Post-hoc Interpretability Analysis

Require: Trained IQNN-CS model ($M_{IQNN-CS}$), Preprocessed Test Data (D_{test}), Set of test instances ($X_{interpret} \subseteq D_{test}$)
Ensure: Interpretability outputs (Attribution Maps, ICAA Matrix, Plots, etc.)
1: Set $M_{IQNN-CS}$ to evaluation mode.
2: **for all** test instance x_{inst} in $X_{interpret}$ **do**
3:   ▷ **1. Gradient-Based Attribution**
4:     For each target class c:
5:         Compute gradient-based attribution $A_c(x_{inst})$ for x_{inst} towards class c using $M_{IQNN-CS}$.
6:         Store or visualize $A_c(x_{inst})$ (e.g., as feature heatmaps).
7:   ▷ **2. Occlusion Analysis**
8:     Perform occlusion analysis on x_{inst} with $M_{IQNN-CS}$ to obtain prediction probability drop curve P_{drop_curve}.
9:         ▷ This involves iteratively masking features of x_{inst} and recording the drop in prediction probability.
10:     Store or visualize P_{drop_curve}.
11:   ▷ **3. Example-Based Attribution (using QNN Activations)**
12:     **if** QNN activations for training data are not precomputed **then**
13:         Extract QNN activations Act_{train} from M_{QNN} for all instances in D_{train}.
14:     **end if**
15:     Extract QNN activation act_{inst} from M_{QNN} for instance x_{inst}.
16:     Calculate cosine similarity scores Sim_{scores} between act_{inst} and all activations in Act_{train}.
17:     Identify influential training examples based on Sim_{scores}.
18:   ▷ **4. Inter-Class Attribution Alignment (ICAA)**
19:     Let $A_0(x_{inst}), A_1(x_{inst}), \ldots$ be the attribution vectors for x_{inst} for each class (obtained from gradient-based attribution).
20:     Calculate $ICAA_{ij} \leftarrow \text{CosineSimilarity}(A_i(x_{inst}), A_j(x_{inst}))$ for all pairs of classes (i, j).
21:     Store the resulting $ICAA_{matrix}$ for x_{inst}.
22: **end for**
23:   ▷ **5. Latent Space Geometry Visualization (on a subset of D_{test})**
24: Select a subset X_{subset} from D_{test}.
25: Extract QNN activations Act_{QNN_subset} from M_{QNN} for all instances in X_{subset}.
26: Apply a dimensionality reduction technique (e.g., t-SNE or PCA) to Act_{QNN_subset}.
27: Plot the reduced-dimension embeddings, coloring them by true class labels.
28: **return** All generated interpretability maps, matrices, plots, and insights.

Table 2. Experimental Configuration for IQNN-CS

Component	Setting
Quantum Framework	PennyLane (default.qubit)
Number of Qubits	6
Quantum Layers	4 (StronglyEntanglingLayers)
Embedding Strategy	AngleEmbedding
Classical Layers	Linear $\rightarrow$ ReLU $\rightarrow$ Linear + Dropout
Preprocessing	StandardScaler, OneHotEncoder, SMOTE
Dimensionality Reduction	PCA (6 components)
Batch Size	16
Optimizer	AdamW
Learning Rate	0.01
Scheduler	StepLR
Loss Function	CrossEntropy (class-weighted)
Epochs	50
Validation Strategy	70-15-15 train-val-test split (stratified)
Random Seed	42 (NumPy + PyTorch)

Table 3. Standard deviation of saliency maps under 20 random Gaussian perturbations for selected test samples.

Sample	Dataset 1 (std)	Dataset 2 (std)	Indecisive?
3	0.0493	0.0651	No
7	0.1570	0.0395	No
12	0.1426	**0.2797**	Yes

Influence of Training Examples on Test Samples: We analyze the influence of training samples on test instance 7 using test-to-train cosine similarity. As shown in Table 4, in Dataset 1, the most influential samples align with the test sample's true class, indicating coherent internal representations. In contrast, Dataset 2 shows high-similarity points from multiple classes, suggesting ambiguity in the learned decision boundary.

Table 4. Cosine similarity between test sample 7 and top influential training examples in both datasets. Dataset 1 shows high similarity to same-class examples, while Dataset 2 yields a mix of same and different-class influences, suggesting less coherent internal representations.

Dataset	Test Sample	Train Index	Label	Cosine Sim.	Observation
1	7	146	2	0.9989	Same class as test
1	7	181	2	0.9976	Same class as test
1	7	160	2	0.9976	Same class as test
1	7	22	2	0.9930	Same class as test
1	7	94	2	0.9416	Same class as test
2	7	503	1	1.0000	Different class
2	7	19	1	1.0000	Different class
2	7	433	1	1.0000	Different class
2	7	588	1	1.0000	Different class
2	7	173	2	1.0000	Same class as test

Prediction Confidence and Attribution Consistency. Figure 7 shows that Dataset 1 had sharper softmax distributions, indicating high-confidence predictions. Dataset 2 yielded broader distributions, consistent with noisy convergence and less reliable model explanations. While the attribution similarity matrix (Fig. 8) reveals well-separated patterns in Dataset 1, while Dataset 2 lacks structure, underscoring inconsistency in explanation reliability.

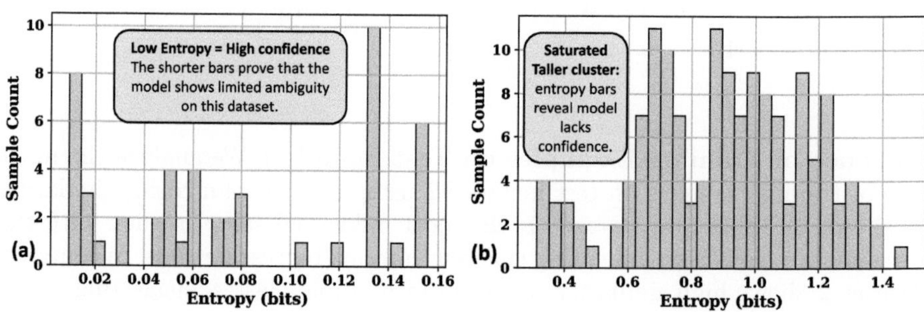

Fig. 7. Softmax entropy distributions. Dataset 1 (a) exhibits lower uncertainty compared to Dataset 2 (b).

Method Evaluation Summary: Table 5 synthesizes the comparative effectiveness of various interpretability techniques across both datasets. The evaluation is based on how well each method provides meaningful, class-consistent, and stable explanations aligned with model behavior.

In Dataset 1, most interpretability methods yielded coherent and high-confidence outputs. Occlusion and ICAA emerged as the most reliable techniques, offering sharply localized and class-discriminative explanations. Gradient-based methods such as saliency maps and integrated gradients also performed well, whereas SmoothGrad shows limited value due to noise sensitivity. In Dataset 2, the utility of nearly all methods degraded, consistent with earlier findings of unstable training dynamics and less confident predictions. ICAA and occlusion retained partial utility but suffered from reduced clarity. Example-based attributions revealed inconsistencies in training-test relationships, and attribution stability flagged specific instances of model indecision. Overall, the summary table highlights that interpretability reliability is closely tied to training convergence quality and dataset structure. Robust interpretability requires not only method design but also stable and semantically meaningful latent representations from the model itself.

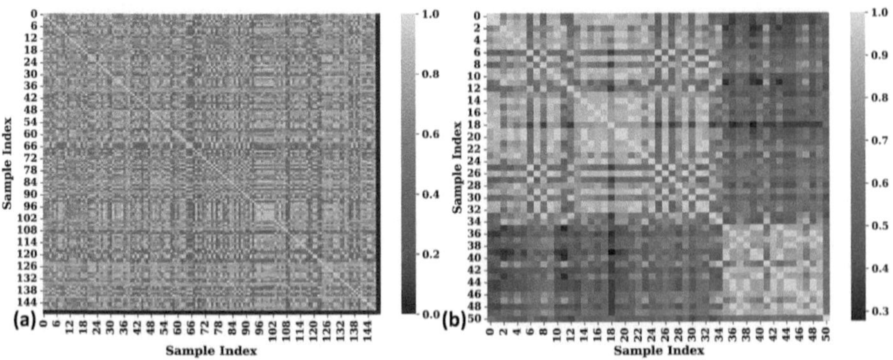

Fig. 8. Attribution similarity matrices across test samples for both Datasets: Dataset 1 (a) and Dataset 2 (b).

Table 5. Interpretability method utility across both datasets.

Method	Dataset 1 Utility	Dataset 2 Utility
Saliency	High	Medium
Gradient × Input	Moderate	Low
Integrated Gradients	High	Medium
SmoothGrad	Low	Low
Occlusion	Very High	Medium
ICAA	Very High	Medium
Example-Based	High	Low
Indecision Detection	Clean (all stable)	Useful (flagged Sample 12)

References

1. Thomas, L., et al.: Credit scoring and its applications. SIAM (2017)
2. Abbas, A., et al.: The power of quantum neural networks. Nat. Comput. Sci. (2021)
3. Innan, N., et al.: Next-generation quantum neural networks: enhancing efficiency, security, and privacy. In: 2025 IEEE 31st International Symposium on On-Line Testing and Robust System Design (IOLTS), pp. 1–4. IEEE (2025)
4. Biamonte, J., et al.: Quantum machine learning. Nature (2017)
5. Innan, N., et al.: LEP-QNN: loan eligibility prediction using quantum neural networks. arXiv preprint arXiv:2412.03158 (2024)
6. Sengar, A., et al.: A comparative analysis of hybrid quantum neural networks in binary credit defaulting tasks. In: 2023 IEEE MIT Undergraduate Research Technology Conference (URTC). IEEE (2023)
7. Choudhary, P.K., et al.: HQNN-FSP: a hybrid classical-quantum neural network for regression-based financial stock market prediction. arXiv preprint arXiv:2503.15403 (2025)
8. Pathak, P., Oad, V., Prajapati, A., Innan, N.: Resource allocation optimization in 5g networks using variational quantum regressor. In: 2024 International Conference on QCNC, pp. 101–105. IEEE (2024)
9. Innan, N., et al.: Financial fraud detection: a comparative study of quantum machine learning models. Int. J. Quantum Inf. (2024)
10. Innan, N., et al.: QFNN-FFD: quantum federated neural network for financial fraud detection. In: 2025 IEEE International Conference on QSW, pp. 41–47. IEEE (2025)
11. Innan, N., et al.: Financial fraud detection using quantum graph neural networks. Quantum Mach. Intell. **6**(1), 7 (2024)
12. Alami, M.E., et al.: Comparative performance analysis of quantum machine learning architectures for credit card fraud detection. arXiv preprint arXiv:2412.19441 (2024)
13. Kashif, M., et al.: Evaluating quantum amplitude estimation for pricing multi-asset basket options. arXiv preprint arXiv:2509.09432 (2025)
14. Innan, N., et al.: Quantum portfolio optimization with expert analysis evaluation. arXiv preprint arXiv:2507.20532 (2025)
15. Mancilla, J., et al.: Empowering credit scoring systems with quantum-enhanced machine learning. arXiv preprint arXiv:2404.00015 (2024)
16. Schetakis, N., et al.: Quantum machine learning for credit scoring. Mathematics **12**(9), 1391 (2024)
17. Zhang, B., et al.: Credit scoring card combination optimization model based on QUBO. In: Proceedings of the International Conference on Machine Learning, Pattern Recognition and Automation Engineering (2024)
18. Chen, C.M., et al.: Quantum optimized cost based feature selection and credit scoring for mobile micro-financing. Comput. Econom. (2024)
19. Pira, L., Ferrie, C.: On the interpretability of quantum neural networks. Quantum Mach. Intell. **6**(2), 52 (2024)
20. Kottahachchi, A.K.D., Khalil, I.: Qrlaxai: quantum representation learning and explainable AI. Quantum Mach. Intell. (2025)
21. Barua, S., et al.: Quxai: explainers for hybrid quantum machine learning models. arXiv preprint arXiv:2505.10167 (2025)
22. Lin, H.-Y., et al.: Quantum gradient class activation map for model interpretability. In: 2024 IEEE Workshop on SiPS, pp. 165–170. IEEE (2024)

23. Credit score classification dataset. https://www.kaggle.com/datasets/sujithmandala/credit-score-classification-dataset
24. Credit score classification dataset. https://www.kaggle.com/datasets/parisrohan/credit-score-classification
25. Bergholm, V., et al.: Pennylane: automatic differentiation of hybrid quantum-classical computations. arXiv preprint arXiv:1811.04968 (2018)

Quantum Geometric Learning: Encoding and Classification in Kendall Shape Spaces

Rasha Friji[2,3]($\boxtimes$) (iD), Mehdi Houas[1], Behjet Boussofara[2], and Mourad Ben Ammar[2]

[1] Talan Group, Paris, France
mehdi.houas@talan.com
[2] Talan Tunisia, Charguia1, Tunisia
{racha.friji,behjet.boussofara,mourad.benammar}@talan.com
[3] Cristal Lab, ENSI University, Manouba, Tunisia

Abstract. Kendall shape spaces provide a powerful geometric framework for modeling shapes independently of rotation, translation, and scale. These spaces, which arise naturally in statistical shape analysis and morphometrics, are structured as quotient Riemannian manifolds, such as complex projective spaces. In parallel, quantum machine learning (QML) has shown promise in tackling high-dimensional data and structured learning problems, yet its extension to non-Euclidean data domains remains largely unexplored. In this work, we propose a theoretical framework that enables QML algorithms to operate on Kendall shape spaces. We define a quantum encoding scheme that maps centered and normalized shape configurations into quantum states via amplitude encoding on the complex projective manifold. Furthermore, we introduce a quantum kernel function grounded in the geodesic distance between shapes, enabling quantum-enhanced classification and clustering algorithms. This approach opens the door to geometric quantum learning on structured manifolds and offers new perspectives for the efficient quantum processing of shape-based data. We discuss the feasibility of implementing this framework on NISQ hardware and highlight its potential for future applications in quantum shape analysis and beyond.

Keywords: QML · Kendall Shape Space · Quantum Kernel · Quantum Encoding

1 Introduction

Quantum machine learning (QML) has emerged as a promising paradigm that aims to leverage quantum computing capabilities to enhance classical learning algorithms in terms of expressivity and computational efficiency [1,2]. Recent advances in QML have shown potential speed-ups in supervised and unsupervised learning tasks, particularly in the context of kernel methods [3], variational circuits [4], and quantum embeddings [5]. However, most QML applications have

F. Barbaresco and F. Gerin (Eds.): QUEST-IS 2025, CCIS 2744, pp. 96–104, 2026.
https://doi.org/10.1007/978-3-032-13855-2_9

been limited to data represented in Euclidean or Hilbert spaces, with little exploration of learning tasks involving non-Euclidean geometric structures.

In parallel, Kendall shape spaces provide a rigorous mathematical formalism for analyzing geometric shapes, especially in biological and computer vision domains [6]. These spaces arise from factoring out translation, rotation, and scale from point configurations, resulting in manifolds such as complex or real projective spaces (e.g., $\mathbb{CP}^{k-2}$ for 2D shapes). The geometric and statistical structure of these manifolds supports tasks such as shape clustering, classification, and regression through intrinsic methods [7,15].

Bridging QML with shape analysis is a compelling yet underexplored direction. Recent works have begun addressing the generalization of machine learning to manifolds [8–11], and interest in quantum geometric representations has grown [12,13], but a quantum framework for learning on shape spaces has not yet been proposed. In this work, we present a theoretical framework that enables quantum learning on Kendall shape spaces. We focus on:

1. Developing an encoding scheme for shape configurations into quantum states based on their projection onto shape manifolds.
2. Defining a quantum kernel that captures the geometric similarity between shapes using quantum state overlaps or geodesic-based distances.

This paper is structured as follows. Section 2 provides background on Kendall shape theory and its manifold structure. Section 3 outlines the proposed quantum encoding scheme and learning model. Section 4 discusses implementation feasibility and theoretical insights. Section 5 presents potential applications and research perspectives.

2   Background

2.1   Kendall Shape Spaces

Kendall shape theory provides a foundational mathematical framework for the statistical analysis of shapes, where a shape is defined as all geometrical information about an object that remains after filtering out location, scale, and rotation [6]. In what follows, k denotes the number of landmark points in the shape configuration. For a configuration of k landmark points in $\mathbb{R}^m$, the pre-shape space is obtained by translating the configuration to the origin and scaling it to unit norm. The resulting pre-shapes lie on a unit hypersphere in $\mathbb{R}^{mk}$. The shape space is then the quotient of this hypersphere under the action of the rotation group $SO(m)$, giving rise to a manifold with non-Euclidean geometry [14].

In the planar case ($m = 2$), the shape space of k-point configurations corresponds to the complex projective space $\mathbb{CP}^{k-2}$, which is a smooth Riemannian manifold equipped with a natural metric structure—the Procrustes distance, derived from geodesic distances on the sphere [7]. This metric allows one to compute intrinsic means, variances, and perform statistical inference in shape spaces, forming the basis of statistical shape analysis [16].

2.2 Applications of Shape Analysis

Kendall shape analysis has been widely applied in biology (e.g., morphological variation in organisms), medicine (e.g., anatomical structure comparison), and computer vision (e.g., gesture and gait recognition) [10,11,17]. These applications rely on understanding shape variability and similarity through intrinsic measures, which remain consistent under transformations that do not affect the shape itself. Traditional machine learning methods have been extended to shape spaces via tangent space approximations, kernel methods, and intrinsic regression models [18]. However, these approaches often face scalability and representation challenges in high-dimensional shape datasets. This motivates the exploration of quantum machine learning paradigms for shape spaces.

2.3 Quantum Learning and Geometric Structures

Recent developments in quantum machine learning have explored the use of quantum-enhanced models for data embedded in structured spaces. For example, quantum kernel methods exploit the inner product structure of Hilbert spaces to capture nonlinear patterns in data [19], and variational quantum circuits can act as universal function approximators in hybrid quantum-classical architectures [20]. While these methods typically assume Euclidean input spaces, there is growing interest in quantum representations of geometric data, such as graphs, manifolds, and topological spaces [13]. Despite this interest, no theoretical framework currently exists for embedding and processing data from Kendall shape spaces within quantum machine learning pipelines. Addressing this gap could open new avenues for shape classification, clustering, and pattern discovery using quantum resources. In the next section, we introduce a novel quantum encoding scheme for Kendall shape configurations and propose learning models adapted to their intrinsic geometry.

3 Theoretical Proposition: Quantum Learning over Kendall Shape Spaces

In this paper, we propose a novel theoretical framework for applying quantum machine learning to data residing in Kendall shape spaces, with a focus on the planar case ($\mathbb{CP}^{k-2}$) for k-landmark shapes. Figure 1 depicts the full framework of the proposed QML framework on Kendall shape space composed of three main processing blocks: 1- the mapping in Kendall shape space of the input landmarks configuration, 2- the encoding quantum state and 3- the quantum classifier. These blocks are described in the following sections.

3.1 Quantum Embedding of Shape Configurations

Let $X \in \mathbb{R}^{2 \times k}$ be a configuration of k landmarks in 2D. We assume that X has been preprocessed to remove translation (centered), scaling (unit norm),

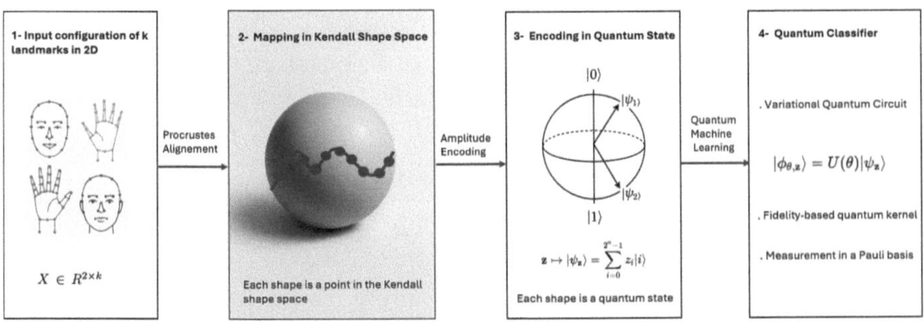

Fig. 1. Illustration of the full workflow of the proposed quantum machine learning framework on Kendall shape spaces. 1- The input configuration of 2D landmarks (representing a face, an action, an organ etc. depending on the use case), 2- Mapping in Kendall shape space, 3- Encoding of each shape as a quantum state using amplitude encoding and 4- Quantum Classifier.

and rotation (Procrustes alignment). This yields a point on the Kendall shape manifold—a nonlinear Riemannian manifold of shapes modulo similarity transformations.

We propose to encode this point into a quantum state via the two following steps: complex representation and quantum state preparation. In the first step, we map the configuration X to a complex vector $z \in \mathbb{C}^k$ such that $\sum_i |z_i|^2 = 1$. In the second step, we normalize the shape vector and loads it into the amplitudes of a n-qubit state using amplitude encoding (Eq. 1).

$$\mathbf{z} \mapsto |\psi_{\mathbf{z}}\rangle = \sum_{i=0}^{2^n - 1} z_i |i\rangle \tag{1}$$

Having encoded shapes as quantum states that respect the geometry of Kendall shape spaces, we can leverage these embeddings to define similarity measures between shapes in the quantum domain. The inner product structure of the Hilbert space naturally leads to the construction of a quantum kernel, which forms the foundation for quantum classification algorithms on shape data.

3.2   Quantum Shape Kernel

We define in Eq. 2 a quantum shape kernel $K_q(X, Y)$, where $|\psi_X\rangle$ and $|\psi_Y\rangle$ are the quantum embeddings of two shape configurations. This kernel corresponds to the fidelity between quantum states and aligns with the **cosine similarity** in Kendall space, which has a geometric interpretation in terms of the angle between complex pre-shapes.

$$K_q(X, Y) = |\langle \psi_X | \psi_Y \rangle|^2 \tag{2}$$

We conjecture that this kernel is positive semi-definite- a crucial property for kernels used in machine learning- and can be implemented via a SWAP test [23] on a quantum device. Moreover, it preserves geodesic distances up to isometric transformations, allowing downstream quantum classifiers to operate in alignment with the intrinsic geometry of the shape manifold.

3.3 Variational Quantum Classifier on Shape Space

We design a variational quantum circuit (VQC) $U(\theta)$, parameterized by a set of angles θ, acting on quantum states $|\psi_X\rangle$. After shapes are encoded into quantum states, the VQC acts as a parametrized quantum model that learns to separate shape classes based on these embeddings. It replaces classical classifiers (like SVM or neural networks) with a trainable quantum model. It applies a sequence of quantum gates with tunable parameters θ, optimized through classical feedback loops. The final measurement output (in a Pauli basis or projection onto specific output states) corresponds to a classification label.

Given that the input quantum states encode shapes while respecting their intrinsic manifold structure, and since the VQC operates directly in this Hilbert space, the model benefits from the nonlinear separation capacity of quantum kernels. This alignment with the shape's geodesic geometry offers better generalization in low-data regimes, especially where classical flattening approaches (e.g., PCA) fail.

4 Feasibility Study

This section evaluates the feasibility of implementing the proposed quantum machine learning framework over Kendall shape spaces, considering both theoretical compatibility and current technological constraints.

4.1 Quantum Encoding Overhead

Encoding shape data into quantum states involves preparing a quantum system into a state $|z\rangle \in \mathbb{C}^k$, where k is the number of landmarks. This can be achieved using amplitude encoding, which requires $\log_2(k)$ qubits, but in practice, state preparation complexity depends on the method. Classical-to-quantum encoding may require $\mathcal{O}(k)$ gate operations approximate or parametrized state preparation circuits can reduce gate depth but may introduce encoding noise. For small k (e.g., $k \leq 16$), encoding is feasible on Noisy Intermediate-Scale Quantum (NISQ devices), with potential for simulation on quantum emulators. So, for low-dimensional shape configurations, such as those common in biological morphometrics and medical imaging, encoding costs are therefore tractable.

4.2 Circuit Architecture and Complexity

The proposed variational quantum classifier (VQC) architecture operates on encoded shape states with several practical considerations. First, the qubit requirement scales logarithmically with the number of landmarks k, needing only $\lceil \log_2(k) \rceil$ qubits; for example, with $k = 8$, just 3 qubits suffice, making it compatible with current quantum hardware. Second, to maintain trainability, moderate-depth ansätze such as hardware-efficient circuits or layered entanglers are employed. Third, while VQCs can suffer from barren plateaus—regions of vanishing gradients that hinder training, especially as circuit depth or input dimensionality increases, the use of structured and geometrically meaningful embeddings, as proposed here, may help mitigate this challenge [21]. Together, these features render the classifier well-suited for prototype implementations on today's quantum devices or simulators.

4.3   Classical vs Quantum Tradeoffs

When considering the integration of quantum models into shape analysis pipelines, it is essential to assess the tradeoffs between classical and quantum approaches across several dimensions. Table 1 summarizes key differences between classical and quantum paradigms for shape analysis. It contrasts their respective strengths in terms of expressivity, computation, generalization in low-data regimes, and compatibility with geometric invariants. Notably, while classical models are well-supported by existing toolchains and scalable architectures, quantum models offer the potential for superior generalization and richer geometric encodings but at the cost of current hardware limitations and algorithmic maturity.

Table 1. Key differences between classical and quantum paradigms for shape analysis.

Criteria	Classical (tangent space methods)	Proposed Quantum Approach
Geometry handling	Linearization introduces approximation	Intrinsic via Hilbert space embedding
Representation	Vectors in tangent space	Rays in complex projective Hilbert space
Scalability	Challenging in high dimensions	Requires qubit-efficient encoding
Interpretability	Established via geodesics and PCA	Requires definition of quantum observables

5   Perpectives and Use Cases

5.1   Perspectives

Leveraging quantum machine learning on Kendall shape spaces opens new avenues for both foundational and applied research. The alignment between Kendall's shape geometry and quantum Hilbert spaces invites exploration of quantum Riemannian geometry, quantum analogues of Procrustes distances, and quantum principal geodesic analysis. Extending fidelity-based kernels to task-specific quantum kernels informed by shape classes or anatomical priors may enhance model expressivity. Quantum generative models, such as quantum GANs or Born machines, could enable shape synthesis with uncertainty quantification and augmentation on shape manifolds.

5.2   Use Cases

This quantum shape learning framework applies broadly where geometric structure is key. In medical imaging and morphometry, it supports shape-based classification from limited data, improving analysis of organ morphology and surgical outcomes. In biomechanics and anthropometry, it enhances robustness in classifying body shapes and gait patterns from noisy landmarks. Robotics benefits via improved shape understanding for grasping and perception tasks. Archaeology and evolutionary biology gain powerful tools for clustering fossil morphologies, reconstructing phylogenies, and restoring fragmented artifacts through invariant quantum shape embeddings.

Conclusion

In this work, we present a theoretical framework integrating quantum machine learning with Kendall shape analysis to develop geometric learning algorithms on shape manifolds. By embedding shape configurations from Kendall's shape space into quantum Hilbert spaces, our approach enables quantum-enhanced shape classification, similarity assessment, and generative modeling. Although theoretical, we discuss feasibility, outline implementation strategies, and illustrate applications. Future work will focus on experimental benchmarking of this framework against classical methods. Overall, this proposal lays the foundation for models that combine geometric invariants with quantum advantages in representation and generalization.

References

1. Biamonte, J., Wittek, P., Pancotti, N., Rebentrost, P., Wiebe, N., Lloyd, S.: Quantum machine learning. Nature **549**(7671), 195–202 (2017). https://doi.org/10.1038/nature23474
2. Schuld, M., Petruccione, F.: Supervised Learning with Quantum Computers. Springer, Cham (2018). https://doi.org/10.1007/978-3-319-96424-9
3. Schuld, M., Bocharov, A., Svore, K.M., Wiebe, N.: Circuit-centric quantum classifiers. Phys. Rev. A **101**(3), 032308 (2021). https://doi.org/10.1103/PhysRevA.101.032308
4. Havlíček, V., Córcoles, A.D., Temme, K., et al.: Supervised learning with quantum-enhanced feature spaces. Nature **567**(7747), 209–212 (2019). https://doi.org/10.1038/s41586-019-0980-2

5. Lloyd, S., Mohseni, M., Rebentrost, P.: Quantum principal component analysis. Nat. Phys. **10**(9), 631–633 (2014). https://doi.org/10.1038/nphys3029

6. Kendall, D.G.: Shape manifolds, Procrustean metrics, and complex projective spaces. Bull. Lond. Math. Soc. **16**(2), 81–121 (1984). https://doi.org/10.1112/blms/16.2.81

7. Dryden, I. L., Mardia, K. V.: Statistical Shape Analysis with Applications in R. Wiley, Chichester (2016). https://doi.org/10.1002/9781119072492

8. Bronstein, M.M., Bruna, J., LeCun, Y., Szlam, A., Vandergheynst, P.: Geometric deep learning: going beyond Euclidean data. IEEE Signal Process. Mag. **34**(4), 18–42 (2017). https://doi.org/10.1109/MSP.2017.2693418

9. Pennec, X.: Intrinsic statistics on Riemannian manifolds: Basic tools for geometric measurements. J. Math. Imaging Vision **65**(1), 81–110 (2018). https://doi.org/10.1007/s10851-006-6228-4

10. Friji, R., Drira, H., Chaieb, F., Kchok, H., Kurtek, S.: Geometric deep neural network using rigid and non-rigid transformations for human action recognition. In: Proceedings of the IEEE/CVF International Conference on Computer Vision (ICCV) 2021, pp. 12611–12620. IEEE, Piscataway (2021). https://doi.org/10.1109/TPAMI.2023.3291663

11. Hosni, N., Ben Amor, B.: A geometric convnet on 3D shape manifold for gait recognition. In: Proceedings of the IEEE/CVF Conference on Computer Vision and Pattern Recognition Workshops, pp. 0–0. IEEE, Piscataway (2020). https://doi.org/10.1109/CVPRW50498.2020.00434

12. Larkin, J., Sornborger, A. T., Cao, Y.: Quantum manifold learning and geometry estimation. arXiv:2301.05707 (2023). https://arxiv.org/abs/2301.05707

13. Gyurik, C., van Vreumingen, D., Dunjko, V.: Structural risk minimization for quantum linear classifiers. In: Quantum 7, 893. Quantum, Vienna (2023). https://doi.org/10.22331/q-2023-01-13-893

14. Le, H., Kendall, D.G.: The Riemannian structure of Euclidean shape spaces: a novel environment for statistics. Ann. Stat. **21**(3), 1225–1271 (1993). https://doi.org/10.1214/aos/1176349259

15. Barbaresco, F.: Information Geometry of Covariance Matrix: Cartan-Siegel Homogeneous Bounded Domains, Mostow/Berger Fibration and Frechet Median. In: Nielsen, F., Bhatia, R. (eds.) Matrix Information Geometry, pp. 199–255. Springer, Berlin (2011). https://doi.org/10.1007/978-3-642-30232-9_9

16. Huckemann, S., Hotz, T., Munk, A.: Intrinsic shape analysis: geodesic PCA for Riemannian manifolds modulo isometric Lie group actions. Stat. Sin. **20**(1), 1–100 (2010)

17. Srivastava, A., Klassen, E.: Functional and Shape Data Analysis. Springer, Cham (2016). https://doi.org/10.1007/978-1-4939-4020-2

18. Sommer, S., Lauze, F., Nielsen, M.: Optimization over geodesics for exact principal geodesic analysis. In: Scale Space and Variational Methods in Computer Vision, LNCS, vol. 6667, pp. 110–121. Springer, Cham (2011). https://doi.org/10.1007/s10444-013-9308-1

19. Schuld, M., Bocharov, A., Svore, K.M., Wiebe, N.: Circuit-centric quantum classifiers. Phys. Rev. A **101**(3), 032308 (2021). https://doi.org/10.1103/PhysRevA.101.032308

20. Benedetti, M., Lloyd, E., Sack, S., Fiorentini, M.: Parameterized quantum circuits as machine learning models. Quan. Sci. Technol. **4**(4), 043001 (2019). https://doi.org/10.1088/2058-9565/ab4eb5

21. McClean, J.R., Boixo, S., Smelyanskiy, V.N., Babbush, R., Neven, H.: Barren plateaus in quantum neural network training landscapes. Nat. Commun. **9**(1), 4812 (2018). https://doi.org/10.1038/s41467-018-07090-4
22. Kent, J.T.: The Fisher-Bingham distribution on the sphere. J. Roy. Stat. Soc.: Ser. B (Methodol.) **44**(1), 71–80 (1982). https://doi.org/10.1111/j.2517-6161.1982.tb01189.x
23. M. A. Nielsen and I. L. Chuang, Quantum Computation and Quantum Information, Cambridge University Press, (2010). (Chapter 9, Exercise 9.10 describes the SWAP test)

Quantum-Enhanced Consensus Clustering Through Quantum Annealing and QAOA

Daniele Franch[1(✉)] , Rui Wang[1,2] , Amer Delilbasic[1,2] ,
Kristel Michielsen[2,3,4] , and Gabriele Cavallaro[1,2,4]

[1] University of Iceland, Reykjavík 102, Iceland
`daf9@hi.is`
[2] Forschungszentrum Jülich, Jülich 52428, Germany
[3] RWTH Aachen University, Aachen 52056, Germany
[4] AIDAS, Jülich 52428, Germany

Abstract. Consensus clustering integrates multiple partitions into a unified solution, enhancing stability across clustering algorithms. However, ensuring global consistency remains computationally demanding. Quadratic Unconstrained Binary Optimization (QUBO) formulations provide a natural framework for this task, making consensus clustering amenable to quantum optimization methods such as quantum annealing and the Quantum Approximate Optimization Algorithm (QAOA).

We extend the established QUBO formulation for correlation clustering with a refinement protocol that allows for improvement of suboptimal solutions. To address current quantum hardware limitations, we implement a subproblem decomposition strategy that iteratively resolves constraint violations on manageable subsets.

Validation on benchmark and real-world datasets demonstrates competitive accuracy compared to established methods.

Keywords: Consensus Clustering · QUBO · Quantum Annealing · QAOA

1 Introduction

Consensus clustering, also known as cluster ensemble or clustering aggregation, is a widely used approach to improve the robustness and stability of unsupervised learning by combining multiple partitions into a single consensus solution [1–3]. Applications range from bioinformatics [4] to document analysis [5], where heterogeneous algorithms often produce divergent results. Many formulations of the problem are combinatorial, and several important ones, such as those based on correlation clustering [6,7], are NP-hard. Classical heuristics can produce reasonable results, but they scale poorly and typically lack performance guarantees.

A natural way to capture consensus clustering is QUBO formulations. These formulations are directly compatible with quantum optimization techniques such as quantum annealing (QA) [8] and QAOA [9]. However, their cubic growth in

F. Barbaresco and F. Gerin (Eds.): QUEST-IS 2025, CCIS 2744, pp. 105–116, 2026.
https://doi.org/10.1007/978-3-032-13855-2_10

transitivity constraints limits scalability on current hardware, and approximate solutions produced by noisy or resource-limited solvers often violate consistency.

In this work, we build on the established QUBO formulation for consensus clustering [7], which encodes pairwise agreement and transitivity constraints in a quadratic framework. Our main contribution is a refinement protocol with monotonicity guarantees that ensures approximate solutions can only improve or remain unchanged, never deteriorate. To address hardware limitations, we employ subproblem decomposition that restricts optimization to subsets of variables involved in constraint violations, iteratively updating until consistency is achieved.

2   QUBO Formulation

We build on the standard QUBO formulation for consensus clustering [7], where clustering is encoded as a pairwise decision problem. Given a dataset of n elements and m base clusterizations, we define s_{ij} and d_{ij} as the counts of how often elements i and j appear in the same or different clusters, respectively, and set $b_{ij} = d_{ij} - s_{ij}$ to capture their disagreement. Consider the upper triangular matrix $X = (x_{ij})$, where x_{ij} denotes a binary variable equal to 1 if i and j are assigned to the same cluster in the consensus, and 0 otherwise. The initial objective function is:

$$H_{\text{pair}}(X) = \sum_{i<j} b_{ij}\, x_{ij}, \tag{1}$$

which favors agreement with the base clusterings.

To ensure consistency in the clustering assignment, we must enforce transitivity: if i and j are clustered together, and j and k are clustered together, then i and k should also be clustered. Violations occur then when $(x_{ij}, x_{ik}, x_{jk}) \in \{(0,1,1),(1,0,1),(1,1,0)\}$. These infeasible triplets are penalized through the introduction of two auxiliary slack variables $s_1^{(i,j,k)}, s_2^{(i,j,k)} \in \{0,1\}$ for each triplet (i,j,k). The penalties are defined as:

$$P_1^{(i,j,k)} = p_1\left(x_{ij} + x_{ik} + x_{jk} + s_1^{(i,j,k)} - 2s_2^{(i,j,k)} - 1\right)^2, \tag{2}$$

$$P_2^{(i,j,k)} = p_2\, s_1^{(i,j,k)} s_2^{(i,j,k)}, \tag{3}$$

with $p_2 > p_1 > t$, where t is chosen large enough to make violations prohibitively costly. The complete QUBO objective is therefore:

$$H(X) = \sum_{i<j} b_{ij}\, x_{ij} + \sum_{i<j<k} \left[P_1^{(i,j,k)} + P_2^{(i,j,k)}\right]. \tag{4}$$

Solutions with no violations correspond to valid clusterings. The formulation scales quadratically with the number of elements and does not require setting the final number of clusters in advance.

2.1   Refinement of Approximate Solutions

Approximate solutions obtained from noisy or resource-limited solvers, or even classical heuristics, may satisfy feasibility constraints but still be suboptimal with respect to the linear objective. To enhance such solutions, we consider the following refinement procedure.

Let $x^{\mathrm{app}} = \{x_{ij}^{\mathrm{app}}\}$ be a feasible approximate assignment (or clustering). We define the *disagreement set*

$$\mathcal{D}(x^{\mathrm{app}}) = \{(i,j) : \mathrm{sign}(b_{ij}) = \mathrm{sign}(2x_{ij}^{\mathrm{app}} - 1)\},$$

which contains the indices where the sign of the coefficients b_{ij} opposes the current assignment x_{ij}^{app}. Equivalently, this set identifies where x^{app} differs from the unconstrained solution $x^{\mathrm{unp}} = \Theta(-B)$, where Θ is the Heaviside step function, obtained by ignoring feasibility constraints.

Given any non-empty subset $S \subseteq \mathcal{D}(x^{\mathrm{app}})$, we define a modified coefficient matrix:

$$b_{ij}^* = \begin{cases} -b_{ij}, & (i,j) \in S, \\ b_{ij}, & \text{otherwise}, \end{cases}$$

denoted collectively as $B^* = \{b_{ij}^*\}$. Let x^{imp} be an optimal solution under the modified coefficients B^*. This procedure encourages the solution to the problem with the modified matrix B^* to align with x^{app}.

We now demonstrate that this refinement is monotonic: the original objective cannot worsen.

Proof. Let x^{imp} be an optimal solution for the problem with the modified matrix B^*. By optimality, we have:

$$\sum_{i<j} b_{ij}^* x_{ij}^{\mathrm{imp}} \leq \sum_{i<j} b_{ij}^* x_{ij}^{\mathrm{app}}. \tag{5}$$

We can then express b_{ij}^* in terms of b_{ij} as $b_{ij}^* = b_{ij} - 2\mathbf{1}_{(i,j)\in S} b_{ij}$, where $\mathbf{1}_{(i,j)\in S}$ is the indicator function. Substituting this expression into Eq. 5 and rearranging the terms, we obtain:

$$\sum_{i<j} b_{ij}(x_{ij}^{\mathrm{app}} - x_{ij}^{\mathrm{imp}}) \geq 2 \sum_{(i,j)\in S} b_{ij}(x_{ij}^{\mathrm{app}} - x_{ij}^{\mathrm{imp}}). \tag{6}$$

We can now rewrite the right-hand side by adding and subtracting the unconstrained solution x^{unp}:

$$2 \sum_{(i,j)\in S} b_{ij}(x_{ij}^{\mathrm{app}} - x_{ij}^{\mathrm{imp}}) = 2 \sum_{(i,j)\in S} b_{ij}\left[(x_{ij}^{\mathrm{app}} - x_{ij}^{\mathrm{unp}}) + (x_{ij}^{\mathrm{unp}} - x_{ij}^{\mathrm{imp}})\right]. \tag{7}$$

By definition, the unconstrained solution takes the value 0 if $b_{ij} > 0$ and 1 if $b_{ij} < 0$ and since $(i,j) \in S$ implies $x_{ij}^{\mathrm{app}} \neq x_{ij}^{\mathrm{unp}}$, the difference $x_{ij}^{\mathrm{app}} - x_{ij}^{\mathrm{unp}}$ is either $+1$ or -1.

Considering both cases, for all $(i, j) \in S$ we have

$$b_{ij}(x_{ij}^{\text{app}} - x_{ij}^{\text{unp}}) = \begin{cases} b_{ij}, & b_{ij} > 0, \\ -b_{ij}, & b_{ij} < 0, \end{cases} = |b_{ij}| \geq 0.$$

The second term on the right-hand side of Eq. 7 satisfies the trivial bound:

$$2 \sum_{(i,j) \in S} b_{ij}(x_{ij}^{\text{unp}} - x_{ij}^{\text{imp}}) \geq -2 \sum_{(i,j) \in S} |b_{ij}|,$$

because each summand is bounded below by $-|b_{ij}|$. Combining these bounds, we conclude that the right-hand side of (6) is non-negative:

$$2 \sum_{(i,j) \in S} b_{ij}(x_{ij}^{\text{app}} - x_{ij}^{\text{imp}}) \geq 0.$$

Substituting back into (6) we obtain:

$$\sum_{i<j} b_{ij}(x_{ij}^{\text{app}} - x_{ij}^{\text{imp}}) \geq 0,$$

or equivalently,

$$\sum_{i<j} b_{ij} x_{ij}^{\text{imp}} \leq \sum_{i<j} b_{ij} x_{ij}^{\text{app}}.$$

This shows that the refinement step never increases the original linear objective. The argument is general for any linear objective with an additive penalty that vanishes on feasible assignments.

3   Methodology

The empirical evaluation of our quantum-enhanced consensus clustering framework pursues three main objectives: first, to empirically validate the monotonicity property established in Sect. 2.1; second, to compare the QUBO-based consensus clustering against both individual clustering methods and established ensemble approaches; and third, to investigate scalability and refinement strategies when hardware limitations prevent solving the full problem.

3.1   Datasets and Experimental Setup

We selected two benchmark datasets for our experiments: the Iris dataset and the Wine dataset, both accessed through scikit-learn [10]. Additionally, we validated our approach on the reduced EuroSAT satellite imagery dataset [11,12] to demonstrate scalability to real-world scenarios. The Iris dataset was used both in its full version and in reduced subsets of 10, 15, 20, and 25 samples where the latter are used for empirically demonstrating the refinement property. The datasets are small enough to allow evaluation under classical consensus clustering frameworks while providing meaningful validation of our approach.

3.2   Base Clustering Ensemble Construction

To construct diverse ensembles of base clusterings, we employed heterogeneous algorithms representing distinct clustering paradigms. For the benchmark datasets (Iris and Wine), we used centroid-based partitions via K-Means [13] with $k \in \{2, 3, 4\}$, hierarchical partitions via Agglomerative Clustering [14] with $k \in \{2, 3, 4\}$, and graph-based partitions via Spectral Clustering [15] for $k \in \{2, 3, 4\}$. Density-based partitions with DBSCAN [16] using neighborhood radius $\varepsilon \in \{0.3, 0.5, 0.7\}$ were additionally included for the Iris dataset. For the EuroSAT dataset, we employed K-Means with $k \in \{5, 8, 14\}$ using different random seeds, Agglomerative Clustering with $k = 10$, and Spectral Clustering with nearest neighbors affinity ($k = 10$).

3.3   Evaluation Metrics and Baseline Comparisons

We assessed clustering quality using both external and internal evaluation metrics. External indices measure agreement with ground-truth labels and include the Adjusted Rand Index (ARI) [17] and Normalized Mutual Information (NMI) [1]. The Silhouette coefficient [18] serves as our internal evaluation metric, assessing cluster quality independently of ground-truth labels. Additionally, we employ the Mean ARI metric, computed as the average ARI between the final consensus clustering and each of the m base clustering methods, to quantify how well the consensus solution captures agreement across all input methods.

We compared the QUBO-based approach against several established consensus clustering methods. For the Cluster-based Similarity Partitioning Algorithm (CSPA) [1] with METIS and the Meta-CLustering Algorithm (MCLA) [1], we used implementations from the consensus clustering package available at [19]. Additionally, we implemented agglomerative and spectral clustering variants of CSPA, Non-negative Matrix Factorization (NMF)-based consensus clustering [20], and the Da-sm and Da-cr methods proposed by Cohen et al. [21].

3.4   QUBO Solution Methods

We solved the QUBO formulation using multiple optimization approaches. For Quantum Annealing, we used D-Wave's Advantage_system4.1 with 3500 reads, keeping all other parameters at their default values. As a classical baseline, we employed Simulated Annealing [22,23]. Solutions obtained with both optimization approaches were identical on our benchmark problems.

To test the QUBO on gate-model quantum hardware, we implemented a QAOA solver in Qiskit. We instantiate a QAOA Ansatz directly from the QUBO cost operator and sweep the circuit depth $p \in \{1, \ldots, 5\}$. Parameters are trained with a multi-start SPSA routine: each batch/round starts from the previous angles and adds at least one random seed (60âĂŞ100 iterations per start), retaining the best expectation. Execution uses Aer in two modes: (i) noiseless `AerSimulator` and (ii) noisy (Aer with the `FakeTorino` noise model), with 4,096 shots unless stated otherwise.

3.5   Scalability and Refinement Protocol

To address hardware scalability limitations, we adopted a submatrix decomposition strategy combined with iterative refinement. The process works as follows: starting from the global bias matrix, we identify transitivity violations across all element triplets. We then extract a subset of 15 elements, chosen as a safe bound for the quantum annealer's capacity, focusing on elements involved in violations.

The subproblem is solved iteratively, as in [24]: violations within the submatrix are detected, penalized, and the problem is resolved until no violations remain in the subset. The resulting assignment updates the global bias matrix by flipping the sign of entries where the submatrix solution conflicts with the global configuration. We repeat this process until convergence, yielding the final solution $\mathbf{x} = \Theta(-\mathbf{B})$. For QAOA, we mirror this refinement: in each round, we partition violating constraints into disjoint batches (5 qubits per batch) that fit a fixed qubit budget, run QAOA per batch, and accept a batch only if its local cost decreases or its violation count drops until no batch is accepted or all violations are cleared.

4   Results

This section presents the empirical evaluation of the QUBO consensus clustering framework across benchmark and real-world datasets. The QUBO-based approach was tested using both quantum annealing and simulated annealing, with both methods producing identical solutions on benchmark datasets. Additionally, QAOA was evaluated specifically on the EuroSAT dataset to demonstrate the approach's compatibility with gate-model quantum hardware.

4.1   Empirical Validation of the Refinement Protocol

To provide empirical evidence that the refinement of an approximate solution is feasible, we constructed four reduced versions of the Iris dataset containing 10, 15, 20, and 25 elements, respectively. For each dataset, four feasible but approximate solutions (i.e., valid assignments without transitivity violations) were generated. For these cases, we applied the refinement strategy described in Sect. 2.1. As shown in Fig. 1, across all datasets every point lies on or below the line $E_i = E_f$, empirically confirming that the refinement procedure may improve but never worsens the energy and may improve the quality of approximate solutions. Many points lie on the diagonal $E_i = E_f$, showing no improvement. This likely occurs because $|D \setminus S|$ is too small, leaving insufficient disagreement pairs available for refinement.

The main purpose of refinement is to simplify the problem by reducing constraint violations, deliberately settling for approximate solutions rather than pursuing optimality. Despite the lack of theoretical guarantees when using subproblem decomposition, we tested this approach on the full Iris dataset to evaluate its practical effectiveness on problems exceeding hardware limits.

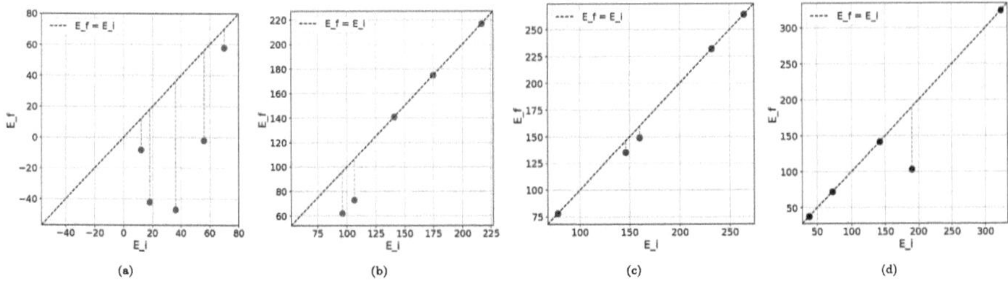

Fig. 1. Final energy versus initial energy for the four reduced Iris datasets: 10 elements (Fig. 1a), 15 elements (Fig. 1b), 20 elements (Fig. 1c), and 25 elements (Fig. 1d). Each point represents an approximate solution refined using the proposed protocol. All points lie on or below the diagonal $E_i = E_f$, showing that the refinement procedure never increases energy and can improve solution quality.

For this experiment, we considered two different starting solutions: the true labels and a random solution close to the optimal one. In both cases, we tested configurations where 50, 100, 200, and 300 pairs from the disagreement set are left unchanged, i.e., pairs in $D \setminus S$, with each larger set containing all pairs from the smaller configurations. This setup allows us to analyze how the number of fixed pairs influences the refinement process and simplifies the problem. The refinement results for both starting solutions are reported in Table 1. Overall, the final energy generally improves as more pairs are left unchanged, though some fluctuations are observed. Starting from the true labels, the procedure successfully reduces the number of subproblems to roughly one-third of those required by the full solution, while a substantial gap in solution quality compared to the complete solution remains. When starting from the threshold-based solution, the refinement behavior is more variable. Notably, for $|D \setminus S| = 50$, the final energy is worse than the initial solution, likely due to the decomposition into subproblems. This suggests that, although the method can still be effective, its performance is not guaranteed in all instances.

Table 1. Refinement results for different starting solutions and numbers of unchanged pairs. E_f denotes the final energy after refinement.

	True Labels			Near-optimal solution		
$|D \setminus S|$	E_f	Violations	Subproblems	E_f	Violations	Subproblems
0	-24101	0	0	-28024	0	0
50	-24101	1798	12	-28002	4310	28
100	-24452	3002	45	-28656	7976	43
200	-24439	4336	39	-28896	14300	105
300	-24624	4274	29	-28024	18102	98
$|D|$	-28963	26452	121	-28963	26452	121

In general, the refinement method's effectiveness depends both on the quality of the initial solution and on the number of pairs left unchanged. It proves more beneficial when starting farther from the optimal solution, while small numbers of fixed pairs offer limited computational savings. The approach remains restricted to smaller datasets until hardware improvements allow scaling to larger problems.

4.2   Comparison with Consensus Clustering Methods

We benchmarked our approach against representative consensus clustering methods as described in Sect. 3.3. For methods requiring the number of clusters k as input, we set k equal to the ground-truth number of clusters. Evaluation metrics include standard external indices (ARI, NMI), the Silhouette score, and the QUBO cost function Eq. (4).

Table 2. Comparison of consensus clustering methods on Iris and Wine datasets.

Dataset	Method	Mean ARI	ARI	NMI	Silhouette	QUBO objective
Iris	CSPA	0.715	0.688	0.705	0.501	-26508
	CSPA Agglomerative	0.689	0.731	0.770	0.554	-30748
	CSPA Spectral	0.689	0.731	0.770	0.554	-30748
	NMF	0.595	0.470	0.661	0.405	-21100
	MCLA	0.759	0.758	0.786	0.542	-30320
	Da-sm	0.688	0.745	0.778	0.553	-30690
	Da-cr	0.689	0.731	0.770	0.554	-30748
	QUBO	0.689	0.731	0.770	0.554	-30748
Wine	CSPA	0.650	0.391	0.400	0.531	-28150
	CSPA Agglomerative	0.665	0.368	0.416	0.565	-34283
	CSPA Spectral	0.665	0.368	0.416	0.565	-34283
	NMF	0.665	0.368	0.416	0.565	-34283
	MCLA	0.703	0.359	0.420	0.557	-30162
	Da-sm	0.665	0.368	0.416	0.565	-34283
	Da-cr	0.665	0.368	0.416	0.565	-34283
	QUBO	0.665	0.368	0.416	0.565	-34283

Table 2 shows that our QUBO solver matches the best consensus baselines on both Iris and Wine. On Iris it achieves strong agreement with the base clusterers and competitive external metrics; on Wine, where methods perform more similarly, it remains on par across criteria. Notably, QUBO consistently attains the lowest value of its own objective, confirming alignment with the intended optimization. The main drawback is runtime from solving many subproblems, which requires multiple iterations. This efficiencyâĂŞoptimality trade-off should lessen as quantum hardware permits larger subproblems, improving scalability and applicability.

4.3   EuroSAT Dataset Validation

We evaluated our approach on a subset of 50 EuroSAT satellite images, which include 10 land-use classes, using feature embeddings from a pre-trained ResNet18 [25].

Noise Robustness of the Batched QAOA. To evaluate the performance under realistic noise conditions, we assess the noise robustness of the batched QAOA by comparing simulations with a realistic gate-level noise model (`FakeTorino`) to an ideal noiseless simulator (`Aer`) for circuit depths $p \in \{1, \ldots, 5\}$. For each batch, we recorded the batch Hamiltonian expectation value E_{batch}, the local energy change ΔE_{local}, and the noise-induced energy inflation, which was quantified as $\Delta E_{\text{noise}} = E_{\text{batch}}^{\text{Torino}} - E_{\text{batch}}^{\text{Noiseless}}$, calculated for each batch at a fixed circuit depth. Remarkably, the acceptance rate for local updates remained identical (0.625) for both the noisy and noiseless simulations across all depths. This stability arises because, as shown in Fig. 2(a), the noise

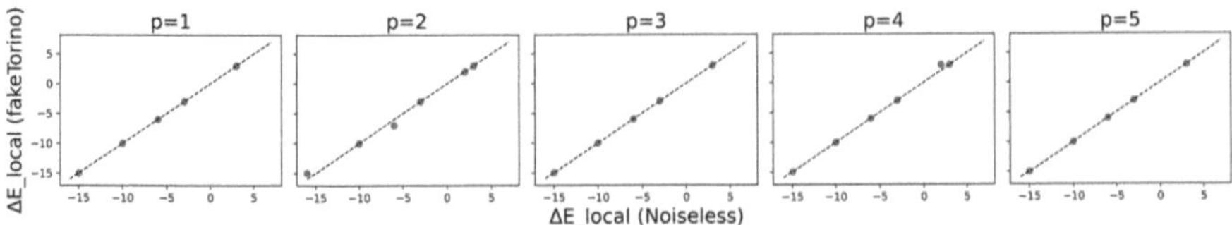

(a) **Local decision stability.** Comparison of ΔE_{local} on the `FakeTorino` backend versus the noiseless simulator for $p = 1, \ldots, 5$.

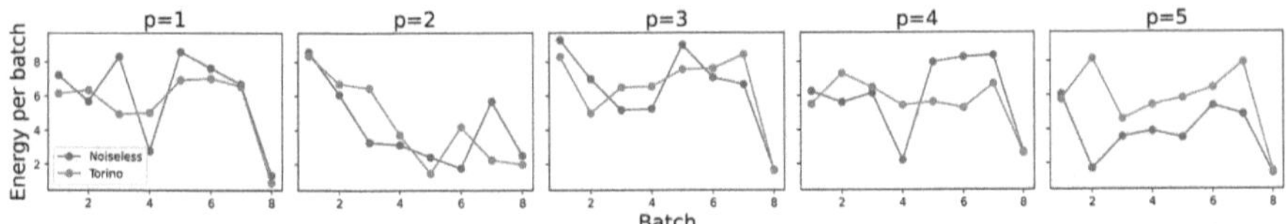

(b) **Batch energy trajectories.** Evolution of the per-batch E_{hat} for the noiseless (blue) and `FakeTorino` (orange) simulations. The curves track closely across batches and depths, with deviations increasing slightly for $p = 5$.

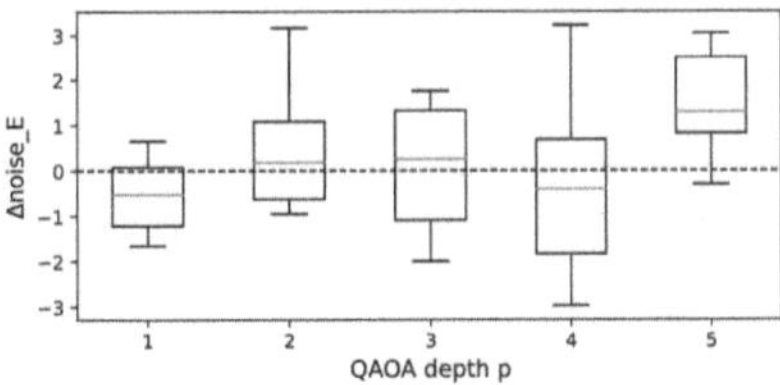

(c) **Noise-induced energy inflation.** Boxplots of the energy difference ΔE_{noise} by depth.

Fig. 2. Impact of simulated gate-level noise on the batched QAOA refinement. The analysis shows that acceptance statistics are close on both backends, as the local energy-based decisions are robust to the effects of noise from the `FakeTorino` model.

model preserves the sign and relative ordering of ΔE_{local}. The energy expectation trajectories in Fig. 2(b) track each other closely, with minor deviations increasing with circuit depth. The resulting energy inflation, ΔE_{noise}, is centered near zero for $p \leq 4$ and exhibits a slight positive shift at $p = 5$ that is insufficient to alter acceptance behavior (Fig. 2(c)). In summary, our numerical simulations demonstrate that for the problem scale and batching scheme considered, the local decisions within the batched QAOA framework are robust to a realistic gate-level noise model. The `FakeTorino` backend introduces a modest, depth-dependent energy inflation that is insufficient to alter the algorithm's acceptance mechanism. However, the listed results are only for one device model and problem size; deeper circuits, fewer shots, or different mixers could increase inflation. Real hardware may add calibration that the FakeTorino model does not capture. Therefore, exploring hardware execution and depth-aware optimizers remains future work.

Clustering Performance on EuroSAT. Table 3 summarizes the clustering performance for classical and quantum-based methods on the 50-image EuroSAT subset. Metrics include the Adjusted Rand Index (ARI), Normalized Mutual Information (NMI), and Silhouette score. For QUBO, solutions were obtained via quantum annealing, while for QAOA, results correspond to batched QAOA refinement with circuit depth $p = 2$.

Table 3. Clustering performance on the EuroSAT subset using classical and quantum-based methods.

Method	ARI	Mean ARI
NMF	0.393	0.654
Agglomerative	0.548	0.731
Spectral	0.393	0.656
QUBO (QA)	0.552	0.727
Batched QAOA	0.553	0.729

The performances of batched QAOA and QUBO are comparable, whereas classical methods perform at a similar or lower level, highlighting the effectiveness of quantum-based optimization for clustering on the EuroSAT subset.

5   Conclusions

This study extends earlier QUBO formulations [7] by introducing a refinement procedure that can improve approximate clustering solutions. While decomposition into subproblems is necessary due to current hardware limits and does not guarantee convergence to the global optimum, the workflow effectively combines quantum exploration with classical refinement.

Experiments show that QUBO-based clustering performs comparably to classical ensemble methods, with refinement ensuring transitivity. Similarly, batched QAOA demonstrates robust local decision-making even under realistic gate-level noise, with modest energy inflation insufficient to alter acceptance behavior. These results indicate that quantum solvers can contribute to clustering pipelines.

Future work will explore alternative penalty formulations, larger datasets, and hardware-specific optimizations, including noise-aware strategies for QAOA.

Acknowledgments. This work is supported by the Icelandic Centre for Research, project *"Hybrid Quantum-Classical Workflows for EO"*, RANNÍS grant number: 2511078-051 (http://en.rannis.is/).

Disclosure of Interests. The authors have no competing interests to declare that are relevant to the content of this article.

References

1. Strehl, A., Ghosh, J.: Cluster ensembles – a knowledge reuse framework for combining multiple partitions. J. Mach. Learn. Res. **3**, 583–617 (2002)
2. Vega-Pons, S., Shulcloper, J.: A survey of clustering ensemble algorithms. Int. J. Pattern Recogn. Artif. Intell. **25**, 337–372 (2011)
3. Meng, F., Tong, X., Wang, Z.: A clustering-ensemble approach based on voting. In: Deng, H., Miao, D., Lei, J., Wang, F.L., eds. In: Artificial Intelligence and Computational Intelligence, pages 421–427, Berlin, Heidelberg, (2011). Springer Berlin Heidelberg
4. Monti, S., Tamayo, P., Mesirov, J., Golub, T.: Consensus clustering: a resampling-based method for class discovery and visualization of gene expression microarray data. Mach. Learn. **52**, 91–118 (2003)
5. Brodley, C.: Solving cluster ensemble problems by bipartite graph partitioning. (2004)
6. Bansal, N., Blum, A., Chawla, S.: Correlation clustering. Mach. Learn. **56**, 89–113 (2004)
7. Charikar, M., Guruswami, V., Wirth, A.: Clustering with qualitative information. J. Comput. Syst. Sci. **71**(3), 360–383 (2005). Learning Theory 2003
8. Finnila, A.B., Gomez, M.A., Sebenik, C., Stenson, C., Doll, J.D.: Quantum annealing: a new method for minimizing multidimensional functions. Chem. Phy. Lett. **219**(5-6), 343–348 (1994)
9. Farhi, E., Goldstone, J., Gutmann, S.: A quantum approximate optimization algorithm (2014)
10. Pedregosa, F., et al.: Scikit-learn: Machine learning in python. J. Mach. Learn. Res. **12**, 2825–2830 (2011)
11. Helber, P., Bischke, B., Dengel, A., Borth, D.: Introducing eurosat: A novel dataset and deep learning benchmark for land use and land cover classification. In: IGARSS 2018-2018 IEEE International Geoscience and Remote Sensing Symposium, pages 204–207. IEEE (2018)
12. Helber, P., Bischke, B., Dengel, A., Borth, D.: Eurosat: A novel dataset and deep learning benchmark for land use and land cover classification. IEEE J. Sel. Top. Appl. Earth Observ. Rem. Sens. (2019)

13. MacQueen, J.: Multivariate observations. In: Proceedings of the 5th Berkeley Symposium on Mathematical Statistics and Probability, volume 1, pages 281–297 (1967)
14. Ward Jr, J.H.: Hierarchical grouping to optimize an objective function. J. American Stat. Assoc. **58**(301), 236–244 (1963)
15. Ng, A., Jordan, M., Weiss, Y.: On spectral clustering: Analysis and an algorithm. In: Dietterich, T., Becker, S., Ghahramani, Z., eds. Adv. Neural Inf. Process. Syst. **14**. MIT Press (2001)
16. Ester, M., Kriegel, H.-P., Sander, J., Xu, X.: A density-based algorithm for discovering clusters in large spatial databases with noise. In: Knowledge Discovery and Data Mining (1996)
17. Hubert, L.J., Arabie, P.: Comparing partitions. J. Classif. **2**, 193–218 (1985)
18. Rousseeuw, P.J.: Silhouettes: a graphical aid to the interpretation and validation of cluster analysis. J. Comput. Appl. Math. **20**, 53–65 (1987)
19. Giecold, G.: Cluster_ensembles (2025). https://github.com/GGiecold-zz/Cluster_Ensembles. 2016. Accessed: September
20. Lee, D., Seung, H.S.: Algorithms for non-negative matrix factorization. In: Leen, T., Dietterich, T., Tresp, V., eds. Adv. Neural Inf. Process. Syst. **13** MIT Press (2000)
21. Cohen, E., Mandal, A., Ushijima-Mwesigwa, H., Roy, A.: Ising-based consensus clustering on specialized hardware. In: Berthold, M.R., Feelders, A., Krempl, G. (eds.) IDA 2020. LNCS, vol. 12080, pp. 106–118. Springer, Cham (2020). https://doi.org/10.1007/978-3-030-44584-3_9
22. Kirkpatrick, S.: C Daniel Gelatt Jr, and Mario P Vecchi. Optimization by simulated annealing. Science **220**(4598), 671–680 (1983)
23. D-Wave Systems Inc. neal: Simulated annealing sampler for dimod. GitHub repository, (2021). Implementation of simulated annealing sampler for general Ising model graphs; used with dimod—see Simulated Annealing Sampler
24. Franch, D., Zardini, E., Blanzieri, E., Pastorello, D.: Consensus ranking by quantum annealing (2025)
25. He, K., Zhang, X., Ren, S., Sun, J.: Deep residual learning for image recognition (2015)

Hybrid Quantum Classical Algorithms: A Cloud On-Demand Viewpoint

Aleksander Wennersteen$^{(\boxtimes)}$ [iD], Kemal Bidzhiev, Mauro D'Arcangelo, Matthieu Moreau [iD], Anton Quelle, Alexandre Dauphin [iD], and Mourad Beji [iD]

Pasqal, 24 Av. Emile Baudot, Palaiseau 91120, France
{aleksander.wennersteen,mourad.beji}@pasqal.com

Abstract. In the last decade, advances in quantum technologies have allowed for the rapid development of industrialized quantum processing units. These new devices exploit the laws of quantum mechanics to perform complex calculations. Quantum processing units require new ways of thinking and programming. In particular, these new algorithms will be hybrid, with part of the computation performed on classical high-performance computing hardware and part on the dedicated quantum hardware. At Pasqal, we have developed a cloud platform hosting a neutral atom quantum processing unit (QPU) operating in the analog paradigm and a series of hybrid quantum classical algorithms that cover applications such as quantum optimization, quantum machine learning and quantum simulation. In this paper, we will show how this platform is used during the execution of real workloads.

1 Introduction

Neutral atom based quantum computers are gaining prominence as a leading Quantum Processing Unit (QPU) modality due to their scalability, programmability, and robustness. At Pasqal, we have developed and deployed cloud-accessible[1] QPUs based on neutral atom architectures that operate in the analog mode [11]. However, a purely quantum computational paradigm is insufficient. This has led to the emergence of hybrid quantum-classical algorithms as a practical and scalable approach.

Hybrid quantum-classical algorithms leverage the strengths of classical high-performance computing (HPC) systems alongside quantum processors. Moreover, integrating QPU and emulator backends through cloud access, offering the choice of backend to the user, allows efficient workflows during the whole development and execution process.

In this paper, we present how Pasqal's cloud platform supports this hybrid paradigm, providing an integrated architecture for algorithm design, benchmarking, and execution. The tools available on the cloud allow the user to:

- prepare sequences that can be sent to the QPU using both traditional and graphical programming SDKs in Pulser and Pulser Studio,

---

[1] Pasqal Cloud Service available at: https://portal.pasqal.cloud.

F. Barbaresco and F. Gerin (Eds.): QUEST-IS 2025, CCIS 2744, pp. 117–125, 2026.
https://doi.org/10.1007/978-3-032-13855-2_11

- benchmark the algorithms on system size from 50 to 100 qubits with a suite of state-of-the art emulators,
- benchmark the resilience of the algorithms to noise with the noisy emulators,
- Monitor your jobs through our all in one platform: Follow the progress of your jobs in real-time, download or visualize their results when they are ready, retry or cancel your jobs when needed, extract logs and add per-time-step observables for emulators to verify the simulations validity,
- easily switch between the emulators and the QPU backends.

After introducing the architecture and the tools, we show the usage of the cloud with a series of applications.

2   Platform Architecture

In this section, we discuss the full architecture of our platform. We describe briefly how the HPC cluster is integrated with cloud services to provide our quantum device emulation, and cloud infrastructure, including orchestration, scheduling, and dispatching of jobs. The emulator which is used herein, which we will refer to as EMU-TN as introduced in [3], and more recently a suite of statevector and tensor network emulators [2].

Orchestration and Scheduling. The infrastructure, illustrated in Fig. 1, is split between public cloud services orchestrated through Kubernetes, a HPC cluster hosted in a private data center running a Slurm scheduler, and the QPU located in Pasqal's datacenter. We next describe their integration for providing quantum computing as a service.

It is possible to run the same workloads on a Kubernetes pod alongside the other services, either as a long-running service responsible for receiving and executing a job or scheduled using Kubernetes compatible schedulers. However, a downside of these approaches is that it would be challenging to integrate the on-premise HPC cluster with the Kubernetes scheduler. Not scheduling the jobs on the cluster, which would mean just running a job execution service in a Kubernetes pod, would put limitations both on how effectively the resources in the cluster are used, and ultimately on the performance of the HPC applications. In this work we use a custom, internal, scheduler for our program, it is the same scheduler as for QPU jobs, to dispatch jobs.

Job Dispatching. To connect our two clusters, we used the Slurm REST API [19] to submit jobs to Slurm from the regular backend service. A "Slurm connector" service was developed, later merged into "ManGo" to have a single scheduler, to act as a bridge between the existing Pasqal cloud Services and the HPC cluster. The service creates a control script on-demand for the job which is submitted to Slurm as the batch script. This script then takes care of executing all the higher-level logic of the job. The service is also responsible for setting up the job with appropriate resources and inserting all necessary information for future

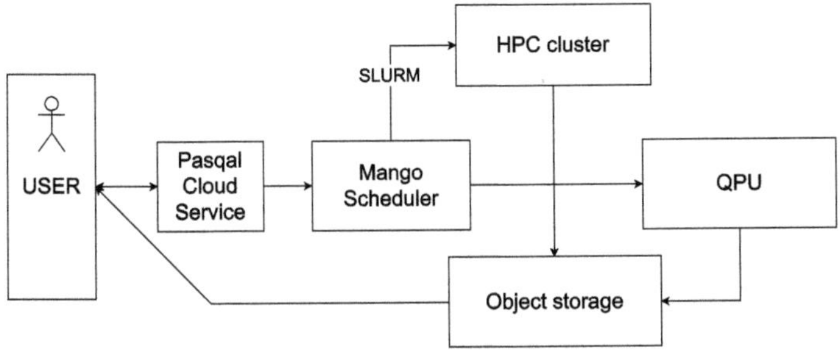

Fig. 1. Architecture of the cloud platform providing quantum computing as a service. The cloud services are deployed on a Kubernetes cluster and use a microservice architecture, each microservice exposing a public REST API to end users. The HPC cluster where the application runs, consists of 10 DGX A100 systems from NVIDIA and uses Slurm, an HPC cluster management and job scheduling system, for resource management and job submission.

communication with the cloud service. An alternative would have been to extend the Kubernetes cluster using the bridge pattern [16]. However, this would not have allowed us to manage the emulators in the same way as the QPU without significant development.

Upon completion of the job the results, as well as the input, are uploaded to an object storage. The full results include the logs from the emulator, bitstrings and upon request arbitrary Pauli-string observables and the quantum state. In order to aid in the exploration of the job, the individual programs can be loaded into Pulser Studio[2], a Pulser graphical user interface version that can visualize the program. For smaller programs (up to 12 atoms) the system can be simulated in the browser. For larger programs Pulser Studio can submit jobs to the cloud platform described.

The internal logic of the job submission endpoint validates the pulse sequence and the emulation configuration. The validation step checks that the request and the underlying data such as the sequence and emulator configuration are correctly formatted, that the requested resources are accessible to the user and, for a real device, that the sequence is physically valid. It then dispatches the request to the appropriate scheduler. In Code Sample 1, we show an example of sending a serialized pulse sequence to Pasqal Cloud Services for execution. To build the sequences we have used **Pulser** [20], an open source package for designing neutral atom QPU sequences. To communicate with cloud services we have used the **PasqalCloud Pulser** extension.

```
1  import numpy as np
2  import pulser
3  from pulser_pasqal import PasqalCloud, EmuMPSBackend
4
5  connection = PasqalCloud(
```

---

[2] Available at: https://pulserstudio.pasqal.cloud/.

```
 6        username=USERNAME, # Your username on the Pasqal Cloud Platform
 7        project_id=PROJECT_ID, # The ID of the billed project
 8        password=PASSWORD, # The password for your account
 9    )
10
11   # Build a device object to describe the specs of our "Fresnel" QPU
12   device = connection.fetch_available_devices()["FRESNEL"]
13
14   # Create a register with 2 atoms with automatic calibration
15   reg = pulser.Register({"q0": (-5, 0), "q1": (5, 0)}).
            with_automatic_layout(device)
16   seq = pulser.Sequence(reg, device)
17   seq.declare_channel("rydberg_global", "rydberg_global")
18   t = seq.declare_variable("t", dtype=int)
19
20   amp_wf = pulser.BlackmanWaveform(t, np.pi)
21   det_wf = pulser.RampWaveform(t, -5, 5)
22   seq.add(pulser.Pulse(amp_wf, det_wf, 0), "rydberg_global")
23
24   # Run the program on a real QPU
25   # backend = pulser.backends.QPUBackend(seq, connection=connection)
26   # Choose the latest emulator, Emu-MPS
27   backend = EmuMPSBackend(seq, connection=connection)
28
29   # Parameteraized job execution with job_params
30   job_params = [
31       {"runs": 100, "variables": {"t": 1000}},
32       {"runs": 50, "variables": {"t": 2000}},
33   ]
34   results = backend.run(job_params=job_params, wait=True)
35   print([job.bitstring_counts for job in results])
```

Code Sample 1. Running QPU or emulation jobs through the cloud service. The cloud connection is configured the program is created using Pulser the chosen backend selected and the final parameter values as inserted before the job is finally executed and the results are retrieved.

3   Applications

In this section, we discuss an application of the framework, motivated by typical tasks that can be performed with a Neutral-Atom Quantum Processor addressing the ground-Rydberg basis [11]. The interactions between atoms causes the quantum system to evolve according to the following Hamiltonian, that is reminiscent of the Ising model Hamiltonian:

$$\hat{H}(t) = \sum_{\text{atom } i} \left(\frac{\Omega(t)}{2}\hat{\sigma}_i^x - \delta(t)\hat{n}_i + \sum_{j<i} \frac{C_6}{|\mathbf{r}_{ij}|^6}\hat{n}_i\hat{n}_j \right), \tag{1}$$

where $\hat{n}_i = |1_i\rangle\langle 1_i| = (\mathbb{I} - \hat{\sigma}_i^z)/2$ is the projector into the one-qubit excited state of the i-th atom, C_6 is a constant that depends on the properties of the targeted excited state, and we have we have set the phase $\phi = 0$.

As shown in (1), the Hamiltonian can be split into two parts: the global single-qubit interactions driven by programmable lasers through the Rabi frequency $\Omega(t)$ and the detuning $\delta(t)$; and the atom-atom interactions that are programmable via the position of the atoms in the register.

Two promising areas of applications to explore are graph-based problems [6–9,12,15] and quantum simulation [1,4,5,10,14,17,24]. These are seen as particularly promising on the neutral atom analog QPUs compared to other qubit modalities and paradigms such as digital superconducting or ion-trap QPUs. The difference comes down to two distinguishing features:

1. The ability to encode graph structures in the register by placing the atoms in the form of the encoded graph, utilizing the atom-atom interactions in the Hamiltonian (1).
2. The natural time-evolution of the Hamiltonian, unlike in the digital case where said Hamiltonian must be Trotterized to an acceptable precision with a possibly large number of gates, each of which will have an error rate.

In order to further democratize quantum computing, along with the cloud platform outlined herein, we are developing open-source software to provide reusable solvers abstracting specific projects, such as Maximum Independent Set (MIS) solver [22] and Quantum Evolution Kernel (QEK) solver [23]. The MIS problem is explored further in the following subsection. For an introduction to QEK see [12]. The aforementioned libraries come with out-of-the-box support for Pasqal's cloud platform.

After the quantum evolution and readout, one obtains a bitstring of 0's and 1's representing excited and ground-state qubits, respectively. To calculate the expectation value of a given observable, many cycles need to be performed in order to generate enough statistics. The output of the emulator provides bitstrings that simulate the actual output of the QPU.

Maximum Independent Set on Unit Disk Graphs. We present a typical use case of neutral atom devices for two-dimensional registers. The possibility of placing atoms in arbitrary 2D configurations allows to solve certain classes of hard graph problems on the QPU. An interesting application is solving the Maximum Independent Set (MIS) problem on Unit Disk (UD) graphs [18].

A graph is an object with nodes and connections between nodes. A UD graph is a graph where two nodes are connected if an only if they are closer than a certain minimal distance. By representing the nodes of a graph with neutral atoms and matching the Rydberg blockade radius with the UD graph minimal distance, one can establish a direct mapping where a connection exists between atoms that are within a blockade distance of each other. The quantum evolution of such a system is naturally restricted to those sectors of the Hilbert space where excitations of connected atoms are forbidden. These configurations translate to independent sets of a graph, *i.e.* subsets of nodes that are not directly connected to each other. Driving the quantum evolution in such a way as to produce as many excitations as possible, one would obtain with high probability as the outcome of a measurement an independent set of high cardinality, representing a good candidate solution to the MIS problem.

The best pulse sequence to find the MIS of a graph is based on adiabatic pulses. As shown in [9], however, for arbitrary graphs it is often necessary to optimize the pulse shape in order to find the MIS with a high enough probability. We present in Fig. 2 a one-parameter family of pulses where the parameter t_c controls at which time the detuning goes from negative to positive. Rather than sweeping over the free parameter sequentially, one can send multiple simultaneous jobs for each value to be tested, and finally select the parameter giving the best result. Parallelizing this task is a direct advantage of the cluster infrastructure, and the cloud service provides a dedicated interface in which all jobs can be organized and retrieved, saving large amounts of time. We perform a sweep over 9 parameters for a pulse of 5 μs on a graph with 30 nodes. The results are shown in Fig. 2, where the optimal zero-crossing for the detuning is found to be at around 1.6 μs.

While the example shown here is centered on pulse parameters, other tasks suited for parallelization are also implementable, e.g. averages over noise and disorder realizations. Being able to quickly set up and run studies of this kind for systems of many qubits opens the door for creative algorithm design that is at the same time compatible with a QPU. By attaching `NoiseModels` to the job submission in Code Sample 1, this submission can easily be extended to also include a study of the implementations noise-robustness by running noisy simulations.

4   Discussion

In this paper, we described the components on the basis of hybrid quantum-classical workflows. We discussed how a cloud on-demand service can make use of a cluster backend to simplify access to the required computational operations. This includes implementing parallelization routines on the algorithm management side, controlled exploitation of cluster resources for heavy computational loads, and efficient solvers.

Developing workflows is much easier with a fully fledged cloud backend to orchestrate the tasks. This enables creating workflows that automatically assess the validity of the submitted pulse sequence for the QPU. Simulating the dynamics from a particular pulse sequence requires attention to tuning and understanding how to set the algorithm parameters. The cloud platform can provide the user suggestions for future hyperparameters based on live monitoring of the simulation. By having emulators on the same platform as QPUs, we also minimize the code changes required to move from emulation to QPU jobs.

A frequent problem when running hybrid algorithms is co-scheduling the quantum and classical resources. Part of this is solved in our platform with "open

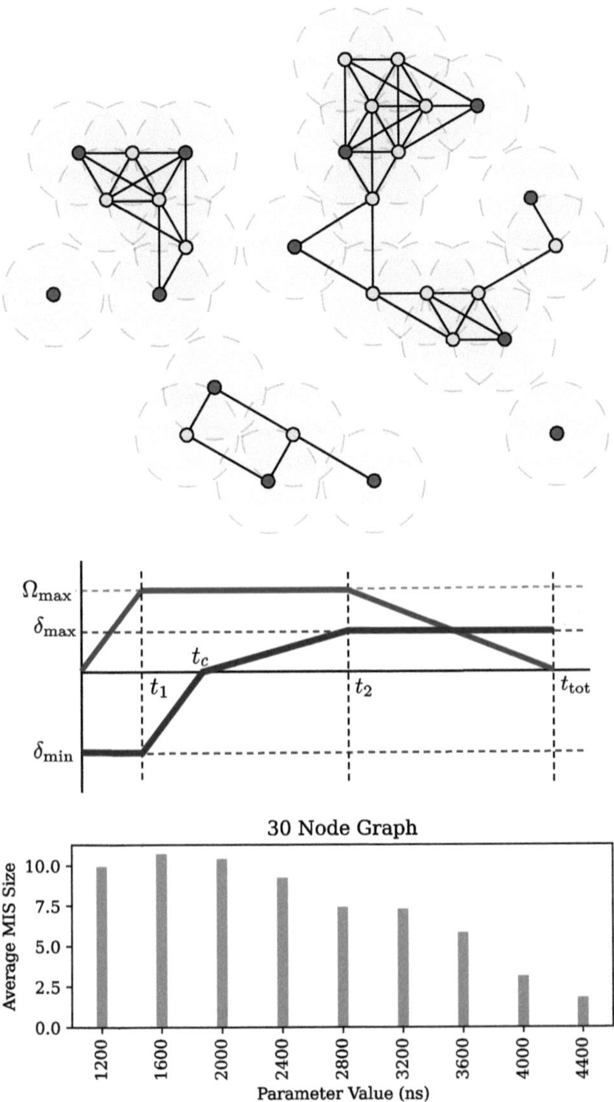

Fig. 2. Parameter sweep. We arrange 30 qubits in 2D (above), where the dashed circles show half the blockade radius and nodes corresponding to a MIS solution are colored in red. The pulse sequence (middle, $t_{\mathrm{tot}} = 5\mu s$) contains a tunable parameter t_c controlling the time when the detuning goes from negative to positive. We sweep in parallel for 9 values of t_c. The results (below) show the average size of the MIS found for each value of t_c. (Color figure online)

batces", where a session is opened with the device granting exclusive access and enabling the user to not have to wait in the queue between jobs. To improve the experience with classical resources we are working on integrating a recent extension allowing Slurm to manage the QPU resources both on-prem, cloud based and emulated [21]. This will make the process more streamlined in the HPC setting. Pasqal is also developing a runtime environment which will primarily integrate with the Slurm integration [25], intended to provide a more seamless experience when developing and executing hybrid programs across different access models.

Using a cloud connection to the QPU might be problematic due to the communication latency [13]. However, the current repetition rate of our QPU platform is on the order of $1\,\mathrm{Hz}$. Given that it takes on the order of 10^4 shots to obtain a sufficiently strong signal from the device, the incurred latency of a cloud platform is negligible. On the other hand, the benefits in terms of flexibility with respect to an on-premise device are, in our opinion, significant.

We believe that the platform presented here can greatly improve reproducibility and benchmarking of research in the developing quantum industry, as well as allowing a broader scope of researchers, engineers, and developers to understand, explore, and design the behavior of quantum systems and quantum devices.

Acknowledgments. We thank Sebastian Grijalva, Mario Dagrada, Anne-Claire Le Henaff, Caroline de Groot, and Alvin Sashala Naik, for valuable conversation and collaboration in related work.

References

1. Bernien, H., et al.: Probing many-body dynamics on a 51-atom quantum simulator. Nature **551**(7682), 579–584 (2017)
2. Bidzhiev, K., Grava, S., le Henaff, P., Mendizabel Pico, M., Merhej, E., Quelle, A.: pasqal emulators (2025). https://github.com/pasqal-io/emulators
3. Bidzhiev, K., et al.: Cloud on-demand emulation of quantum dynamics with tensor networks (2023). https://arxiv.org/abs/2302.05253
4. Bluvstein, D., et al.: Controlling quantum many-body dynamics in driven rydberg atom arrays. Science **371**(6536), 1355–1359 (2021)
5. Browaeys, A., Lahaye, T.: Many-body physics with individually controlled rydberg atoms. Nat. Phys. **16**(2), 132–142 (2020)
6. Cazals, P., et al.: Identifying hard native instances for the maximum independent set problem on neutral atoms quantum processors (2025). https://arxiv.org/abs/2502.04291
7. Cong, I., Choi, S., Lukin, M.D.: Quantum convolutional neural networks. Nat. Phys. **15**(12), 1273–1278 (2019)
8. Dalyac, C., et al.: Graph algorithms with neutral atom quantum processors (2024). https://arxiv.org/abs/2403.11931
9. Ebadi, S., et al.: Quantum optimization of maximum independent set using rydberg atom arrays. Science **376**(6598), 1209–1215 (2022)
10. González-Cuadra, D., et al.: Observation of string breaking on a $(2+1)$ d rydberg quantum simulator. Nature, 1–6 (2025)
11. Henriet, L., et al.: Quantum computing with neutral atoms. Quantum **4**, 327 (2020)

12. Henry, L.P., Thabet, S., Dalyac, C., Henriet, L.: Quantum evolution kernel: Machine learning on graphs with programmable arrays of qubits. Phy. Rev. A **104**(3) (2021). https://doi.org/10.1103/physreva.104.032416
13. Humble, T.S., McCaskey, A., Lyakh, D.I., Gowrishankar, M., Frisch, A., Monz, T.: Quantum computers for high-performance computing. IEEE Micro **41**(5), 15–23 (2021)
14. Julià-Farré, S., Vovrosh, J., Dauphin, A.: Amorphous quantum magnets in a two-dimensional rydberg atom array. Phy. Rev. A **110**(1) (2024). https://doi.org/10.1103/physreva.110.012602
15. Kornjača, M., et al.: Large-scale quantum reservoir learning with an analog quantum computer. arXiv preprint arXiv:2407.02553 (2024)
16. Lublinsky, B., Jennings, E., Spišaková, V.: A kubernetes 'bridge' operator between cloud and external resources (2022). https://doi.org/10.48550/ARXIV.2207.02531
17. Manovitz, T., et al.: Quantum coarsening and collective dynamics on a programmable simulator. Nature **638**(8049), 86–92 (2025)
18. Pichler, H., Wang, S.T., Zhou, L., Choi, S., Lukin, M.D.: Quantum optimization for maximum independent set using rydberg atom arrays (2018). https://doi.org/10.48550/ARXIV.1808.10816
19. Rini, N.: Rest API (2020). https://slurm.schedmd.com/SLUG20/REST_API.pdf. slurm User Group Meeting
20. Silvério, H., et al.: Pulser: an open-source package for the design of pulse sequences in programmable neutral-atom arrays. Quantum **6**, 629 (2022)
21. Sitdikov, I., et al.: Quantum resources in resource management systems (2025). https://arxiv.org/abs/2506.10052
22. Teller, D., Guichard, R., Lahariya, M.: Maximum Independent Set. https://github.com/pasqal-io/maximum-independent-set
23. Teller, D., Guichard, R., Lahariya, M.: Quantum Evolution Kernel. https://github.com/pasqal-io/quantum-evolution-kernel
24. Vovrosh, J., de Hond, J., Julià-Farré, S., Knolle, J., Dauphin, A.: Meson spectroscopy of exotic symmetries of ising criticality in rydberg atom arrays (2025). https://arxiv.org/abs/2506.21299
25. Wennersteen, A., Moreau, M., Nober, A., Mourad, B.: Towards a user-centric HPC-QC environment. Workshops of the International Conference for High Performance Computing, Networking, Storage and Analysis (SC Workshops '25) .https://doi.org/10.1145/3731599.3767549. to appear

Session: 7 Quantum Algorithms, Computing; Simulation – Quantum Machine Learning B

Quantum Neural Networks Under Threat: Modeling Security Risks and Attack Vectors

Nouhaila Innan[1,2(✉)], Walid El Maouaki[1,3], Alberto Marchisio[1,2], and Muhammad Shafique[1,2]

[1] eBRAIN Lab, Division of Engineering, New York University Abu Dhabi (NYUAD), Abu Dhabi, UAE
{nouhaila.innan,we2077,alberto.marchisio,muhammad.shafique}@nyu.edu
[2] Center for Quantum and Topological Systems (CQTS), NYUAD Research Institute, NYUAD, Abu Dhabi, UAE
[3] Quantum Physics and Spintronic Team, LPMC, Faculty of Sciences Ben M'sick, Hassan II University of Casablanca, Casablanca, Morocco

Abstract. Quantum Neural Networks (QNNs) represent a promising frontier in quantum machine learning, offering potential advantages in terms of expressivity and computational efficiency. However, their rapid development has outpaced critical attention to their security. In this work, we position QNN security as a foundational research challenge and present a comprehensive threat model that includes input encoding, circuit compilation, execution, and deployment. We introduce a taxonomy of emerging quantum-specific attack classes, including dataset poisoning, circuit tampering, side-channel leakage, multi-tenant interference, and federated learning threats. Our analysis highlights key gaps, including the lack of quantum-native robustness metrics, hardware variability, and cross-layer vulnerabilities. To address these issues, we outline strategic directions for developing secure-by-design QNN architectures, emphasizing formal verification, quantum-aware defenses, and adversarial benchmarking. This work lays the groundwork for a systematic approach to building trustworthy and resilient quantum learning systems.

Keywords: Quantum Machine Learning · Quantum Neural Networks · Security · Attacks

1   Introduction

Quantum Neural Networks (QNNs) have become a crucial component of Quantum Machine Learning (QML), providing a framework for learning tasks that harness the structure of quantum information [1,2]. These models, typically based on parameterized quantum circuits and hybrid quantum-classical training loops, are being actively explored for applications in fields such as finance, healthcare, materials science, and cryptography [3–13]. As the availability of

F. Barbaresco and F. Gerin (Eds.): QUEST-IS 2025, CCIS 2744, pp. 129–139, 2026.
https://doi.org/10.1007/978-3-032-13855-2_12

quantum hardware expands and variational algorithms mature, QNNs represent viable candidates for integration into critical data-driven systems.

However, the rapid development of QNNs has not been accompanied by a corresponding focus on their security and robustness. In classical Machine Learning (ML), significant effort has been devoted to understanding and mitigating vulnerabilities such as adversarial attacks, data poisoning, and model extraction. These efforts have led to the establishment of well-defined threat models and a diverse range of defense mechanisms. In contrast, the QML community has yet to fully articulate the potential attack surfaces of QNNs or to develop a coherent framework for assessing their security risks.

QNNs introduce several challenges that are fundamentally distinct from their classical counterparts. Phenomena such as superposition, entanglement, and the stochastic nature of quantum hardware execution complicate traditional security assumptions. Furthermore, hardware-specific noise profiles, the irreversibility of measurement, and the no-cloning property of quantum states introduce novel vectors for side-channel attacks and information leakage. These characteristics necessitate a rethinking of existing threat models and highlight the need for quantum-specific definitions of robustness, privacy, and trust.

This paper presents a scientific position on the necessity of treating QNNs security as a foundational research problem, and makes the following contributions:

- A multi-layer threat model covering QNNs vulnerabilities across encoding, compilation, execution, and deployment;
- A taxonomy of quantum-specific attack classes, including poisoning, tampering, side-channels, and cloud-based interference;
- An analysis of core challenges such as hardware variability and the lack of quantum-native robustness metrics;
- Strategic directions for secure-by-design QNNs through verification, benchmarking, and quantum-aware defenses.

2   Background and State of the Art

2.1   Quantum Neural Networks

QNNs refer to a class of models that combine principles from quantum computing with neural network-inspired learning mechanisms. They are typically constructed using parameterized quantum circuits (PQCs), where learnable parameters modulate quantum gates acting on qubits. These parameters are updated iteratively using classical optimization routines, forming what is known as a hybrid quantum-classical learning architecture [14,15].

As shown in Fig. 1, a standard QNN pipeline involves three key stages: (i) encoding classical data into quantum states via an embedding circuit (e.g., angle or amplitude encoding), (ii) applying a sequence of parameterized quantum gates forming the trainable circuit (often referred to as a variational layer), and (iii) performing projective measurements to extract classical output used to compute a loss function. The resulting gradients are estimated, commonly via the parameter-shift rule, and used to update the quantum circuit parameters in a classical outer loop.

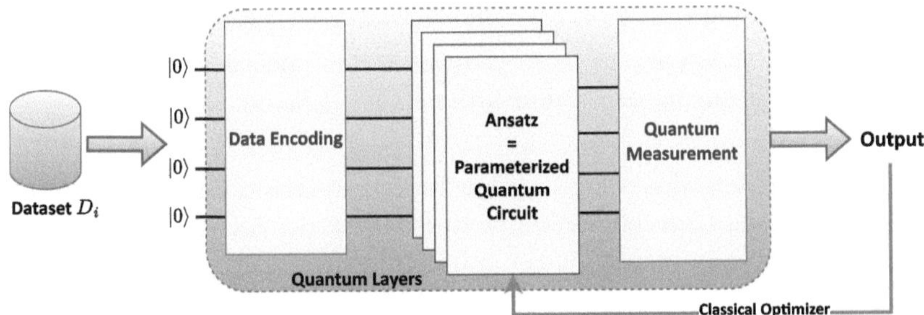

Fig. 1. QNNs pipeline where classical data is encoded into quantum states, processed by a parameterized quantum circuit (Ansatz), measured, and updated via a classical optimizer.

QNNs have been proposed for supervised learning tasks such as classification and regression, as well as generative modeling and reinforcement learning. Notable architectures include quantum convolutional neural networks, quantum graph neural networks, and quantum recurrent neural networks, each adapting structural elements from classical deep learning models into the quantum domain. Many of these models operate within the constraints of Noisy Intermediate-Scale Quantum (NISQ) hardware, which limits the circuit depth and qubit count but allows for practical experimentation on near-term devices [16–19].

Despite their increasing popularity, QNNs lack formal analysis from the perspective of robustness, reliability, and security. Their reliance on fragile quantum states, stochastic measurements, and hardware-dependent behavior introduces non-trivial challenges for stability and trustworthiness [2]. Furthermore, unlike classical models, QNNs do not yet benefit from standardized evaluation criteria for adversarial robustness, differential privacy, or information leakage. As these models are increasingly deployed in sensitive and high-assurance applications, understanding their attack surfaces and developing appropriate threat models becomes an essential direction for QML research.

2.2   Security in Classical ML

Classical ML systems are widely used, but their adoption has revealed various security risks impacting data integrity, confidentiality, and system reliability. This section outlines key threats: adversarial attacks, privacy breaches, hardware-level exploits, reliability issues, and side-channel attacks.

Adversarial Attacks involve crafted inputs designed to mislead models. These perturbations can occur during training, inference, or in physical environments [20].

Privacy Threats include model inversion, which reconstructs input data, membership inference, which identifies whether a sample was in the training set [21], and model extraction, which replicates a model based on its outputs [22].

Hardware-Based Threats target ML computations through faults or malicious alterations. Fault injection introduces errors such as bit flips [23], while hardware Trojans are hidden modifications that activate under specific conditions [24].

Reliability Threats are caused by hardware imperfections. Permanent faults result from defects, transient faults stem from environmental factors [25], and aging effects gradually increase error rates [26].

Side-Channel Attacks extract information from physical signals such as timing, power, or electromagnetic emissions. Power analysis reveals input patterns [27], EM attacks capture signals during computation [28], and cache attacks analyze memory access to infer model behavior [29].

3   Threat Modeling for QNNs

Identifying where adversaries can influence QNNs is key to building robust security models. We outline the main threat surfaces and their corresponding threat classes and defenses (see Fig. 2).

3.1   Threat Surfaces

A robust threat model for QNNs must begin with an analysis of components where adversaries can interact with, observe, or influence the model's behavior.

Input-Level. The input level in QNNs presents unique security challenges. This stage involves converting classical data into quantum states, a process not found in traditional machine learning. This opens new possibilities for attacks. For example, attackers might subtly alter input data in ways that are hard to detect but can lead to incorrect results from the QNN. Additionally, the methods used for encoding can unintentionally reveal information about the original data, making it easier for attackers to reverse-engineer or infer sensitive details. These vulnerabilities are especially concerning in environments where QNNs are accessed through shared or cloud-based platforms. Therefore, it's crucial to carefully design and secure the input encoding processes to protect against potential threats.

Model-Level. At the model level, QNNs face risks from both design exposure and the compilation process. Adversaries may steal or replicate models by analyzing their architecture or behavior, especially when hosted on external platforms. They could also infer sensitive training data by observing how the model processes inputs. Additionally, during compilation, when high-level algorithms are translated for specific hardware, malicious or compromised compilers can secretly alter circuits, harming performance or leaking information. These threats are greater with third-party or cloud-based tools, underscoring the need for secure deployment, trusted compilers, and strict access control.

Execution-Level. During execution, QNNs process data by sending control pulses to manipulate qubits. This phase faces significant security risks if the quantum hardware or environment is compromised. Attackers gaining access could manipulate computations by subtly altering these critical pulses. Modifications such as changing pulse timing or frequency can cause unpredictable qubit behavior, leading to incorrect outputs, data leakage, or hardware damage. These pulse-level attacks are challenging to detect, as they may not appear in higher-level circuit representations. Ensuring the integrity and confidentiality of the execution environment is crucial to mitigating these threats.

3.2   Threat Classes and Defenses

Quantum Dataset Manipulation. Quantum dataset manipulation poisons training data by inserting mislabeled or slightly perturbed quantum states. These manipulations compromise the integrity of quantum training data, causing misclassification and performance degradation during inference [30–32]. Notably, quantum computing properties such as superposition, quantum entanglement capabilities, as well as quantum space dimensionality affect the distinguishability, making poisoned embeddings inherently difficult to detect or correct [33]. Tactics such as label-flipping, clean-label poisoning, and test-time evasion have cut accuracy by up to 92% in simulation and 75% on real hardware [34]. Defenses range from fidelity checks and data anonymization to adversarial training with poisoned samples [35]. Circuit-aware designs, like low-entanglement quanvolutional networks with CRz gates, retain much more accuracy under attack [36].

Model Poisoning. Beyond the data, QNN circuits can be poisoned both during compilation and at the pulse layer. A hostile transpiler may tweak gate angles, add stealth gates, or remap qubits, embedding a backdoor that pushes the model toward chosen errors [37]. Even with a clean gate list, a pulse-mismatch attack can alter the microwave waveform, causing the hardware to apply a different unitary [38]. Compile-time tools such as CertiQ compare the original and transpiled circuits to spot hidden changes [39], while gate-set tomography can flag unusual drops in fidelity that hint at hardware sabotage [40]. Still, the field lacks shared benchmarks and cross-platform tests, so a full defense will require joint hardware-software safeguards.

Model and Data Extraction. QNN models, especially when deployed on cloud quantum services, face threats of intellectual property theft and privacy breaches through extraction attacks. Adversaries may attempt to steal the model (recover its parameters or structure) or extract sensitive data from it by exploiting side-channels and query access. Because quantum computations are physical processes, they can inadvertently leak information via timing, power consumption, or other physical observables. Power-side-channel analysis has been shown to reconstruct full gate sequences for benchmark circuits from a single trace, while timing analysis of cloud job latencies can reveal qubit-mapping decisions and ultimately the circuit topology. Masking and shuffling, mainstays of classical cryptography, are less effective because qubit control pulses are highly structured. Research is now exploring noise injection and measurement staggering as more quantum-native counter-measures [41,42]. Beyond these physical traces, untrusted QNN-as-a-Service platforms also allow query-based model stealing: CopyQNN, Split Co-Teaching, and related studies show that with only thousands of input-output queries an attacker can train a substitute QNN that reproduces 90-99% of the victim's accuracy even under realistic NISQ noise [43–45].

Cloud Multi-tenant Attacks. In shared quantum computing environments, multi-tenancy introduces unique interference threats. When multiple QNN workloads run on the same hardware concurrently, one adversarial job can intentionally degrade the performance of another through crosstalk and other noise injection. Crosstalk is unwanted interaction between qubits or circuits executing in parallel. For example, an attacker can submit a specially crafted circuit with frequent entangling operations (like repeated CNOT gates) that induces excess crosstalk noise on a victim QNN running at the same time [46,47]. This interference can corrupt the victim's qubit states and lead to higher error rates or model mispredictions, effectively a denial-of-service on the QNN's accuracy. Such cross-tenant fault injection attacks have been identified as a serious emerging threat in the NISQ era, as they require no special privileges beyond sharing the hardware. Alongside crosstalk, contention for shared resources (e.g., microwave control channels or readout lines) can similarly be exploited to impair a victim's model. To defend against multi-tenant attacks, cloud providers can incorporate isolation and scheduling strategies, like scheduling conflicting quantum jobs sequentially or allocating them to well-separated qubit groups to minimize cross-coupling [48].

Quantum Federated Learning Attacks. Quantum Federated Learning (QFL) [10] allows multiple parties to collaboratively train QNNs without sharing raw data, but it inherits classical federated learning vulnerabilities, including Byzantine attacks [49]. Malicious clients may submit poisoned quantum models or falsified gradients during aggregation, aiming to degrade accuracy or insert backdoors while hiding their behavior. To counter this, early research adapts robust aggregation methods like Byzantine-resilient averaging to the quantum setting and explores dynamic filtering of outlier updates [50]. Effective QFL security likely requires a combination of these defenses, validation of quantum updates, and secure communication channels to ensure trust in the training process.

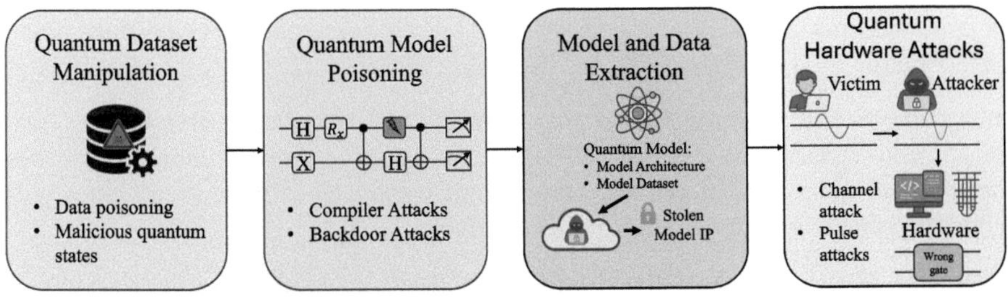

Fig. 2. Overview of Quantum threat surfaces and classes targeting key stages of QNNs.

4   Open Challenges in QNNs Security

As QNNs transition from theoretical prototypes to deployed systems, the lack of foundational security principles presents urgent challenges. Many of these are not just unsolved, they are undefined. Below, we highlight the most pressing barriers that restrain progress toward trustworthy quantum learning:

Absence of Attack Benchmarking Protocols: Unlike classical ML, there is no standardized methodology to evaluate how resilient QNNs are under various threat models. This inhibits fair comparison across defense strategies and limits reproducibility.

Unclear Adversarial Bounds for Quantum States: Classical adversarial attacks rely on well-understood norms to measure perturbations. In contrast, it remains unclear what constitutes a "small" or "stealth" perturbation in Hilbert space, given quantum state collapse and non-determinism.

Entanglement-Induced Complexity: In multi-qubit QNNs, entangled states introduce global dependencies, meaning an attack on one qubit may have non-local effects. This undermines the intuition built from classical model modularity and complicates localized defense design.

Hardware-Layer Fragility and Opacity: Due to vendor-specific implementations, developers often lack full visibility into low-level operations like pulse shaping or measurement timing. This black-box hardware abstraction makes it difficult to verify the integrity of the full model stack.

Tooling Fragmentation: The ecosystem of quantum programming frameworks (Qiskit, PennyLane, Cirq, etc.) offers differing abstractions and interfaces, leading to fragmented security practices. Currently, no platform provides native support for secure QNN development or verification. Addressing these challenges requires not only technical innovation but also community-level consensus on definitions, tools, and shared security benchmarks.

5   Position and Strategic Recommendations

Security must evolve as a co-equal objective alongside accuracy and efficiency in QNNs development. We consider reactive, patch-based approaches insufficient for quantum models, and argue that QML should adopt security-by-design as a core principle. Toward this end, we propose the following strategic priorities:

Integrate Security into Quantum Compiler Infrastructure: Future transpilers and circuit mappers should natively support tamper-detection, cross-checking of optimization passes, and formal verification hooks that align with security policies specified by model developers.

Promote Quantum-Native Resilience Mechanisms: Defenses should capitalize on quantum properties rather than imitate classical ones. For example, decoherence-aware dropout layers or entanglement-based watermarking could serve as native countermeasures for model theft and tampering.

Establish Continuous Validation Pipelines: Just as continuous integration has become essential for software development, QNNs should be evaluated for adversarial robustness and side-channel resistance on a rolling basis during training, deployment, and retraining cycles.

Foster Red-Teaming and Adversarial Benchmarking in QML: Encouraging external audits, red-team evaluations, and challenge-based competitions will accelerate the discovery of vulnerabilities and stimulate robust solution design. This mirrors the success of adversarial ML competitions in the classical domain.

Incentivize Interdisciplinary Collaboration: Quantum security demands expertise across quantum physics, compiler theory, cryptography, and machine learning. Funding bodies and academic institutions should support initiatives that bring these disciplines together under shared research agendas. These actions lay the groundwork for a secure quantum AI ecosystem, one that is proactive, rigorous, and robust against both present and future threats.

6   Conclusion

QNNs present interesting opportunities for advancing ML into the quantum era, but they also introduce vulnerabilities that we are only beginning to understand. In this work, we present a structured view of QNN-specific attack vectors and provide a layered threat taxonomy that highlights new risks arising from quantum data encoding, model compilation, and hardware-level execution. We emphasize that security must not be treated as an afterthought or follow-up phase, but as an integrated component of QNNs design and deployment. The characteristics that give quantum models their properties also introduce new risks that evade traditional defenses. A trustworthy QML future depends on rigorous security modeling, practical mitigation tools, and coordinated community efforts. By initiating this step now, we can build QNNs that are performant, expressive, secure, interpretable, and resilient from the ground up.

Acknowledgments. This work was supported in part by the NYUAD Center for Quantum and Topological Systems (CQTS), funded by Tamkeen under the NYUAD Research Institute grant CG008, and the Center for Cyber Security (CCS), funded by Tamkeen under the NYUAD Research Institute Award G1104.

References

1. Zaman , K., et al.: A survey on quantum machine learning: Current trends, challenges, opportunities, and the road ahead, arXiv preprint arXiv:2310.10315 (2023)
2. Innan, N., et al.: Next-generation quantum neural networks: enhancing efficiency, security, and privacy, In: IOLTS (2025)
3. Innan, N., et al.: Qfnn-ffd: Quantum federated neural network for financial fraud detection, In: QSW, pp. 41–47, IEEE (2025)
4. Innan, N., et al.: Lep-QNN: Loan eligibility prediction using quantum neural networks, arXiv preprint arXiv:2412.03158 (2024)
5. Choudhary, P.K., et al.: Hqnn-FSP: A hybrid classical-quantum neural network for regression-based financial stock market prediction, arXiv preprint arXiv:2503.15403 (2025)
6. Sawaika, A., et al.: A privacy-preserving federated framework with hybrid quantum-enhanced learning for financial fraud detection, arXiv preprint arXiv:2507.22908 (2025)
7. Maouaki, W.E., et al.: Quantum clustering for cybersecurity, In: QCE, vol. 2, pp. 5–10, IEEE (2024)
8. Innan, N., et al.: Fl-qdsnns: Federated learning with quantum dynamic spiking neural networks, arXiv preprint arXiv:2412.02293 (2024)
9. Siddiqui, O.I., et al.: Quantum bayesian networks for machine learning in oil-spill detection, arXiv preprint arXiv:2412.19843 (2024)
10. Innan, N., et al.: Fedqnn: Federated learning using quantum neural networks, In: IJCNN (2024)
11. Dave, K., et al.: Optimizing low-energy carbon IoT systems with quantum algorithms: performance evaluation and noise robustness, IEEE Internet Things J. (2025)
12. Innan, N., et al.: Qnn-vrcs: A quantum neural network for vehicle road cooperation systems, IEEE Trans. Intell. Trans. Syst. (2025)
13. Dave, K., et al.: Sentiqnf: A novel approach to sentiment analysis using quantum algorithms and neuro-fuzzy systems, IEEE Trans. Comput. Soc. Syst. (2025)
14. Zaman, K., et al.: Studying the impact of quantum-specific hyperparameters on hybrid quantum-classical neural networks, In: World Congress in Computer Science, Computer Engineering & Applied Computing, pp. 132–149, Springer (2024)
15. Zaman, K., et al.: A comparative analysis of hybrid-quantum classical neural networks, In: World Congress in Computer Science, Computer Engineering & Applied Computing, pp. 102–115, Springer (2024)
16. Marchisio, A., et al.: Cutting is all you need: Execution of large-scale quantum neural networks on limited-qubit devices, arXiv preprint arXiv:2412.04844 (2024)
17. Kashif, M., et al.: Computational advantage in hybrid quantum neural networks: Myth or reality? arXiv preprint arXiv:2412.04991 (2024)
18. Ahmed, T., et al.: Noisy hqnns: A comprehensive analysis of noise robustness in hybrid quantum neural networks, arXiv preprint arXiv:2505.03378 (2025)

19. Ahmed, T., et al.: Quantum neural networks: A comparative analysis and noise robustness evaluation, arXiv preprint arXiv:2501.14412 (2025)
20. Chakraborty, A., et al.: A survey on adversarial attacks and defences, CAAI Trans. Intell. Technol. (2021)
21. Shokri, R., et al.: Membership inference attacks against machine learning models, In: SP (2017)
22. Tramèr, F., et al.: Stealing machine learning models via prediction APIs, In: USENIX Security (2016)
23. Rakin, A.S., et al.: Bit-flip attack: Crushing neural network with progressive bit search, In: ICCV (2019)
24. Clements, J., Lao, Y.: Hardware trojan attacks on neural networks, arXiv preprint arXiv:1806.05768 (2018)
25. Raghunathan, B., et al.: Cherry-picking: exploiting process variations in dark-silicon homogeneous chip multi-processors, In: DATE (2013)
26. Kang, K., et al.: Nbti induced performance degradation in logic and memory circuits: How effectively can we approach a reliability solution? In: ASP-DAC (2008)
27. Yan, X., et al.: Defense against ml-based power side-channel attacks on DNN accelerators with adversarial attacks, arXiv preprint arXiv:2312.04035 (2023)
28. Méndez Real, M., Salvador, R.: Physical side-channel attacks on embedded neural networks: A survey, Appl. Sci. **11**(15) (2021)
29. Tong, Z., et al.: Cache side-channel attacks detection based on machine learning, In: TrustCom (2020)
30. Lu, S., et al.: Quantum adversarial machine learning, Phy. Rev. Res. (2020)
31. Maouaki, W.E., et al.: Advqunn: A methodology for analyzing the adversarial robustness of quanvolutional neural networks, In: QSW (2024)
32. Maouaki, W.E., et al.: Robqunns: A methodology for robust quanvolutional neural networks against adversarial attacks, In: ICIPC Workshop (2024)
33. Liu, N., Wittek, P.: Vulnerability of quantum classification to adversarial perturbations, Physical Review A **101**(6) (2020)
34. Kundu, S., Ghosh, S.: Adversarial poisoning attack on quantum machine learning models, arXiv preprint arXiv:2411.14412 (2024)
35. Maouaki, W.E., et al.: Qfal: Quantum federated adversarial learning, arXiv:2502.21171 (2025)
36. Maouaki, W.E., et al.: Designing robust quantum neural networks via optimized circuit metrics, Adv. Quan. Technol. (2025)
37. Chu, C., et al.: Qtrojan: A circuit backdoor against quantum neural networks, In: ICASSP (2023)
38. Xu, C., et al.: Classification of quantum computer fault injection attacks, arXiv preprint arXiv:2309.05478 (2023)
39. Shi, Y., et al.: Certiq: a mostly-automated verification of a realistic quantum compiler, arXiv preprint arXiv:1908.08963 (2019)
40. Proctor, T., et al.: Detecting and tracking drift in quantum information processors, Nature communications **11**(1) (2020)
41. Erata, F., et al.: Quantum circuit reconstruction from power side-channel attacks on quantum computer controllers, arXiv preprint arXiv:2401.15869 (2024)
42. Lu, C., et al.: Quantum leak: Timing side-channel attacks on cloud-based quantum services, arXiv preprint arXiv:2401.01521 (2024)
43. Fu, Z., et al.: Copyqnn: Quantum neural network extraction attack under varying quantum noise, arXiv preprint arXiv:2504.00366 (2025)

44. Kundu, S., et al.: Evaluating efficacy of model stealing attacks and defenses on quantum neural networks, In: Proceedings of the Great Lakes Symposium on VLSI 2024 (2024)
45. Fu, Z., Chen, F.: Quantum neural network extraction attack via split co-teaching, arXiv preprint arXiv:2409.02207 (2024)
46. Ash-Saki, A., et al.: Analysis of crosstalk in nisq devices and security implications in multi-programming regime, In: ISLPED (2020)
47. Harper, B., et al.: Crosstalk attacks and defence in a shared quantum computing environment, arXiv preprint arXiv:2402.02753 (2024)
48. Murali, P., et al.: Software mitigation of crosstalk on noisy intermediate-scale quantum computers, In: ASPLOS (2020)
49. Xia, Q., et al.: Defending against byzantine attacks in quantum federated learning, In: MSN (2021)
50. Zhang, Y., et al.: Federated learning with quantum secure aggregation, arXiv preprint arXiv:2207.07444 (2022)

Machine Learning-Driven Quantum Hacking of CHSH-Based QKD: Exploiting Entropy Vulnerabilities in Self-testing Protocols

Hubert Kołcz[1]([⊠]) (iD), Tushar Pandey[2] (iD), and Yug Shah[3] (iD)

[1] Warsaw University of Technology, Doctoral School, Pl. Politechniki 1,
00-661 Warsaw, Poland
`hubert.kolcz@ieee.org`
[2] Department of Mathematics, Texas A&M University, 400 Bizzell Street,
College Station, TX 77843, USA
[3] Department of Computer Science, University of Toronto, 27 King's College Circle,
Toronto, ON M5S 1A1, Canada

Abstract. Despite the mathematical excellence of CHSH-based QKD, their engineering possibilities compromise their security foundations, creating space for the so-called *Quantum Hacking*. Several real-world quantum hacking techniques have been successfully demonstrated against commercial QKD systems in 2025, utilising methods such as Phase Remapping Attack, Trojan Horse Attack, Time-Shift Attack, and Detector Blinding Attack.

This work introduces a new method to exploit weak Random Number Generators (RNGs) in QKD systems using quantum RNG (qRNG) and machine learning. We demonstrate how compromised RNGs enable side-channel attacks by monitoring entropy and external factors (e.g., temperature). Our framework integrates:

- **Multi-modal RNG Analysis:** A 12-qubit qGAN quantifies distribution similarity via relative entropy (KL divergence < 0.1) and discriminator loss (3.7–17), validated on Rigetti Aspen-M-3 (80 qubits) and IonQ Aria-1 (25 qubits) hardware.
- **Bias Detection:** Markov chain-enhanced logistic regression identifies hardware-induced biases (59% '1' frequency in compromised RNGs vs. 54% in certified devices).
- **Real-Time Monitoring:** A neural network classifier (58.7% accuracy) discriminates quantum-generated randomness from classical simulations using 100-bit entropy profiles.

Experimental results show CHSH scores correlate with RNG quality (Rigetti: 0.8036, IonQ: 0.8362), while gate fidelity (IonQ: 99.4% vs. Rigetti: 93.6%) impacts certifiable randomness. Combining device-independent CHSH validation with machine learning, this framework detects attacks such as phase remapping and detector blinding through entropy deviations.

Keywords: quantum key distribution · self testing · quantum hacking · random number generator · machine learning

1   Introduction

The Bell and Clauser-Horne-Shimony-Holt (CHSH) inequalities test quantum mechanics against local hidden variable theories but differ fundamentally in experimental feasibility. The original Bell inequality requires perfect anti-correlation (identical outcomes when measurement settings align) and ideal detectors with 100% efficiency. Mathematically, for correlation functions $E(\theta_A, \theta_B)$, the Bell inequality for settings θ_A, θ'_A and θ_B, θ'_B is:

$$|E(\theta_A, \theta_B) + E(\theta_A, \theta'_B) + E(\theta'_A, \theta_B) - E(\theta'_A, \theta'_B)| \leq 2, \tag{1}$$

which demands visibility $V > 97.4\%$ to violate classically, a threshold unattainable in real-world setups due to noise [5]. In contrast, the CHSH inequality tolerates imperfections, requiring only statistical correlations across four different settings. For outcomes $A, A' \in \{\pm 1\}$ and $B, B' \in \{\pm 1\}$, the CHSH value is:

$$S = \langle AB \rangle + \langle AB' \rangle + \langle A'B \rangle - \langle A'B' \rangle, \tag{2}$$

with classical bound $S \leq 2$ and quantum maximum $S = 2\sqrt{2}$. Experimentally, CHSH violations require detector efficiency $\eta > \frac{2}{1+\sqrt{2}} \approx 82.8\%$ [6], achievable with entangled states like $|\phi^+\rangle = \frac{1}{\sqrt{2}}(|00\rangle + |11\rangle)$. Real-world implementations achieve $S \approx 2.42$ for electron spins [8], far exceeding the classical limits. This robustness makes CHSH pivotal for QKD. Entanglement-based protocols (e.g. E91, DI-QKD) distribute pairs with correlations $E(\theta_A, \theta_B) = -\cos(\theta_A - \theta_B)$, yielding $S > 2$ to certify security. Prepare-and-measure protocols (e.g. BB84) lack this, relying on orthogonal state indistinguishability.

Although the mathematical rigor of the original Bell inequality makes its practical adoption almost impossible, the CHSH approach is preferred not only for its experimental feasibility but also for its built-in self-testing capabilities, which are critical for real-world quantum security. This feature is essential for the mathematical formulation of device-independent QKD (DI-QKD), where security persists even if the hardware (e.g., photon sources, detectors) cannot be fully trusted. By contrast, non-entanglement-based QKD protocols inherently assume device trustworthiness, limiting their applicability in adversarial scenarios.

Despite the theoretical rigor of entanglement-based QKDs, real-world systems remain vulnerable to quantum hacking attacks, exploiting imperfections in hardware or protocol implementation. In particular, compromised RNGs can introduce biases or correlations that undermine security, even in device-independent settings. This motivates the need for dynamic, ML-driven verification of randomness and entropy in operational QKD systems through a penetration test approach, focused on qRNGs, which exploit intrinsic quantum uncertainty to produce true randomness, forming the foundation of secure QKD protocols. While established qRNGs are well-studied, emerging platforms (e.g., KwanTeach [3], low-cost home-based quantum computers [4]) require robust verification protocols.

2 Related Work

Attacks such as phase-remapping, Trojan horse, time-shift, and detector blinding have all been demonstrated against commercial QKD hardware, with some studies reporting successful key extraction in over 60% of targeted attacks under specific laboratory conditions [16]. For example, phase-remapping attacks manipulate phase modulators in bidirectional systems to alter correlation measurements [12], while Trojan horse attacks use injected light to probe internal device settings via back-reflections [13]. Time-shift attacks exploit mismatches in single-photon detector response times to bias measurement outcomes [14], and detector-blinding attacks use intense light to force detectors into a controllable classical regime, enabling full key extraction by an adversary [15]. These sophisticated strategies, often successful due to engineering imperfections rather than theoretical flaws, highlight the need for robust, implementation-aware countermeasures that can operate in real-time [7,17].

To validate the randomness of qRNGs underpinning QKD, traditional statistical test batteries such as NIST SP 800-22, Dieharder, and TestU01 have long been the standard. These suites apply a range of hypothesis tests—examining frequency, runs, autocorrelation, and more to ensure output unpredictability. However, these methods are fundamentally static: they are designed to detect gross statistical anomalies in large, stationary datasets, and are typically run only during device certification or at long intervals. As a result, they can miss subtle regressions or emerging vulnerabilities in certified devices, especially those induced by environmental drift, aging, or targeted quantum hacking attacks. For example, NIST SP 800-22 is known to have high false positive rates and limited sensitivity to temporal or low-order correlations, while Dieharder and TestU01 may not identify biases that appear only under certain operational conditions or attacks [9,10].

In the age of artificial intelligence, there is a growing recognition that static batteries are insufficient for the dynamic threat landscape facing modern QKD systems. Recent research has turned to ML and QML as more adaptive, real-time solutions. Neural network classifiers have been developed to distinguish quantum-generated randomness from classical or compromised sources, achieving high accuracy in attack detection and device identification. QGANs provide a powerful tool for quantifying distributional shifts in RNG outputs using metrics such as relative entropy to enable device-agnostic validation of quantum randomness. These dynamic, data-driven approaches can monitor entropy in real time, detect subtle anomalies missed by static tests, and provide actionable feedback for device tuning and certification [11].

Despite these advances, most existing frameworks focus on isolated aspects of the problem, either targeting specific attacks or validating randomness in isolation. Few solutions holistically address the interplay between RNG vulnerabilities, entropy monitoring, and adaptive protocol-level responses. Our work aims to bridge this gap by combining CHSH-based self-testing with ML-driven, real-time entropy analysis, enabling both attack detection and hardware-agnostic certification. This unified approach advances the state of the art beyond isolated countermeasures, supporting the secure and scalable deployment of QKD technologies in realistic, adversarial environments.

3   Methodology

We integrate quantum GANs and classical machine learning to verify qRNG outputs on IBM Qiskit simulators, Rigetti Aspen-M-3, and IonQ Aria-1 platforms [11]. Each device produced 2000 samples of 100-bit entries, classified as in Table 1. For each sample, the Bell value for trial i and the CHSH game value are:

$$J_i = \begin{cases} 1, & \text{if } x_i \oplus y_i = a_i b_i \\ 0, & \text{otherwise} \end{cases} \qquad J = \frac{1}{n} \sum_{i=1}^{n} J_i - \frac{3}{4}$$

where $\frac{3}{4}$ is the classical threshold.

Table 1. Benchmarking metrics for tested platforms [1].

	IBM Qiskit	Rigetti Aspen-M-3	IonQ Aria-1
Qubit Tech.	Simulation	Superconducting	Trapped Ion
Qubit Count	Flexible	80	25
CHSH Score	1.0	0.8036	0.8362
2-Qubit Fidelity (%)	100	93.6	99.4
Connectivity	Full	Local	All-to-all

3.1   Data Generation and Feature Engineering

Quantum circuits prepared multi-qubit superpositions (e.g., Hadamard gates), with both small (2, 4 qubits) and large (up to 100 qubits) circuits. Reference datasets from classical PRNGs were included. Feature extraction (Python) computed mean, variance, run-length, autocorrelation, Markov transitions, entropy measures (Shannon, min-entropy, KL divergence) and per-qubit statistics.

3.2   Machine Learning Model Architectures

Neural Network Classifier. A PyTorch feedforward network with two hidden layers (30 and 20 neurons, ReLU, dropout 0.2) and a softmax output was used for device/source classification.

Markov Chain-Enhanced Logistic Regression. Logistic regression was trained on Markov transition and run-length features, capturing sequential dependencies.

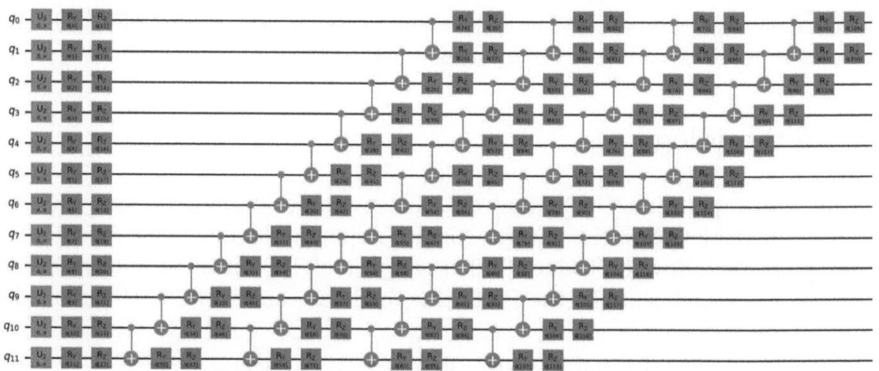

Fig. 1. Variational Quantum Eigensolver within qGAN model [2].

Quantum GAN (qGAN). A 12-qubit quantum generator (Qiskit/Pennylane) and classical discriminator were trained adversarially; the generator uses parameterized quantum gates, optimized via a VQE subroutine. The discriminator is a classical neural network (Fig. 1).

4 Experimental Results

Table 2 presents a comparative summary of the main quantum-classical verification methods. The baseline DoraHacks model, which serves as a reference for random guessing, achieves an accuracy of 54%. Both the neural network (NN) and logistic regression models show incremental improvements, with the best NN configuration (batch size 8, 4:1 split, L1 regularization) reaching 58.67% accuracy and a KL divergence of 0.75. Logistic regression, enhanced with Markov chain features, achieves 56.1% accuracy, demonstrating its effectiveness in detecting sequential dependencies and hardware-induced bit biases. Notably, compromised RNGs exhibited a 59% frequency of '1's, compared to 54% in certified devices, confirming the model's sensitivity to such biases.

The quantum GAN (qGAN) approach provides a more nuanced assessment of distribution similarity. KL divergence values for qGAN comparisons range from 3.7 (Set 1 vs 2) to 17 (Set 2 vs 3), with corresponding p-values of 0.01, indicating statistically significant differences between device outputs. The qGAN's ability to capture higher-order statistical structure is further reflected in the discriminator loss metrics, which provide a robust, device-agnostic measure for qRNG validation.

4.1 Entropy and Hardware Analysis

Figure 2 shows the entropy analysis for the evaluated quantum hardware platforms. IonQ Aria-1, despite having fewer qubits than Rigetti Aspen-M-3, demonstrates higher CHSH game scores (0.8362 vs 0.8036) and superior two-qubit gate

Table 2. Performance comparison of quantum-classical verification methods. The hybrid approach shows improvement over baseline DoraHacks models.

Method	Accuracy (%)	KL Divergence	p-value
Baseline (Dorahacks)	54.00	–	–
NN (batch=16, split=3:1)	54.80	1.1	–
Logistic (split=7:3)	56.10	–	–
NN (batch=8, split=4:1, L1 Reg.)	58.67	0.75	–
qGAN (Set 1 vs 2)	–	3.7	0.01
qGAN (Set 2 vs 3)	–	17	0.01
qGAN (Set 1 vs 3)	–	16	0.01

fidelity (99.4% vs 93.6%). This translates into higher entropy and more certifiable randomness in the generated bitstreams. IBM Qiskit simulators, while idealized, serve as an upper bound for entropy and randomness quality, but do not capture the practical noise and bias present in physical devices.

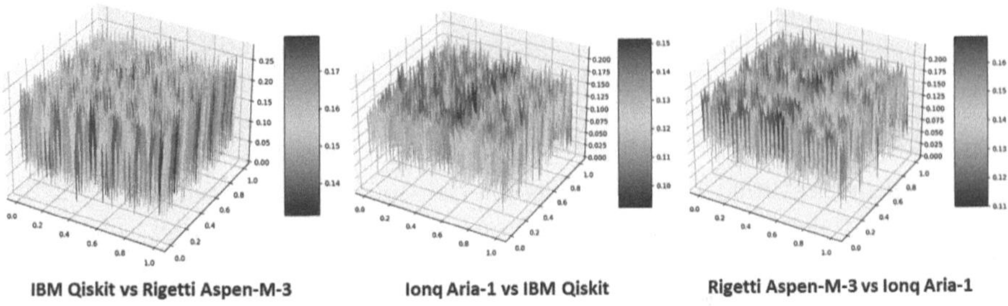

Fig. 2. Entropy analysis for evaluated quantum hardware platforms.

The entropy analysis highlights that higher CHSH scores and gate fidelities correlate with an improved certifiable randomness. This is consistent with theoretical predictions: as noise and hardware imperfections increase, both entropy and the ability to violate the CHSH inequality decrease, compromising the security guarantees of device-independent QKD.

4.2 qGAN Training Dynamics and Novel Metrics

A key contribution of this work is the introduction of qGAN training dynamics as a novel metric for qRNG validation. Figure 3 presents representative learning curves from qGAN training. The stability and convergence behavior of generator/discriminator losses and relative entropy serve as quantitative indicators of data randomness. Rapid convergence and low steady-state KL divergence are characteristic of high-quality quantum randomness, while unstable or divergent curves suggest noise, bias, or classical contamination.

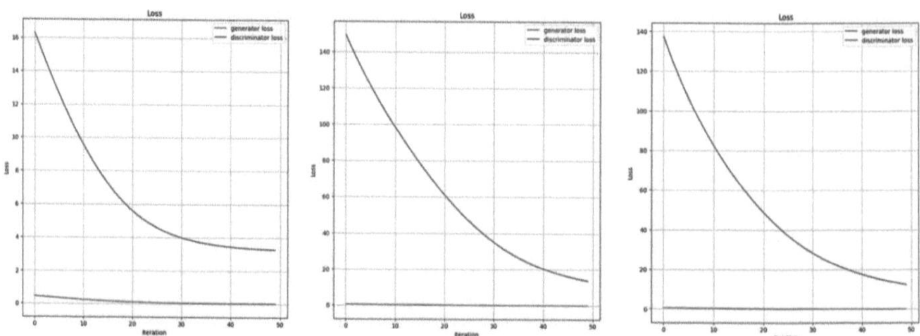

Fig. 3. Representative learning curves from qGAN training. The stability and convergence behavior of generator/discriminator losses and relative entropy serve as quantitative indicators of data randomness.

4.3   Case Study: Application to a Short-Range QKD System

To validate our framework in a practical setting, we applied our entropy monitoring methodology to a simulated real-world scenario based on a short-range QKD system architecture for a metro access network [18]. We generated an additional 500 samples of 100-bit keys using accessible cloud-based QPUs, labeling this dataset as "non-certified" to emulate randomness from a potentially untrusted, low-cost device.

Our neural network classifier successfully distinguished the "non-certified" data from our certified IonQ and Rigetti datasets with an accuracy of 62.3%, primarily by flagging subtle periodic biases and lower overall entropy. The qGAN analysis corroborated this, showing a high KL divergence of 19.5 between the non-certified data and the IonQ reference set. This case study demonstrates that our methodology can serve as an effective real-time verification layer to detect the integration of unvetted or compromised hardware into a QKD network.

4.4   Interpretation and Implications

The observed differences between hardware platforms underscore the importance of entropy analysis and device benchmarking in practical QKD deployments. IonQ's higher fidelity and CHSH performance, for example, translate directly into improved randomness quality, while classical simulators and lower-fidelity devices exhibit predictable biases that can be exploited by adversaries. The integration of qGAN training dynamics as a validation metric opens new avenues for adaptive device tuning and certification. By monitoring generator/discriminator loss and KL divergence in real time, it is possible to detect regressions, guide hardware improvements, and enhance the security of quantum random number generation pipelines. This approach is particularly valuable for emerging quantum platforms and uncertified devices, where traditional static test batteries may fail to capture dynamic or context-dependent vulnerabilities.

5   Discussion

Our results demonstrate that machine learning models, especially when combined with device-independent CHSH validation, can detect subtle entropy deviations caused by compromised RNGs or environmental factors. The observed 59% frequency of '1's in compromised RNGs, while based on the current dataset, points to a statistically significant bias that our logistic regression model effectively captured. Although the dataset size is a limitation in this preliminary study, this finding strongly suggests a hardware- or software-level vulnerability that a conventional statistical test might miss without a large sample size. This underscores the value of ML models trained to identify specific, subtle patterns of non-randomness.

To build upon these findings, future work should focus on expanding the dataset by generating more samples from a wider variety of certified and non-certified cloud QPUs. This would not only enhance the statistical power of our models but also improve their generalization capabilities for identifying diverse RNG failure modes. Ultimately, the classifier and qGAN-based metrics can challenge certification, guide device tuning, and be integrated into "randomness extractor" pipelines to enhance compliance with NIST's randomness standards and mitigate the risk of quantum hacking.

6   Conclusion

This study bridges the gap between theoretical device-independent security and engineering realities in QKD. By integrating quantum GANs and ML-based entropy monitoring, we provide a robust framework for detecting and mitigating entropy-based attacks. Our approach supports the scalable deployment of secure QKD systems, guiding both certification and adaptive operation under realistic noise conditions.

Acknowledgments. We thank Yale University and DoraHacks for organizing the YQuantum 2024 hackathon, as well as prof. Teodor Buchner and prof. Jerzy Balicki from Warsaw University of Technology for consultations on RNG and qGAN.

Artifact Description (AD)

This paper is accompanied by a computational artifact available at: https://github.com/hubertkolcz/NoiseVsRandomness. The repository contains all code, data, and instructions necessary to reproduce the main results. Specifically, it provides:

- Raw and processed datasets used for training and evaluating the machine learning models, including quantum random number generator outputs from IBMQ simulators, Rigetti Aspen-M-3, and IonQ Aria-1.

- Python scripts for data preprocessing, feature extraction, and entropy analysis.
- Implementations of the neural network classifier, Markov chain-enhanced logistic regression, and quantum GAN (qGAN) models as described in the methodology.
- Jupyter notebooks that reproduce the main figures and tables from the paper, including CHSH score analysis, entropy plots, and learning curves.

The artifact is intended to facilitate full transparency and reproducibility of the computational experiments reported in this work. All code dependencies and hardware requirements are documented in the repository.

Artifact Evaluation (AE)

The artifact was evaluated for functional completeness, reproducibility, and documentation quality. All scripts and notebooks were tested on a standard Python 3.10 environment with the required packages listed in the IPython notebooks. The main results—including neural network classification accuracy, qGAN training dynamics, and entropy analysis—can be reproduced on a standard workstation (16GB RAM, 4 CPU cores) within a few hours.

The repository includes sample datasets for quick testing, as well as instructions for downloading or generating the full datasets used in the study. The code is modular and well-documented, enabling users to adapt the models to new quantum hardware or alternative random number sources. The artifact has been independently validated by the authors and is suitable for further research and benchmarking in quantum randomness verification and quantum hacking detection.

References

1. DoraHacks: Global Quantum Randomness Generator Competition Dataset (2023)
2. Zoufal, C., Lucchi, A., Woerner, S.: Quantum generative adversarial networks for learning and loading random distributions. npj Quantum Inf. **5**, 103 (2019)
3. KWANTEACH Project: KwanTeach quantum platform. https://kwanteach.org (2024)
4. Kołcz, H., Przybyła, G., Rukat, P., Łabaj, F., Maruszak, P.: Low-Cost, Home-Made Quantum Computer. Warsaw University of Technology, Warsaw, Poland (2025)
5. Brunner, N., Cavalcanti, D., Pironio, S., Scarani, V., Wehner, S.: Bell nonlocality. Rev. Mod. Phys. **86**, 419–478 (2014)
6. Hensen, B., et al.: Loophole-free bell inequality violation using electron spins separated by 1.3 kilometres. Nature **526**, 682–686 (2015)
7. Zapatero, V., et al.: Advances in device-independent quantum key distribution. npj Quantum Inf. **9**, 10 (2023)
8. Rosenfeld, W., et al.: Event-ready bell test using entangled atoms simultaneously closing detection and locality loopholes. Phys. Rev. Lett. **119**, 010402 (2017)

9. Saarinen, M.J.O.: SP 800-22 and GM/T 0005-2012 Tests: Clearly Obsolete, Possibly Harmful. Cryptology ePrint Archive, Report 2022/169 (2022)
10. L'Ecuyer, P., Simard, R.: TestU01: A C library for empirical testing of random number generators. ACM Trans. Math. Softw. **33**(4), Article 22 (2007)
11. Kołcz, H., Pandey, T., Shah, Y., et al.: Verification of qRNG with qGAN and Classification Models. QUACC+ CTP PAS, Warsaw, Poland (2025)
12. Fung, C.-H.F., Qi, B., Tamaki, K., Lo, H.-K.: Phase-remapping attack in practical quantum-key-distribution systems. Phys. Rev. A **75**, 032314 (2007)
13. Gisin, N., Fasel, S., Kraus, B., Zbinden, H., Ribordy, G.: Trojan-horse attacks on quantum-key-distribution systems. Phys. Rev. A **73**, 022320 (2006)
14. Qi, B., Fung, C.-H.F., Lo, H.-K., Ma, X.: Time-shift attack in practical quantum cryptosystems. Quantum Inf. Comput. **7**(1), 73–82 (2007)
15. Lydersen, L., Wiechers, C., Wittmann, C., Elser, D., Skaar, J., Makarov, V.: Hacking commercial quantum cryptography systems by tailored bright illumination. Nat. Photonics **4**, 686–689 (2010)
16. Xu, F., Ma, X., Zhang, Q., Lo, H.-K., Pan, J.-W.: Secure quantum key distribution with realistic devices. Rev. Mod. Phys. **92**, 025002 (2020)
17. Zhou, Z., et al.: Real-time monitoring and protection of superconducting nanowire single-photon detectors against blinding attacks. Phys. Rev. Appl. **19**, 014027 (2023)
18. Ha, N.T.T., Van, D.T., Thang, V.V., Luyen, H.D.: A low-cost and compact quantum key distribution system for metro access network. J. Eur. Opt. Soc.-Rapid Publ. **17**, 1 (2021)

EniQmA - Enabling Hybrid Quantum Applications

Dimitris Badounas[1], Marvin Bechtold[2], Martin Beisel[2], Patricia Bickert[3], Arcesio Castañeda Medina[4], Michael Fromm[1], Alexander Geng[4], Ilie-Daniel Gheorghe-Pop[5]([✉]), Florian Girtler[6], Cristian Grozea[5], Dominik Heldwein[7], Michael Holzki[3], Matthias Kabel[4], Colin Kai-Uwe Becker[5], Sophia Lahs[3], Fernando Lima[4], Théo Lisart-Liebermann[4], Darya Martyniuk[5], Merlin Mengel[1], Ali Moghiseh[4], Andreas Müller[3], Joe Rixon[6], Marcel Seelbach Benkner[1], Sebastian Senge[7], Nikolay Tcholtchev[5,8], Felix Truger[2], Juris Ulmanis[6], Sebastian Wagner[9], Benjamin Weder[2], and Armin Wolf[5]

[1] eleQtron GmbH, Heeserstraße 5, 57072 Siegen, Germany
{dimitris.badounas,Michael.Fromm,merlin.mengel,
marcel.seelbach}@eleqtron.com

[2] Institute of Architecture of Application Systems, University of Stuttgart, Universitätsstraße 38, Stuttgart, Germany
{marvin.bechtold,martin.beisel,truger,
benjamin.weder}@iaas.uni-stuttgart.de

[3] DB Systel GmbH, Jürgen-Ponto-Platz 1, 60329 Frankfurt am Main, Germany
{patricia.bickert,michael.holzki,sophia.lahs,
andreas.ze.mueller}@deutschebahn.com

[4] Fraunhofer ITWM, Fraunhofer-Platz 1, 67663 Kaiserslautern, Germany
{arcesio.castaneda.medina,alexander.geng,matthias.kabel,fernando.lima,
theo.lisart,ali.moghiseh}@itwm.fraunhofer.de

[5] Fraunhofer FOKUS, Kaiserin-Augusta-Allee 31, 10589 Berlin, Germany
{ilie-daniel.gheorghe-pop,cristian.grozea,darya.martyniuk,
nikolay.tcholtchev}@fokus.fraunhofer.de

[6] Alpine Quantum Technologies GmbH, Technikerstraße 17/1, 6020 Innsbruck, Austria
{florian.girtler,joe.rixon,juris.ulmanis}@aqt.eu

[7] Accenture GmbH, Kaistr. 20, 40221 Düsseldorf, Germany
{dominik.heldwein,sebastian.senge}@accenture.com

[8] RheinMain University of Applied Sciences, Unter den Eichen 5, 65195 Wiesbaden, Germany

[9] Kipu Quantum GmbH, Greifswalderstrasse 212, 10405 Berlin, Germany
sebastian.wagner@kipu-quantum.com
https://www.hs-rm.de

Abstract. The transition of quantum computing into practical applications depends on the successful combination of quantum algorithms with existing classical software systems. The resources for creating hybrid applications are currently fragmented, requiring teamwork efforts across various disciplines. Moreover, there are currently no clear predefined methods, standards, or solutions for these complex operations. We introduce hereby EniQmA which stands as a quantum software

F. Barbaresco and F. Gerin (Eds.): QUEST-IS 2025, CCIS 2744, pp. 150–163, 2026.
https://doi.org/10.1007/978-3-032-13855-2_14

engineering framework for unifying this development process as well as deployment and operation of hybrid quantum-classical applications. EniQmA integrates agile workflows with domain-specific tool-chains along with quantum DevOps practices to enable connecting theoretical quantum algorithms with scalable industrial applications. The framework delivers structured process models together with quality standards and cohesive development tools to produce reliable and sustainable quantum software. Through its low-code/no-code capabilities, EniQmA makes quantum application development accessible to domain experts who have minimal quantum expertise or enable specialists to provide cutting-edge approaches in quantum computing. This framework is showcased in three real-world industrial scenarios: sustainability in aircraft building, anomaly detection in production, as well as railway transportation. EniQmA creates an essential initial step towards standardizing quantum software engineering, which speeds up the journey to practical quantum advantage in industrial applications.

Keywords: Quantum Computing · Hybrid Applications · Software Development Framework · Quantum DevOps

1 Introduction

The development of quantum applications with industrial significance has so far required heterogeneous teams of experts from different domains. An interdisciplinary approach comprising quantum physics, algorithm development, industry specialization and IT operations is required, among others.

The EniQmA [1] framework attempts to streamline this often complex process by creating structured, workflow-based processes for efficient quantum software engineering. These processes and corresponding tool-chains are further connected to hybrid environments, managed through one or multiple orchestration systems while a monitoring system provides live-data to ensure smooth operations. Modern Quantum DevOps [2] practices process a hybrid application multiple rounds through a testing and benchmarking framework to ensure successful execution on quantum backends both during development as well as during operations sub-cycles. Ultimately these developments will be used to support hybrid quantum applications. Such development software systems are essential due to multiple barriers of entry into the field, especially where algorithm designers are not from the physics field, and other contributors come from outside computer science. Fragments and algorithm design at the low-level should be build by quantum-inclined engineers, while the full vertical integration of solutions is usually built by computer scientists and programmers. This split widens and complicates quantum software research, where innovative algorithms and low-level procedures take a lot of time to emerge to possible industrial applications. This problem is what we are aiming to tackle through our work.

2   Related Work

2.1   Hybrid Quantum-Classical Routines

Hybrid quantum-classical routines, often of the form of Variational Quantum Algorithms (VQAs), are promising approaches for achieving near-term quantum advantage [3]. These algorithms use an iterative method to evaluate an objective function on quantum hardware, followed by fine-tuning of the outcome using a classical optimization routine. This approach is particularly beneficial for applications like quantum chemistry [4], combinatorial problems [5], simulations [6], quantum machine learning (QML) [7] and numerical methods [8].

The effectiveness of these applications is significantly impacted by the classical routine ability to find a satisfactory approximation of the global optimum. The non-convex nature of VQA optimization landscape and noisy gradient sampling undermine theoretical convergence guarantees. Successful VQA approaches rely on several factors like proper workflow design, good parametric initial conditions (warm-starting), hyperparameter fine-tuning of the classical optimizer and problem-specific ansatz design. In addition, circuit-cutting methods to fit larger problems onto the current QC scale is considered. In the next section, we go over some of those elements, then we proceed to show software concepts that allow us to bind them together.

2.2   Hybrid Quantum Algorithms and Workflows

Quantum applications require orchestration of both classical and quantum programs. Workflows are a well-established approach for integrating heterogeneous tasks. Quantum Modeling Extension (QuantME) [9] was introduced to facilitate standardized modeling of quantum workflows using BMPN[1], offering custom-tailored constructs for recurring tasks and graphical notation for better understanding. QuantME also includes a dedicated algorithm for portability, transforming QuantME modeling constructs into natively supported workflow language constructs using reusable process fragments [10]. By modeling and executing quantum applications as workflows[2] they can inherently benefit from their advantages, e.g., robustness, scalability, and transactional processing. Those workflow tools are developed with quantum algorithm designers in mind, as it is likely that a collection of trainability-improvement techniques would be needed to be deployed simultaneously for the success of VQA-types of algorithms [33].

Warm-starting techniques can help overcome issues that negatively impact the execution of quantum algorithms, we show in [11] that issues such as barren plateaus, local optima, and the scarcity of quantum devices can be significantly lowered.

We introduce "convex-hull" cost-to-value analysis tools for selecting gradient based, second order gradient and natural gradient optimizers. The problem of

---

[1] BPMN,https://camunda.com/products/camunda-bpm/bpmn-engine.2023, (accessed 17.09.2025).

[2] Quantum Worfklow Modelling Demonstration (accessed 17.09.2025), https://www.youtube.com/watch?v=U9kJvRw3WjY.

Barren plateaus are tackled through the introduction of novel SP-BFGS, an hybrid first and second order method allowing for better optimization landscape exploration [13]. This method stabilizes the Broyden–Fletcher–Goldfarb–Shanno algorithm (BFGS) to non-convex landscapes, as well as leverages multi-policy bayesian estimation for hyper-parameters fine-tuning from ground noise, capable of absorbing high shot-noise, and diminishing Quantum Processing Unit (QPU) calls.

Then, we explored quantum-classical routine improvements on the quantum side: we increase parallelization opportunities within the reconfigurable ADAPT-QAOA [16], an Adaptive Quantum Approximate Optimization Algorithm algorithm for optimization problems [14], improving operator selection through Clifford-Point approximations [12] and low-rank stabilizer decomposition, delegating non-quantum logic to classical HPC resources.

Finally, we investigated circuit-cutting procedures in [15], providing an approach to execute quantum circuits that surpass the width and depth constraints of current quantum devices. This strategy involves the fragmentation of the quantum circuit into several smaller subcircuits, each with a reduced number of qubits and gates, making them compatible with existing devices. The final measurement output of the original circuit can then be reconstructed through a classical combination of the results derived from these subcircuits.

Leveraging those multiple, often sophisticated techniques simultaneously, especially when expecting further compilation and transpilation to run programs on real hardware is a complex software engineering problem. We are exploring our solution in the next section.

3   EniQmA Framework: Foundations and Architecture

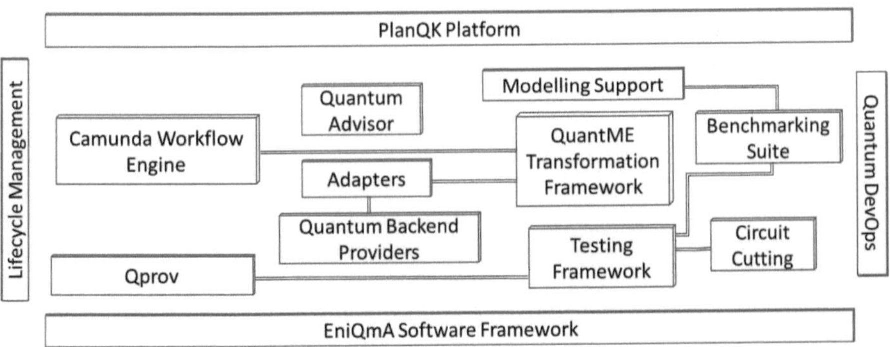

Fig. 1. The EniQmA Architecture.

As a novel and exotic domain, quantum computing did not have a very large supporting community of specialists compared to other popular scientific research fields. This has changed in the past years as quantum computing gained notoriety via worldwide dissemination[3]. The issue remains for many new companies

---

[3] SCOPUS quantum computing number of publications per year trend graph, https:// tinyurl.com/QCpubTrend18-24 (accessed 17.09.2025).

tackling the emerging field of quantum technologies is a severe lack of quantum specialists. EniQmA aims to make it easier for new professionals to enter the field, for existing programmers to switch towards quantum application development as well as to provide tools for existing quantum developers to enhance their productivity.

The proposed architecture in Fig. 1 accommodates both quantum specialists and less experienced practitioners. Through the graphical workflow modeler, non-experts can compose and deploy hybrid applications via drag-and-drop components[4] Experts, by contrast, continue working in their preferred IDEs, into which the quantum debugger, unit-testing framework for hybrid code, automated circuit-cutting and a standardized benchmarking suite are provided as plug-in extensions.

4   EniQmA Quantum Application Development Workflow

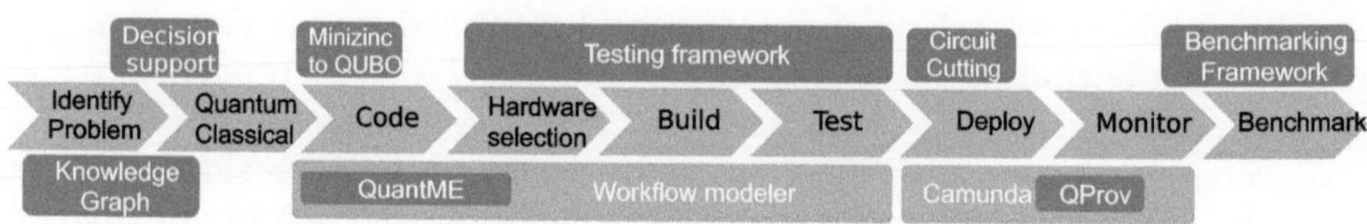

Fig. 2. The EniQmA Development Flow.

The EniQmA framework for hybrid quantum–classical application development provides the workflow shown in Fig. 2 that spans problem identification to deployment and monitoring[5]. A knowledge-driven Quantum Advisor first classifies subproblems by suitability for quantum or classical processing. Then using the Knowledge Graph the appropriate available EniQmA tools and algorithms are identified. When the problem can be expressed as Quadratic Unconstrained Binary Optimization (QUBO), a MiniZinc-to-QUBO converter automatically maps high-level constraint models onto QUBO instances. Users then assemble hybrid workflows in a BPMN editor using QuantME, leveraging built-in circuit-cutting and hardware-selection modules, and deploy them using the Camunda engine. Via the Testing Framework users are able to identify compatible quantum backends or adjust the runtime workload using the Circuit Cutting Service. During execution, QProv continuously captures hardware calibrations, circuit metadata, and performance metrics to enable detailed analysis.

The project partners develop tools for quantum software engineering systematization, expanding consulting services, product development business models, and PoC implementations. The tools within the EniQmA framework support

---

[4] Quantum Worfklow Modelling Demonstration (accessed 17.09.2025), https://www.youtube.com/watch?v=U9kJvRw3WjY.

[5] All software components described in Artifact Description Annex.

the application development process as depicted in Fig. 2. Hardware provider partners are at the same time developing their capacity to compute the newly developed application. Algorithmic developments strengthen hybrid runtime co-development and establish a reference for hybrid runtimes.

4.1 Platform and Backend Providers

Quantum applications require quantum computing hardware resources, which are limited due to the early stage of the industry. Within this work, we exploited the following ressources:

PLANQK. The PLANQK platform simplifies the development of hybrid quantum applications by providing unified access to various quantum hardware backends and simulators [17]. This allows developers to test quantum algorithms and deploy production workloads on actual quantum hardware. PLANQK also integrates classical computing resources like CPUs and GPUs, allowing for easy integration of classical and quantum computing. It supports the integration of services hosted externally in cloud environments or private data centers, enhancing flexibility and interoperability. While the use-case applications covered within the EniQmA Framework might seem limited to combinatorial optimisation and QML, its' integration with the PLANQK platform[6] expands the current application range while at the same time ensures future use-cases to be made available via the platform offered services.

AQT. To enable more practical use of the EniQmA framework with AQT's trapped-ion quantum computers, an integration stack has been developed using a dedicated cloud backend called ARNICA [18]. Users can submit jobs for processing and retrieve results through a web API[7], providing information about the status and capabilities of the quantum computers. Circuits must be defined exclusively in the available gate set or transpiled beforehand. The cloud service is asynchronous, returning a job ID when a job is submitted.

EleQtron Backend. Founded in 2020 as a spin-off at the University of Siegen quantum optics group, eleQtron uses trapped ions in the MAgnetic Gradient Induced Coupling (MAGIC) scheme [19] combined with well-proven microwave control of individual qubits. This setup allows to take advantage of the native trapped-ion all-to-all connectivity at very low crosstalk error. Users can submit QASM circuits to the eleQtron quantum backend via an API using token-based authentication[8].

---

[6] PlanQK Platform, https://planqk.de/en/ (accessed 17.09.2025).

[7] AQT Arnica backend https://arnica.aqt.eu/api/v1/docs (accessed 17.09.2025).

[8] EleQtron homepage https://eleqtron.com/en/ (accessed 17.09.2025).

5   Use Cases

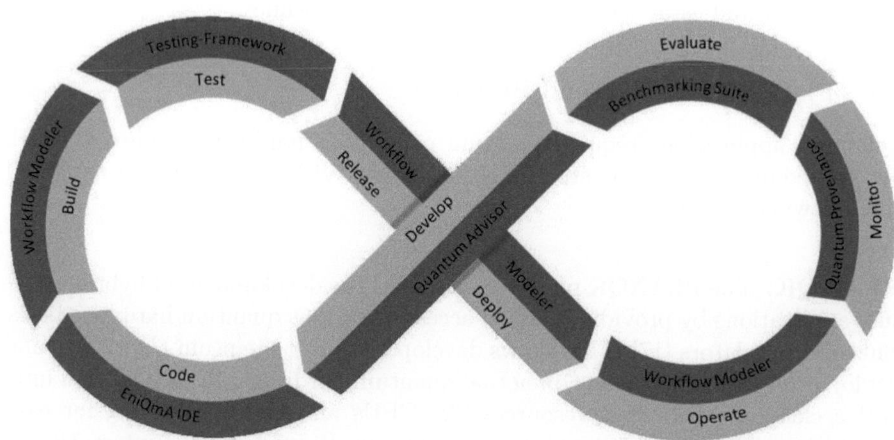

Fig. 3. The EniQmA tools mapped to the quantum DevOps cycle.

Continuously applying the processes and workflows to industry-relevant use cases represents a pivotal procedure for validating the developed tools and assessing the practicality of the targeted engineering approach with respect to the design objectives. These are aimed at unifying and expediting the development of hybrid quantum applications. This internal feedback loop within the EniQmA framework, shown in Fig. 3, ensures a sustained influx of improvement suggestions that can be seamlessly integrated in an agile manner. Consequently, this process also empowers the industrial consortium partners to construct quantum proof-of-concept applications.

5.1   Sustainability in Aircraft Construction and Operation

Modern aircraft construction relies on the use of composite materials for performance, energy efficiency, and sustainability. Numerical simulation can accelerate the design process, but stress analysis of complex composites is time-consuming and computationally demanding. This research proposes using quantum models to build surrogate models trained on DNS data [20], aiming to represent effective material behavior with low inference time. The study aims to address open questions [7] and benchmark performance [21] on real data and quantum devices, fostering efficiency, precision, and sustainability in aircraft construction.

5.2   Anomaly Detection in Production

Digitalization in production processes requires real-time data for anomaly detection, making Machine Learning (ML) a viable solution. Quantum Machine Learning (QML) offers superior results with reduced data requirements, but challenges [22] arise due to Noisy Intermediate-Scale Quantum (NISQ) computing [23,24]. Hybrid quantum computing, combining traditional and quantum layers, and quantum support vector machines (QSVMs) [26] are being explored for anomaly detection. Thus, recent research [25] proposed the variational quantum linear solver enhanced QSVM (VQLS-enhanced QSVM). Experimental results indicate that the algorithm successfully identified anomalies in real data, making it a suitable support vector classifier for anomaly detection in production.

5.3 Service for Quantum-Based Capacity Optimization

Digitalization in transport has emphasized process optimization, especially in route planning and rolling stock optimization. Quantum computers offer a solution through quantum annealing as well as hybrid approaches. Combining problem decomposition, reduced data models, and quantum computers allows for optimal solutions in a short time interval, enabling real-world applications in train fleet management [27].

6 Limitations and Future Work

This framework, while promising, also reveals key challenges for interdisciplinary teams. Error rates on NISQ hardware remain the primary bottleneck, though the QProv system helps mitigate this by tracking performance across different quantum backends. As hardware improves, the framework's modular design ensures its benchmarks will evolve to measure true quantum advantage across more application domains. Even though integration between so many components could be effectively done, larger scale, fully integrated benchmarks remain to be executed and evaluated.

7 Conclusion

Validated through industrial use cases in sustainability, anomaly detection, and optimization, covering methods in QML and optimization, EniQmA steers towards practical use of multiple experts tackling the complexities in deploying industrial applications of quantum computing. Its modular design ensures adaptability as quantum technology evolves, and flexibility for quantum algorithms designers to plug in the most recent advances in the field. For interdisciplinary teams, the framework lowers adoption barriers through standardized interfaces and decision-support tools.

Looking ahead, EniQmA's focus on reproducibility and collaboration will remain crucial for translating quantum potential into real-world impact. Future work will expand benchmarking on specific problems fitting to the industrial needs, expanding research on problem-specific modules (of the likes of warm-starting, optimizers, and ansatz design) while fostering community contributions through the knowledge graph. By balancing technical depth with usability,

EniQmA advances quantum software engineering while serving diverse stakeholders across research and industry.

Acknowledgment. This work was funded by the Federal Ministry for Economic Affairs and Climate Action (German: Bundesministerium für Wirtschaft und Klimaschutz) under the project funding number 01MQ22007A. The authors are responsible for the content of this publication.

Disclosure of Interests. The authors have no competing interests.

A   Artifact Appendix

A.1   Abstract

This appendix contains the description of software modules mentioned in the paper as well as providing more details about the services used. No experimentation results were disseminated within this paper therefore there is no necessity for information pertaining to replication of such experiments.

A.2   Artifact Check-List (Meta-information)

- **Algorithm: Variational Quantum Algorithms (VQAs), Circuit cutting algorithms**
- **Program: Circuit Cutter, Quantum Adviser, QProv, Quantum Workflow Modeller, Quantum Testing Framework, Quantum Benchmarking Suite, Benchmarking methods for automatic otimization/synthesis of PQCs, Quantum-Classical splitting Service, MiniZinc-to-QUBO Converter, Knowledge Graph**
- **Publicly available?: Yes**
- **Workflow framework used?: Camunda Workflow Engine, PlanQK Platform, Quokka Ecosystem**

A.3   Description

Quantum Adviser. Quantum computing is a promising paradigm that can solve various problems more efficiently than classical computers. However, it requires the integration of new tools and techniques into the software lifecycle [31, 32] and imposes a new programming paradigm on developers. Businesses need to determine which application parts might benefit from quantum computers [33] and decide which parts to implement as quantum or classical applications [34]. A quantum adviser system aims to assist stakeholders in determining which quantum and classical computing technologies are best suited for their specific scenarios. The system consists of three main components: a knowledge base, an interactive questionnaire, and an analysis component. Via the questionnaire the user describes the problem, algorithm-specifics and existing solution. The tool then compares the best quantum and classical algorithms for the given use case using the knowledge base. The advisor also provides decision support for the selection of warm-starting techniques.

QProv. Quantum computers are error-prone and have changing characteristics, making it crucial to collect and store long-term provenance data to improve reproducibility and detect errors [35]. The provenance system, QProv [36], was developed to overcome this issue by automatically retrieving data from quantum hardware vendors' APIs and executing calibration circuits. The collected data can be divided into four categories: executed quantum circuits, used quantum computers, and the overall performance of quantum computers. This data can be used to mitigate readout errors and improve the execution results of quantum circuits.

Camunda Workflow Engine. The Camunda Workflow Engine [37] is a modern workflow engine that executes BPMN workflows. It navigates through the model, determining activities ready to be launched and determining the execution workflow [38]. The engine also needs knowledge about the implementation of activities, such as classical or quantum programs, and how to invoke them. It also handles errors and allows users to specify custom execution flows in case of errors. The engine is designed to handle distributed and heterogeneous execution environments.

Quokka Ecosystem. The Quokka Ecosystem [30] is a service-based platform for building quantum applications using microservices. It comprises seven services, including the quantum circuit generator, circuit execution, error mitigation, objective evaluation, optimization, circuit-cutting, and warm-starting. These services unify access to Quokka, enabling the generation of quantum circuit fragments, execution of quantum circuits, error mitigation, objective evaluation, optimization, and pre-computation of quantum circuit parameters and initial states.

Testing Framework. The modern software engineering process for hybrid quantum computing applications requires a seamless co-development of classical and quantum computing code. Testing the developed application is crucial to ensure it is error-free and can be executed on the target hardware. The testing framework implements a unit testing module for hybrid quantum-classical codebases, targeting Python applications with the *Qiskit* SDK and *Qrisp* programming language. The testing tool can be used independently, but the best functionality is achieved through synergies within the provided ecosystem.

Benchmarking Suite. The benchmarking suite can be called from the testing framework for running suitable benchmarks or retrieving results from previously performed benchmarks stored in the *benchmarking database* or the *provenance repository*. Additionally, the testing framework can call the *circuit cutting tool* when backend decision, validation or evaluation fails. The circuit cutting tool [15] thus can feed back potentially cut circuits to the testing tool and stays active for necessary postprocessing tasks that in general are mandatory for reconstructing the initial probability distributions.

Benchmarking Methods for Automatic Optimization/synthesis of PQCs. AI advancements have prompted interest in quantum circuit optimization and generation [39–42]. Quantum Architecture Search (QAS)) [43] is a subfield that uses AI and heuristic search techniques to generate and optimize parametrized quantum circuits. However, evaluating QAS methods is hindered by a lack of standardized search spaces and high computational costs. SQuASH[9], a surrogate-based benchmark, aims to speed up experimentation, enable broader comparisons, support various QAS strategies, and provide pretrained surrogate models for each search space.

Quantum-Classical Splitting Service. Classical simulation of quantum circuits is crucial for the validation of quantum hardware and testing of quantum algorithms [45]. Stabilizer circuits are particularly interesting as they can be efficiently simulated on classical hardware [44]. A procedure has been developed to parse ZX diagrams and identify a cutoff border between Clifford and non-Clifford sections of a given circuit [46]. This technique can efficiently process the Clifford section via classical means, allowing for further simplifications. The process can also accelerate statevector simulations for split circuits.

Knowledge Graph. The EniQmA knowledge graph (KG) is an extension of the PLANQK KG [47,48], providing a structured representation of quantum software tools and benchmarks. It is structured in two layers, using the Resource Description Framework (RDF) and the PLANQK platform. The graph enables automated information reasoning and serves as a reliable knowledge source for large language models, enhancing tool discovery and collaboration in the quantum computing research community.

MiniZinc-to-QUBO. The MiniZinc-to-QUBO converter is a crucial component in EniQmA, allowing for a high-level specification of optimization problems for quantum computers, particularly on quantum annealers. MiniZinc [49], a free and open-source constraint modeling language, is used to interface with several solvers, such as Google tools[10], Gecode[11], and Chuffed[12]. It converts the problem from MiniZinc format to a strict subset of MiniZinc called FlatZinc, which is then parsed by solvers. The module implements QUBO (Quadratic Unconstrained Binary Optimization) problems, which involve turning the industrial problem definition into a matrix of real numbers that encodes the strengths of qubit couplings in a quantum annealer. This automatic converter avoids the need for industry users to learn the internals of quantum annealing and find adequate QUBO models equivalent to the initial problems they need to solve. The converter addresses problems with integer variables and rational-valued variables

---

with fixed precision. A prototypical implementation of the MiniZinc-to-QUBO converter is available as a web service[13], not only supporting the conversion from MiniZinc to QUBO but also the interpretation of QUBO solutions, i.e. 0/1 vectors, as solutions of the original MiniZinc problems. The solution output is formatted as specified in the MiniZinc program.

References

1. EniQmA Homepage. https://www.eniqma-quantum.de/. Accessed 17 Sept 2025
2. Gheorghe-Pop, I.-D., et al.: Quantum DevOps: towards reliable and applicable NISQ quantum computing. In: IEEE Globecom Workshops. IEEE (2020). https://doi.org/10.1109/GCWkshps50303.2020.9367411
3. Rudolph, M.S., et al.: Synergistic pretraining of parametrized quantum circuits via tensor networks. Nat. Commun. **14**(1) (2023). ISSN: 2041-1723. https://doi.org/10.1038/s41467-023-43908-6
4. Guo, S., et al.: Experimental quantum computational chemistry with optimized unitary coupled cluster ansatz. Nat. Phys. (2024). ISSN: 1745-2481. https://doi.org/10.1038/s41567-024-02530-z
5. Abbas, A., et al.: Quantum Optimization: Potential, Challenges, and the Path Forward (2023). arXiv:2312.02279 [quant-ph]
6. Fauseweh, B.: Quantum many-body simulations on digital quantum computers: state-of-the-art and future challenges. Nat. Commun. **15** (2024). https://doi.org/10.1038/s41467-024-46402-9
7. Abbas, A., et al.: The power of quantum neural networks. Nat. Comput. Sci. **1**(6), 403–409 (2021). ISSN: 2662-8457. https://doi.org/10.1038/s43588-021-00084-1
8. Bravo-Prieto, C., et al.: Variational quantum linear solver. Quantum **7**, 1188 (2023). ISSN: 2521-327X. https://doi.org/10.22331/q-2023-11-22-1188
9. Beisel, M., et al.: QuantME4VQA: modeling and executing variational quantum algorithms using workflows. SciTePress, pp. 306–315 (2023)
10. Eberle, H., Unger, T., Leymann, F.: Process Fragments, pp. 398– 405. Springer (2009)
11. Truger, F., et al.: Warm-Starting and Quantum Computing: A Systematic Mapping Study, arXiv:2303.06133 (2023)
12. Cheng, M.H., et al.: Clifford circuit initialization for variational quantum algorithms. Phys. Rev. A **111**(6), 062413 (2025). https://link.aps.org/doi/10.1103/PhysRevA.111.062413
13. Lisart-Liebermann, T., Medina, A.C.: Preconditioning Natural and Second Order Gradient Descent in Quantum Optimization: A Performance Benchmark, arXiv:2504.16518 [cs.CE] (2025)
14. Lisart-Liebermann, T., Medina, A.C.: Clifford Accelerated Adaptive QAOA, arXiv:2508.16443 [quant-ph] (2025). https://arxiv.org/abs/2508.16443
15. Bechtold, M., et al.: Cutting a wire with non-maximally entangled states. In: 2024 IEEE International Parallel and Distributed Processing Symposium Workshops (IPDPSW), pp. 1136–1145. IEEE (2024). http://dx.doi.org/10.1109/ipdpsw63119.2024.00185
16. Zhu, L., et al.: An adaptive quantum approximate optimization algorithm for solving combinatorial problems on a quantum computer. arXiv preprint arXiv:2005.10258 (2022). https://arxiv.org/abs/2005.10258

---

[13] https://minizinc2qubo.fokus.fraunhofer.de/.

17. Falkenthal, M., et al.: PlanQK - platform and ecosystem for quantum applications. Kunstliche Intelligenz **38**(4), 371–377 (2024). https://doi.org/10.1007/S13218-024-00865-6

18. Frisch, A., et al.: Trapped-ion quantum computing. In: Exman, I., et al. (ed.) Quantum Software: Aspects of Theory and System Design, pp. 251–283. Springer, Cham (2024). https://doi.org/10.1007/978-3-031-64136-710

19. Nagies, S., et al.: The role of higher-order terms in trapped-ion quantum computing with magnetic gradient induced coupling (2024). https://doi.org/10.1548550/ARXIV.2409.10498

20. Montanaro, A., Pallister, S.: Quantum algorithms and the finite element method. Phys. Rev. A **93**(3), 032324 (2016)

21. Ali, M., Kabel, M.: Performance study of variational quantum algorithms for solving the poisson equation on a quantum computer. Phys. Rev. Appl. **20**(1), 014054 (2023)

22. Cerezo, M., et al.: Variational quantum algorithms. Nat. Rev. Phys. **3**(9), 625–644 (2021). ISSN: 2522-5820. https://doi.org/10.1038/s42254-021-00348-9

23. Geng, A., et al.: Improved FRQI on superconducting processors and its restrictions in the NISQ era. Quantum Inf. Process. **22**(2), 104 (2023). https://doi.org/10.1007/s11128-023-03838-0

24. Geng, A., et al.: Hybrid quantum transfer learning for crack image classification on NISQ hardware. AIP Conf. Proc. (2025). https://doi.org/10.1063/5.0246500

25. Yi, J., et al.: Variational Quantum Linear Solver enhanced Quantum Support Vector Machine (2023). arXiv:2309.07770 [quant-ph]

26. Rebentrost, P., Mohseni, M., Lloyd, S.: Quantum support vector machine for big data classification. Phys. Rev. Lett. **113**(13), 130503 (2014)

27. Bickert, P., et al.: Optimising rolling stock planning including maintenance with constraint programming and quantum annealing. arXiv preprint arXiv:2109.07212 (2021)

28. Truger, F., Barzen, J., Leymann, F., Obst, J.: Warm-Starting the VQE with Approximate Complex Amplitude Encoding, arXiv:2402.17378 [quant-ph] (2024). https://arxiv.org/abs/2402.17378

29. Truger, F., et al.: Selection and optimization of hyperparameters in warm-started quantum optimization for the MaxCut problem. Electronics **11**(7) (2022). https://www.mdpi.com/2079-9292/11/7/1033/pdf

30. Beisel, M., et al.: Quokka: a service ecosystem for workflow-based execution of variational quantum algorithms. In: Service-Oriented Computing – ICSOC 2022 Workshops, pp. 369–373. Springer (2023)

31. Weder, B., et al.: Integrating quantum computing into workflow modeling and execution. In: Proceedings of the 13th IEEE/ACM International Conference on Utility and Cloud Computing (UCC), pp. 279–291. IEEE (2020)

32. Weder, B., et al.: Quantum Software Development Lifecycle. In: Serrano, M.A., Perez-Castillo, R., Piattini, M. (eds.) Quantum Software Engineering, pp. 61–83. Springer (2022)

33. Vietz, D., et al.: An exploratory study on the challenges of engineering quantum applications in the cloud. In: Proceedings of the 2nd Quantum Software Engineering and Technology Workshop (Q-SET 2021) co-located with IEEE International Conference on Quantum Computing and Engineering (QCE21), pp. 1–12. CEUR Workshop Proceedings (2021). http://ceur-ws.org/Vol-3008/paper1.pdf

34. Vietz, D., et al.: On decision support for quantum application developers: categorization, comparison, and analysis of existing technologies. In: Computational

Science – ICCS 2021, pp. 127–141. Springer (2021). https://doi.org/10.1007/978-3-030-77980-110

35. Tannu, S.S., Qureshi, M.K.: Not all qubits are created equal: a case for variability-aware policies for NISQ-era quantum computers. In: Proceedings of the 24th International Conference on Architectural Support for Programming Languages and Operating Systems, pp. 987–999 (2019)

36. Weder, B., et al.: Provenance-preserving analysis and rewrite of quantum workflows for hybrid quantum algorithms. SN Comput. Sci. **4**(233), 1–19 (2023)

37. Camunda. Camunda BPMN Workflow Engine (2023). https://camunda.com/products/camunda-bpm/bpmn-engine

38. Leymann, F., Roller, D.: Production Workflow: Concepts and Techniques. Prentice Hall PTR (2000)

39. Paradis, A., et al.: Synthetiq: fast and versatile quantum circuit synthesis. In: Proceedings of the ACM on Programming Languages 8.OOPSLA1, pp. 55–82 (2024)

40. Nakaji, K., et al.: The generative quantum eigensolver (GQE) and its application for ground state search. arXiv preprint arXiv:2401.09253 (2024)

41. Patel, Y.J., et al.: Curriculum reinforcement learning for quantum architecture search under hardware errors. arXiv preprint arXiv:2402.03500 (2024)

42. Tang, W., et al.: AlphaRouter: quantum circuit routing with reinforcement learning and tree search. In: 2024 IEEE International Conference on Quantum Computing and Engineering (QCE), vol. 1, pp. 930–940. IEEE (2024)

43. Martyniuk, D., Jung, J., Paschke, A.: Quantum architecture search: a survey. In: 2024 IEEE International Conference on Quantum Computing and Engineering (QCE), vol. 01, pp. 1695–1706 (2024). https://doi.org/10.1109/QCE60285.2024.00198

44. Aaronson, S., Gottesman, D.: Improved simulation of stabilizer circuits. Phys. Rev. A **70**(5) (2004). ISSN: 1094-1622. https://doi.org/10.1103/physreva.70.052328

45. Xu, X., et al.: A Herculean task: classical simulation of quantum computers (2023). arXiv:2302.08880 [quant-ph]. https://arxiv.org/abs/2302.08880

46. Lima, F., Medina, A.C.: Clifford and Non-Clifford Splitting in Quantum Circuits: Applications and ZX-Calculus Detection Procedure (2025). arXiv:2504.16004 [quant-ph]. https://arxiv.org/abs/2504.16004

47. Martyniuk, D., Jung, J., Paschke, A.: Quantum architecture search: a survey. In: 2024 IEEE International Conference on Quantum Computing and Engineering (QCE), vol. 01, pp. 1695– 1706 (2024). https://doi.org/10.1109/QCE60285.2024.00198

48. Martyniuk, D., et al.: An analysis of ontological entities to represent knowledge on quantum computing algorithms and implementations. In: Conference on Digital Curation Technologies (2021)

49. Nethercote, N., et al.: MiniZinc: towards a standard CP modelling language. In: International Conference on Principles and Practice of Constraint Programming, pp. 529–543. Springer (2007)

Session: 8 Quantum Algorithms, Computing; Simulation – Physics and Engineering Applications

Numerical Experiments Using Block-Diagonalization Technique for Solving Poisson's Equation

Chetra Mang[1]($\boxtimes$), Sunheang Ty[1], Axel TahmasebiMoradi[1], Renaud Vilmart[2], and Rim Kaddah[1]

[1] IRT SystemX, Bd Thomas Gobert, 91120 Palaiseau, France
chetra.mang@irt-systemx.fr
[2] Université Paris-Saclay, Inria, CNRS, ENS Paris-Saclay, Laboratoire Méthodes Formelles, 91190 Gif-sur-Yvette, France

Abstract. This work presents numerical experiments aimed at verifying solutions of Poisson's equation using two existing methodologies. First, block-diagonalization is employed to block-encode the matrix derived from Poisson's equation through the finite difference method (FDM), significantly improving computational complexity from N to $\log(N)$, where N is the matrix size. Second, the Quantum Singular Value Transformation (QSVT) algorithm is applied to invert the matrix. However, while block-diagonalization improves the complexity in N, QSVT introduces a bottleneck due to its linear dependency on the condition number κ, which grows exponentially with N, posing challenges for large-scale problems. As far as we know, this is the first numerical experiments solving problems with matrix size $N = 1024$ and condition number $\kappa = 500000$; the largest matrix size and condition number from existing works are 16 and < 100, respectively.

Keywords: Block diagonalization · Quantum Singular Value Transformation · Partial differential equation · Finite difference method

1 Introduction

Partial Differential Equations (PDEs) are essential for modeling various physical phenomena. Analytical solutions are often infeasible for complex problems, prompting the use of numerical methods. Discretization techniques, including Finite Difference Method (FDM), Finite Element Method (FEM), and Finite Volume Method (FVM), approximate continuous PDEs into solvable algebraic equations ($Ku = f$). While powerful, these methods demand significant computational resources for large-scale applications. Quantum computation holds great promise for large-scale problems where classical resources, such as memory, are insufficient. It has the potential to offer faster computation for solving discretized PDEs, overcoming the bottlenecks of traditional methods. PDEs can be addressed through quantum computation either continuously, using quantum

F. Barbaresco and F. Gerin (Eds.): QUEST-IS 2025, CCIS 2744, pp. 167–174, 2026.
https://doi.org/10.1007/978-3-032-13855-2_15

algorithms such as those outlined in [3] or discretely using the Quantum Linear System Solver (QLSS) [4,9].

Quantum approaches to solving linear systems, such as QLSS, offer computational advantages, particularly in the efficient handling of large systems. These algorithms rely on well-conditioned and sparse matrices. These include the HHL algorithm [7], Quantum Singular Value Transformation (QSVT) [9,10], and Quantum Variational Linear Solver (QVLS) [12,19].

Several studies highlight the potential of quantum computing in solving linear system of equations and differential equations. One such study, presented in [8], combines Quantum Singular Value Transformation (QSVT) with iterative refinement in a hybrid quantum-classical framework. In the experiments, the size of the problem was set to $N = 16$. This approach used QSVT for low-precision initial solutions, refined by classical processors to achieve high accuracy, reducing quantum resource demands through low-degree polynomial approximations. A similar iterative improvement is demonstrated in [16] for HHL algorithm, with simulation result for 4×4 matrices. HHL algorithm was applied in [22] to several scientific applications, including power grid management and heat transfer. Experimentally, variants of HHL algorithm demonstrated on actual gate-based quantum computers solving problems of size 4×4 [11,15,20], while another quantum linear solver algorithm based on adiabatic quantum computing is demonstrated in [18].

This article explores solving Poisson's equation using block-diagonalization as described in [17], a refinement of block-encoding, to simplify quantum circuits with minimal ancilla qubits and basic gates, ensuring almost double-logarithmic depth efficiency. Additionally, instead of using the convention implementation of the inverse function in QSVT, we propose to find its Chebyshev polynomial series approximation via an optimization method. Numerical experiments were performed on matrices of varying sizes and condition number up to $N = 1024$ and $\kappa = 500000$, demonstrating the quantum linear algorithm's capabilities and improved efficiency via the block-diagonalization method. Note that the matrix size and condition number of existing numerical experiments are only up to 16 and < 100, respectively.

2   Description of the Problems

We consider a Poisson equation in a 1-dimensional space $\Omega = [0, 1]$ with a homogeneous Dirichlet boundary condition $u = 0$ on the boundary $\partial\Omega$.

$$-\Delta u(x) = f(x) \text{ for } x \in \Omega \tag{1}$$

where $f(x)$ is a given smooth function and $u(x)$ is the solution to Eq. (1).

We discretize the problem using the finite difference method into $N + 2$ grid points with a grid size of $h = \frac{1}{N+2}$. Then, we derive a linear system of equations of size $N = 2^n$:

$$\frac{1}{h^2}\begin{pmatrix} 2 & -1 & & & & & & & 0 \\ -1 & 2 & -1 & & & & & & \\ & -1 & 2 & & & & & & \\ & & & \ddots & & & & & \\ & & & & 2 & -1 & & & \\ & & & & -1 & 2 & -1 & & \\ & & & & & -1 & 2 & -1 & \\ 0 & & & & & & & -1 & 2 \end{pmatrix} \cdot \begin{pmatrix} u_0 \\ u_1 \\ u_2 \\ u_3 \\ \vdots \\ u_{N-3} \\ u_{N-2} \\ u_{N-1} \end{pmatrix} = \begin{pmatrix} f_0 \\ f_1 \\ f_2 \\ f_3 \\ \vdots \\ f_{N-3} \\ f_{N-2} \\ f_{N-1} \end{pmatrix} \tag{2}$$

$$\underbrace{\hphantom{K}}_{K} \qquad \underbrace{\hphantom{|u\rangle}}_{|u\rangle} \qquad \underbrace{\hphantom{|f\rangle}}_{|f\rangle}$$

3   Methodologies

Block Encoding - Quantum operations are represented by unitary matrices. Block encoding is one of the techniques that allows one to embed a non-unitary matrix into a unitary matrix [6]. The matrix K can be block-encoded into a unitary matrix U:

$$U = \begin{pmatrix} K & * \\ * & * \end{pmatrix} \tag{3}$$

While there are systematic ways to block encode a matrix K, it is paramount for the efficiency of the QLSS that the encoding be done somewhat efficiently, so as to not constitute a bottleneck.

Block Diagonalization - The block-diagonalization technique can be used to efficiently decompose the matrix K into a linear combination of matrices; each of which can be block-encoded with quantum circuits. The detailed decomposition of the matrix K and its quantum circuit can be found in [17]. The complexity in depth of this block-encoding is bounded by $\mathcal{O}(log^2(log(N)))$ with the use of $\mathcal{O}(1)$ ancilla.

Linear System Solver - We use the QSVT method to approximate the inverse of the matrix K. To do so, the inverse function $1/x$ is approximated by a polynomial series. It is then possible to find a series of angles that can be used inside a particular routine to implement the polynomial. There are several techniques to approximate the inverse function, including the truncated Chebyshev polynomial series with degree $D = \lceil \kappa \log(\frac{\kappa}{\epsilon}) \rceil$ to approximate the Child's function $\frac{1-(1-x^2)^b}{x}$ [1] where $b = \lceil \kappa^2 log(\frac{\kappa}{\epsilon}) \rceil$, and the Remez's algorithm [5]. As the condition number κ increases, the required number of terms in the Chebyshev series also grows. However, the influence of the error ϵ is relatively minor, as it only appears logarithmically in the formula as $\log(1/\epsilon)$. To be practical for this preliminary numerical experiment, we propose to find the approximated odd Chebyshev polynomial series via the optimization of Chebyshev polynomial series coefficients with the following cost function:

$$C(w) = \left\| \sum_{i \in \{0,...,d\}} w_i cos((2i+1)\theta) - \frac{1}{2\kappa cos(\theta)} \right\|_{L^2} \tag{4}$$

where $w = \{w_0, ..., w_d\}^T$ and $\theta \in [0, acos(\frac{1}{\kappa})]$.

The proposed technique could significantly reduce the Chebyshev polynomial series terms to reach a certain precision.

State Preparation f - Complexity of arbitrary state preparation of any vector is bounded by $\mathcal{O}(2^n)$ [21]. Here, we aim to generate some arbitrary vectors f with an $\mathcal{O}(1)$ complexity to save resources for the linear system solving part. In this case, we generate the vector f by applying the gate $R_Y(\theta_i)$ on each qubit $i \in \{0, ..., n-1\}$ with θ_i randomly chosen between $[0, \pi]$.

4   Numerical Experiments

4.1   Numerical Experiment Setup

To perform numerical experiments, the block diagonalization to block encode the matrix K, and the QSVT method to invert the matrix have been implemented using the Pennylane simulator platform [13]. To employ QSVT, we first apply the optimization technique to determine the approximated Chebyshev polynomial series, and then use this polynomial with pyqsp [14] to find the phase factors to be used in the QSVT routine.

To efficiently employ the QSVT method, we generate a dictionary of phase factors corresponding to different values of the condition number κ from 20 to 500000 with a tolerance of 1% error.

First, for a given vector f, we solve the problem with different matrix sizes 2^n for $n = \{2, 3, 5, 7, 10\}$ to numerically analyze the precision of the solution via QSVT. Then, we solve the problem with matrix size $N = 32$ for different randomly generated vectors f.

4.2   Numerical Results

QSVT Numerical Analysis - We present in Table 1 the relationship of Chebyshev polynomial degree with respect to the condition number using the optimization function in 4, and the Child's function [1]. For $\kappa = 7500$, the polynomial degree using the optimization technique is 13 times smaller than that of the Child's function while its approximation error is still satisfied. For $\kappa = 500000$, the approximation error through the optimization technique is not fulfilled due the lack of polynomial terms; however, its degree is nearly 400 times smaller than that using the Child's function with the same error. Even if the optimization technique can achieve a very small degree compared to the Child's function, it is still a heuristic method. We cannot always obtain the desired error.

Table 1. Chebyshev polynomial degree via optimization technique and Child's function

κ	20	50	450	7500	500000
ϵ		0.009	0.007	0.004	0.006	0.062
Degree via optimization technique	50	132	1380	8160	22334
Degree via Child's function:$\kappa log(\frac{\kappa}{\epsilon})$	154	443	5233	105289	7967887

We investigate (Fig. 1) the evolution of the condition number κ in terms of the matrix size. κ rapidly increases as the matrix size increases. For the matrix size of $N = 1024$, the condition number is relatively large $\kappa \approx 400000$. This shows that the QSVT method is not suitable for solving large matrices from discretized PDE problems (at least without preconditioning). Figure 2 also illustrates the evolution of the relative error of the quantum solution with respect to the reference (classical) solution. For the matrix size $N = 1024$, the relative error is largely deviated from the tolerance because of the target error of the approximated Chebyshev polynomial series using the optimization technique. It should be noted that we did not reach a sufficient number of terms in the series to achieve the desired level of error.

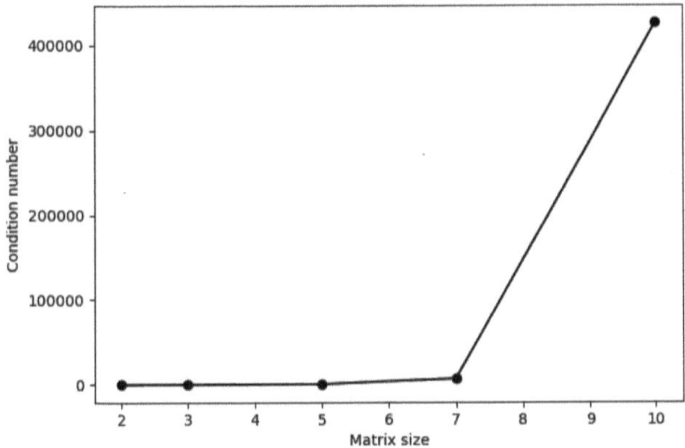

Fig. 1. Condition number κ for different matrix size 2^n for $n = \{2, 3, 5, 7, 10\}$.

PDE Solutions - We present in Fig. 3 the relative error for the problem of matrix size $N = 32$ with different vectors f. Among the solutions corresponding to 50 randomly generated vectors f, 90% of the solutions provide an error less than the tolerance. Nevertheless, the other 10% of the solutions are only 0.25% deviated from the tolerance. Figure 4 illustrates the comparison between the reference and quantum solutions for the problem with matrix size of $N = 32$. The overall behavior of the quantum solution is well fit to the reference solution.

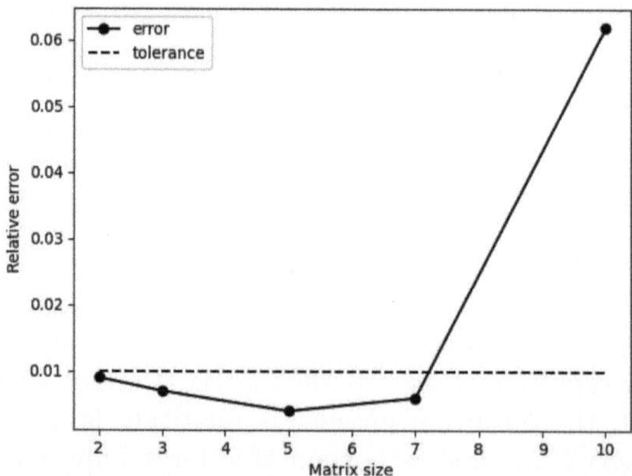

Fig. 2. Relative error ϵ for different matrix size 2^n for $n = \{2, 3, 5, 7, 10\}$.

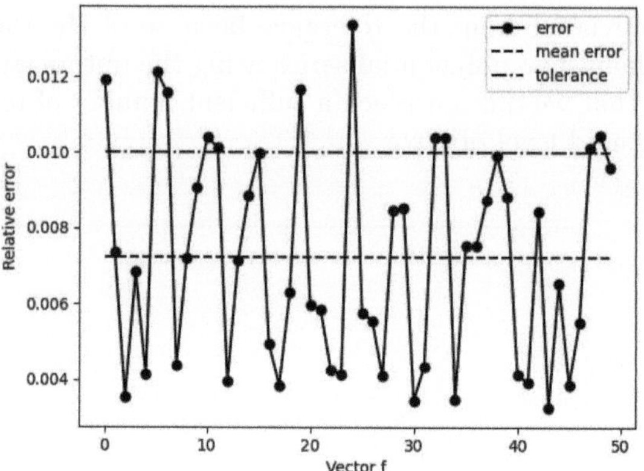

Fig. 3. Relative errors for the matrix size $N = 32$ for different vectors f.

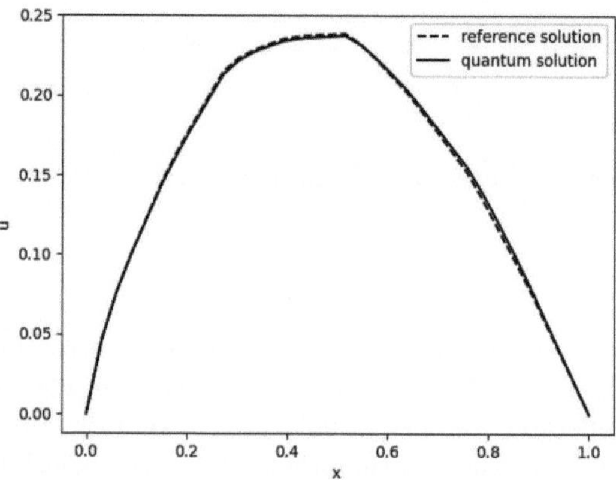

Fig. 4. Reference and quantum solution for the matrix size $N = 32$ for a given vector f.

5   Conclusions and Discussions

We conducted numerical experiments of Poisson's equation by solving the associated linear system of equations using block-diagonalization and the Quantum Singular Value Transformation (QSVT) solver. These preliminary numerical experiments reveal certain limitations when applying the QSVT solver to linear systems derived from Poisson's equation. Specifically, for problems of large size, the condition number of the resulting matrix becomes exponentially large. Hence, the complexity of QSVT grows exponentially as well. To address this issue, the mixed-precision classical-quantum algorithm proposed in [8] offers a promising solution by reducing the number of Chebyshev terms needed through enhanced solution precision. Additionally, leveraging efficient quantum preconditioning techniques, such as those outlined in [2], could significantly mitigate the condition number.

References

1. Childs, A.M., Kothari, R., Somma, R.D.: Quantum algorithm for systems of linear equations with exponentially improved dependence on precision. SIAM J. Comput. **46**(6), 1920–1950 (2017)
2. Clader, B.D., Jacobs, B.C., Sprouse, C.R.: Preconditioned quantum linear system algorithm. Phys. Rev. Lett. **110**(25), 250504 (2013)
3. Dalzell, A.M., et al.: Quantum algorithms: a survey of applications and end-to-end complexities. arXiv preprint arXiv:2310.03011 (2023)
4. Dervovic, D., Herbster, M., Mountney, P., Severini, S., Usher, N., Wossnig, L.: Quantum linear systems algorithms: a primer. arXiv preprint arXiv:1802.08227 (2018)
5. Dong, Y., Meng, X., Whaley, K.B., Lin, L.: Efficient phase-factor evaluation in quantum signal processing. Phys. Rev. A **103**(4), 042419 (2021)
6. Gilyén, A., Su, Y., Low, G.H., Wiebe, N.: Quantum singular value transformation and beyond: exponential improvements for quantum matrix arithmetics. In: Proceedings of the 51st Annual ACM SIGACT Symposium on Theory of Computing, pp. 193–204 (2019)
7. Harrow, A.W., Hassidim, A., Lloyd, S.: Quantum algorithm for linear systems of equations. Phys. Rev. Lett. **103**(15), 150502 (2009)
8. Koska, O., Baboulin, M., Gazda, A.: A mixed-precision quantum-classical algorithm for solving linear systems. arXiv preprint arXiv:2502.02212 (2025)
9. Low, G.H., Chuang, I.L.: Optimal hamiltonian simulation by quantum signal processing. Phys. Rev. Lett. **118**(1), 010501 (2017)
10. Low, G.H., Chuang, I.L.: Hamiltonian simulation by qubitization. Quantum **3**, 163 (2019)
11. Morgan, J., Ghysels, E., Mohammadbagherpoor, H.: An enhanced hybrid hhl algorithm. Phys. Lett. A **532**, 130181 (2025)
12. Patil, H., Wang, Y., Krstić, P.S.: Variational quantum linear solver with a dynamic ansatz. Phys. Rev. A **105**(1), 012423 (2022)
13. Pennylane: Pennylane (2025). https://pennylane.ai/
14. pyqsp: pyqsp (2025). https://github.com/ichuang/pyqsp?tab=readme-ov-file

15. Sævarsson, B., Chatzivasileiadis, S., Jóhannsson, H., Østergaard, J.: Quantum computing for power flow algorithms: testing on real quantum computers. arXiv preprint arXiv:2204.14028 (2022)
16. Saito, Y., Lee, X., Cai, D., Asai, N.: An iterative improvement method for hhl algorithm for solving linear system of equations. arXiv preprint arXiv:2108.07744 (2021)
17. Ty, S., Vilmart, R., TahmasebiMoradi, A., Mang, C.: Double-logarithmic depth block-encodings of simple finite difference method's matrices (2024). https://arxiv.org/abs/2410.05241
18. Wen, J., Kong, X., Wei, S., Wang, B., Xin, T., Long, G.: Experimental realization of quantum algorithms for a linear system inspired by adiabatic quantum computing. Phys. Rev. A $99(1)$, 012320 (2019)
19. Xu, X., Sun, J., Endo, S., Li, Y., Benjamin, S.C., Yuan, X.: Variational algorithms for linear algebra. Sci. Bull. $66(21)$, 2181–2188 (2021)
20. Yalovetzky, R., Minssen, P., Herman, D., Pistoia, M.: Solving linear systems on quantum hardware with hybrid hhl++. Sci. Rep. $14(1)$, 20610 (2024)
21. Zhang, X.M., Li, T., Yuan, X.: Quantum state preparation with optimal circuit depth: implementations and applications. Phys. Rev. Lett. $129(23)$, 230504 (2022)
22. Zheng, M., et al.: An early investigation of the hhl quantum linear solver for scientific applications. arXiv preprint arXiv:2404.19067 (2024)

A Hybrid Quantum-Classical Algorithm for Fault Tree Analysis Via Finding Vertex Separators

Pedro Henrique Pons Fiorentin[1] , Mohamed Hibti[2(✉)] , and Rola Saidi[2]

[1] Amazon, Seattle, USA
[2] EDF Lab Paris-Saclay, PERICLES, Paris, France
`{mohamed.hibti,rola.saidi}@edf.fr`

Abstract. Fault Tree Analysis (FTA) is one of the main problems for reliability studies, in particular for the probabilistic safety assessment of complex systems such as aircrafts or nuclear power plants. The extraction of Minimal Cut Sets (MCS) from a fault tree can be reduced to the problem of finding vertex separators in S-T directed acyclic graphs (S-T DAGs). Prior work introduced a quantum algorithm to identify these s-t vertex separators, but faces scalability issues in larger instances. In this paper, we present a novel hybrid quantum-classical algorithm to solve the problem of st-connectivity using a divide and quantum strategy. Large sâĂŞt DAG instances are partitioned into many subgraphs that fit the qubit capacity of current quantum hardware; each subgraph is solved on a quantum device, yielding partial solutions. These solutions are then recombined classically to produce all global vertex separators of the original s-t DAG, which are mapped back to the MCS of the original fault tree. The results showed that we were able to solve larger instances that have not been solved using a classical algorithm previously. We also quantify the effect of realistic noise by comparing results from noisy versus noise-free simulation, and report on prototype runs using IBM's publicly available quantum hardware.

Keywords: S-t connectivity · Fault-Tree Analysis · Minimal Cut Sets · Divide and Quantum · Hybrid Quantum Classical Algorithm

1 Introduction

Probabilistic Safety Assessment [14] is a systematic approach used to evaluate the safety and reliability of complex engineering systems, particularly in high-risk industries such as nuclear power, aerospace, chemical processing, and transportation. It is a risk-based analysis method that quantifies the probability of hazardous events and their potential consequences.

Fault Tree Analysis (FTA) is the main method for addressing PSA. It has been used since 1961 for visual displaying, analyzing and evaluating failure paths in a system [13]. Its objective is to identify all the combinations of basic events

F. Barbaresco and F. Gerin (Eds.): QUEST-IS 2025, CCIS 2744, pp. 175–185, 2026.
https://doi.org/10.1007/978-3-032-13855-2_16

that lead to an undesired event, called the failure event, and sum up their probabilities. There are many algorithms that were used for FTA within commercial or academic software applications; based on Boolean fusion over modules with approximation procedures for large instances (Riskspectrum, CAFTA, XFTA [15]). Others use Binary Decision Diagram (BDD)-like algorithms [11,16] or/and approximations mixed with BDD or Ternary Decision Diagram (TDD) algorithms [12]. These algorithms are now challenged by the size of the instances and the frequent occurrence of non-rare events due to the implementation of external hazards such as seism.

Some quantum approaches have been investigated for the MCS problem, which mainly in the gate-based paradigm. One method uses a Quantum Fault Tree (QFT) model [1], where the FT is encoded into a quantum circuit, and applies Quantum Amplitude Amplification (QAA) to enhance the probability of sampling minimal cut sets [2]. Another approach reformulates the problem into a Boolean Satisfiablitiy problem and uses a Grover-based solver, with divide-and-conquer strategies [3]. A third method frames the problem as a vertex separator problem (VSP) in a directed graph [10]. In [18], it is shown that a quantum query algorithm can outperform any classical algrorithm for VSP. While promising, these methods remain limited by the small number of qubits and high noise levels of today's Noisy Intermediate-Scale Quantum (NISQ) computers.

Consequently, this work explores a hybrid quantum-classical algorithm for the MCS problem, in an attempt to tackle the number of qubit limitations.

2   FTA as the Vertex Separator Problem

Boolean formula evaluation reduces to st-connectivity problems [5,6]. In [9], the problem of enumerating the cut-sets in a FT's boolean formula is converted to the VSP by representing the FT as a graph; the problem is then solved with a quantum algorithm, which is referred to as Quantum-VSP. In the VSP, the goal is to find all combinations of vertices that, if removed from the graph, disconnect a source vertex **S** from a terminal vertex **T**. These separators are equivalent to the cut-sets of the fault tree. The VSP is NP-Hard [7].

More formally, given a directed acyclic graph (DAG) $G = (V, E)$, where V is the set of vertices and E is the set of directed edges, and two distinguished vertices S (the source) and T (the target) in V, enumerate all minimal vertex separators $\mathcal{C} = \{C_1, C_2, \ldots, C_k\}$ for G, such that:

1. Each C_i is a minimal vertex set that, when removed from G, disconnects all directed paths from S to T in the resulting subgraph $G'_i = (V \setminus C_i, E'_i)$.
2. No proper subset of C_i is a separator, i.e., C_i is minimal in disconnecting S and T.

In [10], the proposed quantum algorithm applies what we call movement oracles for the VSP to incrementally create a state with the superposition of all vertex separators that are not necessarily minimal. Each vertex v has an associated movement oracle that "moves" it to its successors $Succ(v)$. The runtime of the associated quantum circuit is linear, using $O(n + m)$ quantum gates and $n + 2$ qubits, where n is the number of vertices and m the number of edges in the DAG. In order to extract only the minimal separtors, a naive post-processing filtering step with complexity $O(N^2)$ is required, where N is the number of cuts. It consists of going through the list of separators and removing those that include other separators.

3   Hybrid Quantum VSP Algorithm

The approach of [10] to solve VSP problem requires a linear number of qubits to process an S-T graph, which prohibits the algorithm from scaling in current quantum computers with limited numbers of qubits. The question that arises from this is: can this approach be explored under the constraints of lower qubit availability?

This section introduces a hybrid quantum-classical approach for applying the VSP algorithm to subgraphs whose size is bounded by the number of available qubits N_q, enabling the construction of a global solution from smaller, tractable subgraphs.

The hybrid algorithm follows a divide and conquer approach, where the subdivided problems are processed by quantum computers and the global solution is found by a classical computer by joining subsolutions, as seen in Fig. 1. The question is then: can we build the MCS for the whole graph from these subsolutions and how computationally expensive is it? To answer the question of whether it is possible or not, the intuition of how to join subsolutions is explained in Sect. 4.

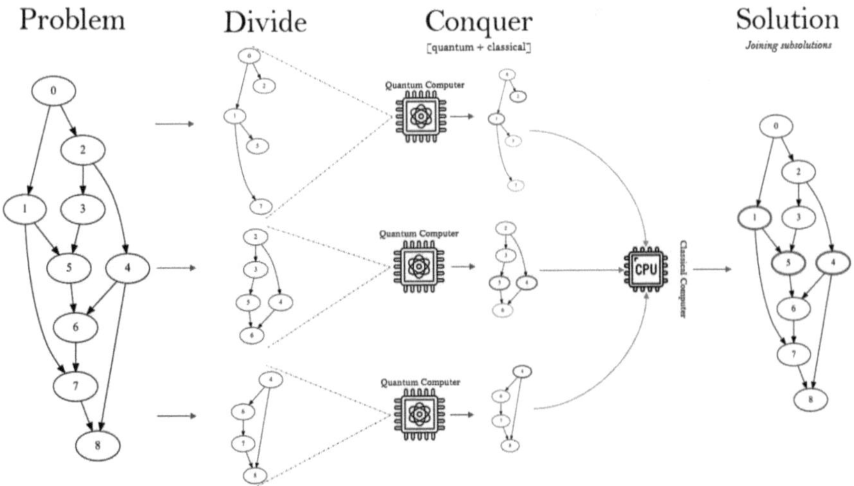

Fig. 1. Flow of hybrid VSP algorithm

4   Intuition of the Algorithm

To introduce the logic of the hybrid algorithm, we first present the basic intuition for it through examples of simple graphs. Consider the first example in Fig. 2, an S-T graph with two vertex-disjoint paths. In this scenario, the minimal separators for this graph can be obtained by performing a cartesian product of sets A and B. This fundamental principle remains consistent for any number of parallel, vertex-disjoint paths that could be added. If a third vertex-disjoint path C were added, the MCS would be $A \times B \times C$. When looking at an S-T graph with only a single path such as in Fig. 3, each element within set A represents an MCS in itself. From an algorithmic perspective, this corresponds to executing a Cartesian product between set A and an empty set. To extend this logic to our method, first let G be a DAG, and let v be a vertex in G. We define $S_G(v)$ as the subgraph of G that includes all vertices which topologically succeed vertex v. In other words, $S_G(v)$ consists of all vertices in G that are reachable from v through a directed path.

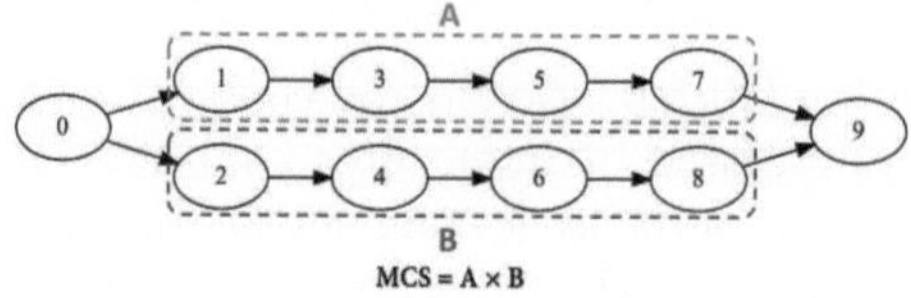

Fig. 2. S-T DAG with two vertex-disjoint paths.

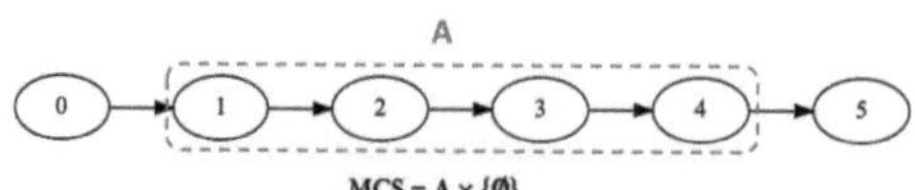

Fig. 3. Simple S-T DAG.

Having this in mind, we reexamine the graph $G = S_G(0)$ in Fig. 2. It becomes clear that the minimal separators of G, i.e. $MCS(S_G(0))$, are equivalent to the cartesian product of the minimal separators of the subgraphs $S_G(1)$ and $S_G(2)$, or $MCS(S_G(1)) \times MCS(S_G(2))$. More generally, the MCS for a source vertex s can be obtained by the cartesian product between the minimal separators of the subgraphs induced by each of $succ(s)$; we define this intermediate set of MCS as $MCS(succ(s))$.

In the pseudocode below, this operation of cartesian products between different sets of MC is defined through a FOLD as $MCS(v) := FOLD(\times, \{\emptyset\}, MCS(succ(v)))$. The FOLD is applied by iteratively folding, or accumulating, the sets of minimal separators using the cartesian product ($\times$) as the combining operator. The folding for the graph in Fig. 2 would be as follows:

1. Starting with a set $\{\emptyset\}$ and the first set $\{\{1\}, \{3\}, \{5\}, \{7\}\}$, we perform the cartesian product: $\{\emptyset\} \times \{\{1\}, \{3\}, \{5\}, \{7\}\} = \{\{1\}, \{3\}, \{5\}, \{7\}\}$.
2. To finish the folding, we take the result from step 1 and perform a cartesian product with the next set: $\{\{1\}, \{3\}, \{5\}, \{7\}\} \times \{\{2\}, \{4\}, \{6\}, \{8\}\}$.

The hybrid algorithm builds upon this logic, generating solutions through the exploration of local separators. Through this intuition, a global solution can be achieved, even if it includes non-minimal MCS that require subsequent filtering. This holds true as long as the algorithm operates in a topological sequence.

Algorithm 1: HYBRIDCUTSEARCH

Input: G - Directed S-T graph
         S_G - Source vertex of G
         N_q - Number of qubits available
         $quantum\_VSP$ - Quantum subroutine to enumerate MCS
Output: $CUTS$ - Separators of a graph

1   $CUTS\_DICT := \{\}$
2   $Q := [S_G]$
3   $processed\_nodes := \{\}$
4   **while** Q *is not empty* **do**
5      $S \leftarrow Q.pop(0)$
6   G', $processed\_v \leftarrow buildSubgraph(G, S, N_q)$
7      $processed\_nodes := processed\_nodes \cup processed\_v \cup S$
      `// Build dictionary with cuts starting from each subgraph source`
8      $CUTS\_DICT[S] \leftarrow quantum\_VSP(G')$
9

      `// Add unprocessed vertices to Q. An unprocessed vertex is a`
      `vertex whose successors were not included in the subgraph or`
      `whose degree is >= 2.`
10     $Q := Q + \{G' \setminus processed\_nodes\}$
11 **end**
12

      `// Now process the cuts in reverse topological order`
13 **for** S *in* $CUTS\_DICT.reversed()$ **do**
14     $new\_cuts := set()$
15   **for** C *in* $CUTS\_DICT[S]$ **do**
16        $cuts\_to\_cross := \{v \cup CUTS\_DICT[v] : v \in C\}$
17        $new\_cuts := FOLD(\times, \{\emptyset\}, cuts\_to\_cross)$
18   **end**
19     $CUTS\_DICT[S] := new\_cuts$
20 **end**
21 **return** $CUTS\_DICT[S_G]$

Algorithm 2: BUILDSUBGRAPH

Input: $\mathbb{G} = (\mathbb{V}, \mathbb{E})$ - Original graph
$\quad\quad\quad s$ - Source of the subgraph
$\quad\quad\quad N_q$ - Number of qubits available
Output: G' - Subgraph of G
$\quad\quad\quad\quad \mathbb{P}$ - Fully processed vertices

1 Initialize an empty subgraph $\mathbb{G}' = (\mathbb{V}' = \emptyset, \mathbb{E}' = \emptyset)$
2 Initialize queue $Q = [s]$
3
4 **while** $|Q| > 0$ *and* $|\mathbb{V}'| < N_q$ **do**
5 $\quad\quad v \leftarrow Q.pop(0)$
6 $\quad\quad \mathbb{V}' = \mathbb{V}' \cup \{v\}$
7 $\quad\quad$ **if** $|\mathbb{V}'| + |Succ(v)| < N_q$ **then**
8 $\quad\quad\quad\quad$ **if** $inDeg(v) = 1$ **then**
9 $\quad\quad\quad\quad\quad\quad \mathbb{P} := \mathbb{P} \cup \{v\}$
10 $\quad\quad\quad\quad Q \mathrel{+}= Succ(v)$
11 **end**
12
13 $E' = \{(a,b) \in \mathbb{E} : a \in \mathbb{V}' \wedge b \in \mathbb{V}'\}$
14 Return $\mathbb{G}' = (\mathbb{V}', \mathbb{E}')$, $\mathbb{P}$

---

5   Evaluation of the Method

This section covers the benchmarks executed to empirically evaluate the performance of the hybrid VSP method. The quantum subroutines execution were simulated with the library Qiskit assuming small devices of 6 qubits to guarantee correct results, as currently available quantum computers have too high of a noise rate.

Two kinds of benchmarks were executed, the first to evaluate the limits and running times of the novel method, and the second to compare its performance to another classical method. The goals of the benchmarks were to test the algorithm on randomly generated graphs of increasing size and analyze its performance, and compare the algorithm's running time against a recent classical method from the literature.

All graph instances generated were S-T DAGs of arbitrary size. We limit all vertices to have a maximum of 4 outgoing edges, following a standard in PSA models which often include 4 levels of redundancy at most. Figure 4 illustrates an instance that was evaluated and is tractable by the hybrid method under 30 min: To contextualize this performance, we compare our method to a recursive cutset enumerating algorithm published in 2012 [4]. This method's time complexity was not reported, but favorable benchmarks were shown in the paper. The classical method's performance is first plotted by itself in Fig. 5 and then compared with the hybrid method in Fig. 6.

The same method of instance generation was followed. The method evaluation was stopped once running times began surpassing 10 min, hence the cutoff

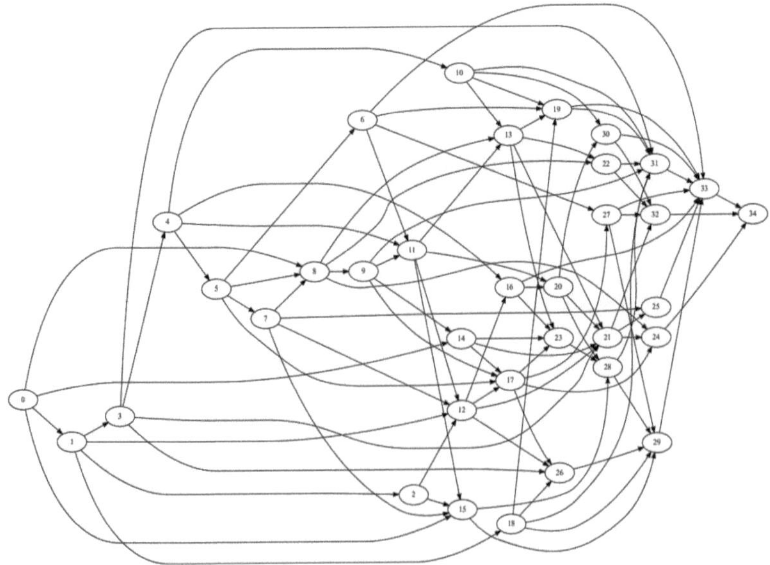

Fig. 4. Randomly generated instance with 34 vertices that have been solved with our hybrid algorithm but could not be solved with the classical one.

point at 15 vertices. An exponential trend can also be seen for the recursive algorithm with a very predictable performance, but the method does not work well for larger instances:

The classical algorithm has a quickly degrading performance once the instance size hits 13 vertices, which our method can still easily handle. This weak performance is naturally not convincing, and the large gap between the methods was particularly surprising (Fig. 6).

We remark that future work should include new benchmarks for the S-T DAG VSP problem with the hybrid algorithm and algorithms from the literature. Although research was done to find different classical implementations, to the best to our knowledge there were no other readily available implementations that could be used for comparisons in this benchmark.

To the best of our knowledge, these benchmarks show that this new hybrid algorithm is capable of dealing with relatively complex instances given the exponential nature of the problem. Additionally, to enhance this algorithm's practical applicability, a potential direction of future work is to implement it in a more performant programming language than Python.

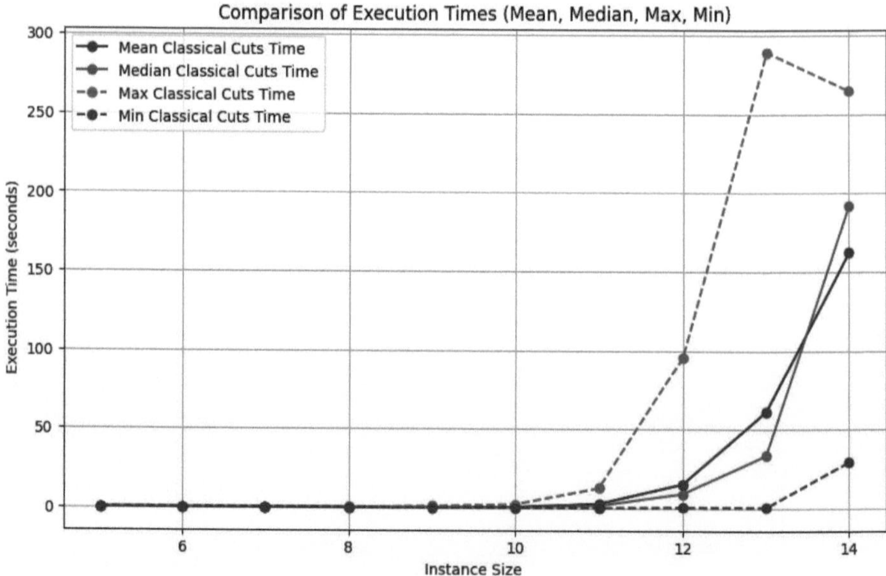

Fig. 5. Performance of classical recursive algorithm over increasing instance sizes.

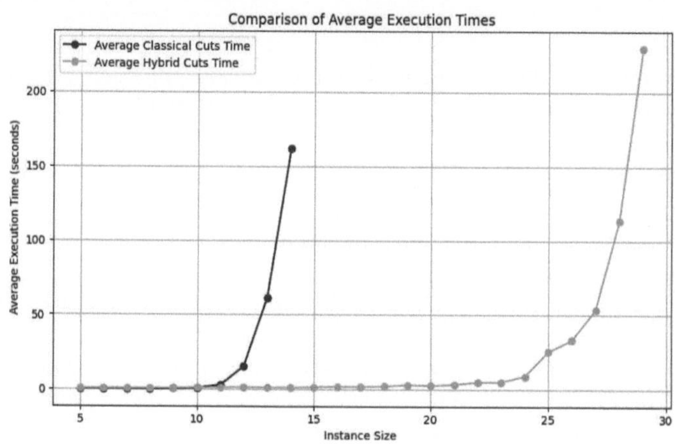

Fig. 6. Comparison of the hybrid algorithm running time against a recursive classical method.

6 Performance Assessment on Noisy Intermediate-Scale Quantum Devices (NISQ)

In the previous section, only Qiskit's classical simulators were used for the execution of the quantum subroutine to avoid the high error rates in quantum computers. However, in this section, we provide an evaluation of the algorithm on current Noisy Intermediate-Scale Quantum (NISQ) devices.

The experiments were performed using the version 1.2.0 of the IBM Qiskit framework. The tests utilized Qiskit's noisy simulator (FakeSherbrooke), as well as the quantum computers IBM Sherbrooke (127 qubits) and IBM Brisbane (127 qubits). To better assess the quality of the results from real quantum hardware, the Hellinger distance [17] was used as a common performance metric. Moreover, error mitigation techniques, specifically the Mthree(M3)[1] IBM's measurement errors mitigation tool [19] were used with only very slight improvement (Fig. 7).

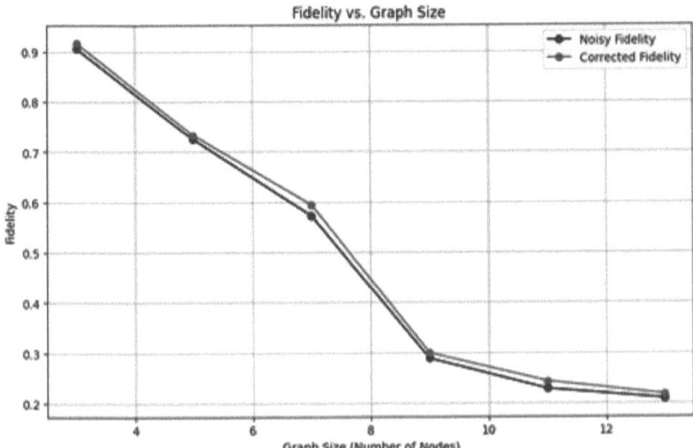

Fig. 7. Fidelity for MCS algorithm in function of graph sizes before and after Error Mitigation

7 Conclusion

This work's main contribution lies in proposing a novel hybrid framework for the S-T Vertex Separator Problem, which can make use of any new quantum subroutines that solve VSP. This hybrid framework enables the use small-scale quantum devices for solving PSA problems. By comparing this new method against a recent classical method for minimal cutset enumeration, there is a performance advantage with the hybrid algorithm. We remark that these results against a single method should be seen with caution and conclude that more benchmarks should be performed against different approaches from the literature.

All things considered, this hybrid framework serves as a foundation upon which subsequent improvements can be made, ultimately leading to more efficient computation in PSA. While practical implementation on current quantum hardware remains a challenge due to noise levels and decoherence, this work underscores the potential of quantum computing for PSA. Future work may focus on discovering more efficient quantum subroutines for the vertex separator problem and improving the filtering processes for minimal separator identification.

---

[1] https://qiskit-extensions.github.io/mthree/.

References

1. San Martín Silva, G., Parhizkar, T., López Droguett, E.: Quantum Fault Trees, Department of Civil & Environmental Engineering, University of California, Los Angeles, and The B. John Garrick Institute for the Risk Sciences, UCLA
2. San Martín Silva, G., López Droguett, E.: Quantum Fault Trees and Minimal Cut Sets Identification, The B. John Garrick Institute for the Risk Sciences, Center for Reliability Science and Engineering, Department of Civil and Environmental Engineering, UCLA
3. Rennela, M., Brand, S., Laarman, A., Dunjko, V.: Hybrid divide-and-conquer approach for tree search algorithms, arXiv preprint arXiv:2007.07040 (2021)
4. Benaddy, M., Wakrim, M.: Cutset enumerating and network reliability computing by a new recursive algorithm and inclusion exclusion principle. Int. J. Comput. Appl. 0975–8887 (2012)
5. Hibti, M.: What if we revisit evaluation of PSA models with network algorithms? ANS PSA 2013 International Topical Meeting on Probabilistic Safety Assessment and Analysis (2013)
6. Jeffery, S., Kimmel, S.: Quantum algorithms for graph connectivity and formula evaluation, arXiv, 1704.00765, (2019). https://arxiv.org/abs/1704.00765
7. Rendl, F., Sotirov, R.: The min-cut and vertex separator problem. Comput. Optim. Appl. **69**(1), 159–187 (2017)
8. Ton, K., Dieter, K.: Listing all minimal separators of a graph. SIAM J. Comput. **27**, 605–613 (1998)
9. Zaiou, A.: Quantum machine learning approaches for graphs and sequences: application to nuclear safety assessment. PhD thesis, 2022. Thèse de doctorat dirigée par Mateï, Basarab Informatique Paris, 13 2022
10. Zaiou, A., Bennani, Y., Hibti, M., Matei, B.: Quantum approach for vertex separator problem in directed graphs. In: Maglogiannis, I., Iliadis, L., Macintyre, J., Cortez, P. (eds.) Artificial Intelligence Applications and Innovations, pp. 495–506, Springer, Cham (2022)
11. Jung, W.S.: ZBDD algorithm features for an efficient probabilistic safety assessment. Nuclear Engineering and Design (2009). ISSN 00295493. https://doi.org/10.1016/j.nucengdes.2009.05.005
12. Jung, W.S.: Zero-suppressed ternary decision diagram algorithm for solving noncoherent fault trees in probabilistic safety assessment of nuclear power plants. Nuclear Eng. Technol. **56**(6), 2092–2098 (2024). ISSN 1738-5733. https://doi.org/10.1016/j.net.2024.01.017, https://www.sciencedirect.com/science/article/pii/S1738573324000184
13. Ericson, C.A.: Fault tree analysis: a history. In: 17th International System Safety Conference (1999)
14. USAEC. Reactor safety study: An assessment of accident risks in the U.S. Commercial Nuclear Power Plants Report WASH 1400 1975. U.S. Atomic Energy Commission, Washington D.C. Eyrolles (1975)
15. Rauzy, A.: An Open-PSA fault tree engine. PSAM/ESREL 2011 (2012)
16. Arboost Technologies. Aralia User Manual. Arboost (2004)
17. Dasgupta, S., Humble, T.S.: Characterizing the reproducibility of noisy quantum circuits, Entropy, vol. 24, num 2, (2022). MDPI https://doi.org/10.3390/e24020244

18. Jeffery, S., Kimmel, S., Piedrafita, A.: Quantum algorithm for path-edge sampling. In: Fawzi, O., Walter, M. (eds.) In: 18th Conference on the Theory of Quantum Computation, Communication and Cryptography TQC (2023)
19. Nation, P.D., Kang, H., Sundaresan, N., Gambetta, J.M.: Scalable mitigation of measurement errors on quantum computers. PRX Quantum **2**(4), 040326 (2021). https://doi.org/10.1103/PRXQuantum.2.040326

Monte Carlo Particle Transport on Quantum Computers

Noe Olivier[✉] and Michel Nowak

cortAIx Labs, Thales Research and Technology, 91120 Palaiseau, France
`{noe.olivier,michel.nowak}@thalesgroup.com`

Abstract. Monte Carlo particle transport codes are well established on classical hardware and are considered as the reference tool for nuclear applications. In a growing number of domains, the design of algorithms is progressively shifting towards the field of quantum computing, where theoretical speedups over their classical counterparts are expected. In some of these domains, Monte Carlo methods have already been converted to a quantum computing friendly setup where the expected and observed gain in complexity is quadratic. In this work, we address the particle transport problem in view of a first implementation on these architectures. We propose a quantum algorithm to model the particle transport based on discrete-time quantum walks and compare our approach to Monte Carlo and deterministic classical numerical schemes. The proof-of-concept algorithm is applied to a 10-qubit small-size problem by simulating its behavior on both ideal and noisy quantum computers based on the qiskit framework in view of reproducing classical results.

Keywords: Monte Carlo · Particle transport · Quantum walks · Quantum algorithm

1 Introduction

Monte Carlo particle transport codes [1–3] are well established within numerous fields of research. They are used as reference calculations in the field of nuclear energy, and more generally for radiation protection. Continuous developments on classical computers still succeed at improving the convergence rate of these codes, in particular by leveraging GPU capabilities [4]. However running those codes on HPC clusters remains costly and resource demanding.

As quantum computers are emerging, researchers develop and implement algorithms that can leverage these architectures in a more and more efficient way. Boosted by promising theoretical speedups, these developments are now guided by real-world applications in many fields. Quantum friendly Monte Carlo methods have already been proposed in a wide range of fields, starting from finance [5,6], to theoretical [7] and practical integration [8], with the motivation of a quadratic speedup.

© The Author(s), under exclusive license to Springer Nature Switzerland AG 2026
F. Barbaresco and F. Gerin (Eds.): QUEST-IS 2025, CCIS 2744, pp. 186–195, 2026.
https://doi.org/10.1007/978-3-032-13855-2_17

In this work, we aim to explore the possibility of computing particle transport on quantum computers. In Sect. 2, we give a brief introduction to the particle transport problem and present associated classical methods. In Sect. 3, we describe our proposition for a quantum algorithm to address the particle transport problem in heterogeneous media. Finally, in Sect. 4 we present and discuss the numerical results obtained from ideal and noisy simulations on small-size problems and compare them to classical ones.

2   Particle Transport

2.1   Problem Definition

In this paper, we address the particle transport problem, which is described by the transport equation. Let's consider an infinitesimal region of phase space positioned at r, of direction Ω and energy E. The exact transport equation that governs the angular flux ψ can be expressed as in page 47 of [9]

$$\Omega \cdot \nabla \psi(r, \Omega, E) + \Sigma_t(r, E)\psi(r, \Omega, E) =$$
$$\int_{4\pi} \Sigma_s(r, \Omega' \to \Omega, E' \to E)\psi(r, \Omega', E')dE'd^2\Omega' + S_0(r, \Omega', E) \quad (1)$$

where Σ_t and Σ_s are respectively the total and scattering cross sections of the medium, and S_0 the source. It can be simplified to the diffusion equation [9], with Σ_a the absorption cross section and D the diffusion coefficient:

$$-\nabla \cdot D(r)\nabla\phi(r) + \Sigma_a(r)\phi(r) = S_0(r) \quad (2)$$

and governs the scalar flux ϕ.

2.2   Monte Carlo Method

Consider a probability distribution that carries information about the system of interest, from which we can sample trajectories. Let $\mathbb{T}$ be the set of possible trajectories. A batch of Monte Carlo sampling consists in generating a set of trajectories $\mathcal{S} \in \mathbb{T}^N$ where N is the number of trajectories sampled.

In order to sample such trajectories, we first initialize particles at specified source points [9]. Each particle then experiments an exponential flight before a next collision or absorption. The flight length ρ between two collisions is sampled from $\pi_{\text{flight}}(\rho) = \Sigma_t e^{-\Sigma_t \rho}$.

We consider that particles are either absorbed or scattered. The scattering phenomenon is composed of two steps: the selection of the isotope on which the particle collides, and the computation of the outgoing scattering angle. For simplicity, we consider that the scattering is isotropic and that the materials are homogenized into macroscopic cross sections. The probability of scattering is characteristic to the medium and is given by $p_s = \Sigma_s/\Sigma_t$.

These steps are repeated until the particle gets absorbed by the material. We can score the flux by counting the number of collisions in a region of interest

$$\phi_{\mathcal{S}}(\mathbf{r}) = \frac{1}{N} \sum_{\mathcal{T} \in \mathcal{S}} \sum_{\mathbf{r}' \in \mathcal{T}} \frac{\mathbb{1}_r(\mathbf{r}')}{\Sigma_t}. \tag{3}$$

In complex radiation shielding problems, the time to solution for such calculations can reach several days on a supercomputer and might be reduced considerably by applying relevant quantum computing tools.

2.3   Finite Differences Method

Deterministic solvers, along with simplifications of the model, are often used in order to achieve a faster result than the classical Monte Carlo method. To this end, we consider Eq. 2 and replace the gradient of the flux by its finite differences approximation. We follow Sect. 3.9.2 of [10] to derive the numerical scheme and implement it in multiple dimensions. The advantage of the deterministic approach is that we get the response in a reasonable amount of time and with reasonable computer resources. However, it requires the end-user of the code to be aware of the simplifications made to the transport equation.

3   Quantum Approach

The Monte Carlo method remains the reference in the field of particle transport, thus making an investigation of a strategy to implement it on a quantum computer highly valuable. In this section, we propose an extension of particle transport numerical methods to the field of quantum computing. This extension comes with various simplifications and intends to model as faithfully as possible the dynamics of the transport with quantum computing tools.

Let's consider the problem of particle transport in a heterogeneous domain represented in Fig. 1 where both media have the same total cross section $\Sigma_t = 1\text{cm}^{-1}$ but differ in their scattering cross section: $\Sigma_s = 0.1\text{cm}^{-1}$ for the obstacle (center), $\Sigma_s = 0.9\text{cm}^{-1}$ for the propagation medium (outside). Although the graphs are two-dimensional for the purpose of this paper, the approach can be generalized by considering that the domain in the z-direction is unbounded.

Generally, a particle transport code relies on numerous calls to the geometry in order to account for the position of the materials, boundary conditions and local materials characteristics. In particular, the probabilities of interaction vary depending on the position of a particle, and on the isotopes present at this location.

In that regard, we propose in this paper to rely on the coin-based discrete-time quantum walk formalism [11,12] on a grid to simulate the particle transport problem.

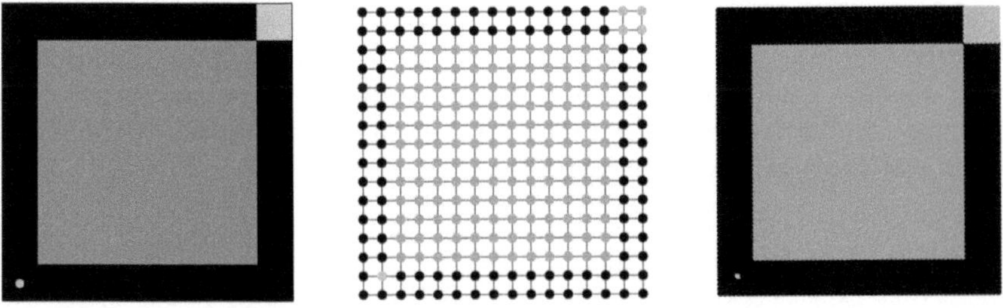

(a) Domain geometry   (b) 16 × 16 nodes, 8 qubits (c) 64 × 64 nodes, 12 qubit

Fig. 1. Domain geometry and its discretized versions. The geometry studied (a) is discretized (b,c) according to the size of the position qubits registers: $2^{n_x} \times 2^{n_y}$. The different colors correspond to different media (grey and dark blue) or help identify the source (green) and the detector (cyan). (Color figure online)

3.1   General Encoding

The modeling of the diffusion of the particles across the domain consists in a coin-based discrete-time quantum walk. The general structure of the proposed algorithm is given in Fig. 2.

The time evolution of the process is discretized in a number of iterations, each of which is composed of two operations. The coin operator acts on a coin register and defines directions of propagation to be applied. The shift operator modifies the position register according to the state of the coin register.

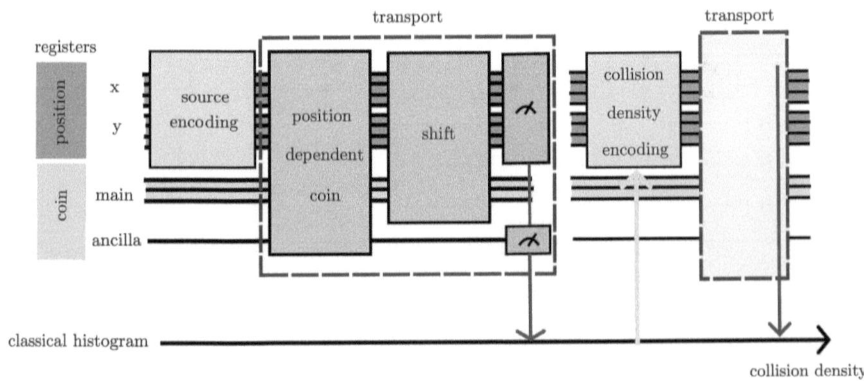

Fig. 2. Diagram of the quantum circuit for the measured quantum walk. The source encoding block consists in a single layer of X-gates for a punctual source, while the collision density encoding relies on the method from [13].

Let's define a position quantum register of $n_{\text{pos}} = n_x + n_y$ qubits, allowing the state representation of all the nodes of a two-dimensional grid of dimensions $2^{n_x} \times$

2^{n_y} as depicted in Fig. 1b. The quantum state corresponding to a single node is described by $|q_{n_{\mathrm{pos}}-1} \cdots q_1, q_0\rangle = |x_{n_x-1} \cdots x_1 x_0 y_{n_y-1} \cdots y_1 y_0\rangle$. This position register encodes the distribution of particles across the geometry in the states amplitudes. Similarly, we define a coin register that encodes the distribution of directions of propagation of the particles.

3.2 Algorithm Description

Source Initialization. As for the classical methods, the first step of the algorithm is the initialization of the particle trajectories. This is done by preparing a uniform superposition state over the position states that matches the source nodes. Initially, we consider a punctual source, efficiently implemented with a single layer of Pauli X gates.

Position-Dependent Coin Operator. Most works on coin-based quantum walks consider global coins, such as the Hadamard or Grover coins, which act on the coin register independently from the state of the qubits in the position register. However, the probabilities of interaction within the geometry are local and the behavior of the particles depends strongly on their location. Therefore, we choose to rely on a local coin operator, in Fig. 3, in order to transport the particles according to the local cross sections. We consider the "naive" position-dependent coin operator introduced by [14] for our implementation despite its exponential scaling in the number of position qubits, due to the definition of one coin operator per position state. This paper also presents another implementation of a local coin operator which allows a linear-depth circuit at the cost of an exponential number of ancillary qubits.

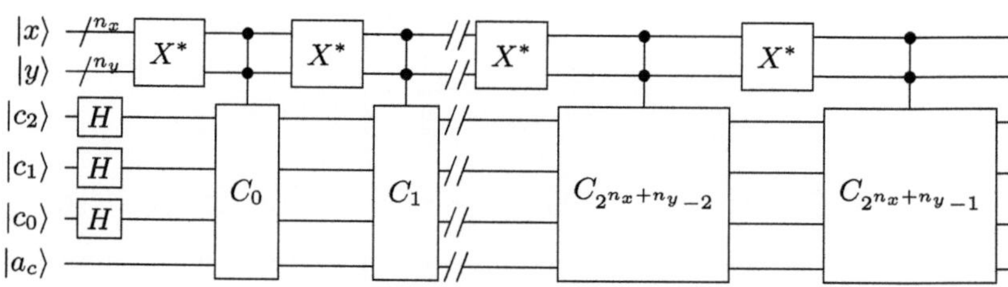

Fig. 3. Quantum circuit for the position-dependent coin, as proposed in [14]. X^* are layers of Pauli X gates to ensure $|x, y\rangle = |i\rangle$ and apply C_i. H gates are required to perform the desired operation described by the matrix $C_{|x,y\rangle}$.

In our algorithm, the coin register is composed of three qubits to account for the four possible directions of propagation and the absorption process using the convention described in Table 1. As a first simplification in our coin operator design, we choose to focus on macroscopic cross sections such that we average the cross sections locally over all isotopes. Let's consider both scattering and absorption effects and their associated local cross sections $\Sigma_s^{x,y}$ and $\Sigma_a^{x,y}$ respectively.

For simplicity, let's omit the indices x, y in the following notations. The coin operator is designed to result in the coin state $|\circlearrowleft\rangle$ with probability $p_a = \Sigma_a/(\Sigma_a + \Sigma_s)$ and in either of the four other states with probability $p_{s,i} = \Sigma_{s,i}/(\Sigma_a + \Sigma_s)$ respectively, where $i \in \{\rightarrow, \uparrow, \leftarrow, \downarrow\}$. Numerical results are obtained assuming that all scattering directions occur with the same probability $p_{s,i} = p_s/4$ with $p_s = \Sigma_s/(\Sigma_a + \Sigma_s)$.

Table 1. Coin-states qubit representation.

Coin state $|c_2 c_1 c_0\rangle$	Directions
$|100\rangle$	$|\rightarrow\rangle$
$|110\rangle$	$|\uparrow\rangle$
$|101\rangle$	$|\leftarrow\rangle$
$|111\rangle$	$|\downarrow\rangle$
$|0\cdot\cdot\rangle$	$|\circlearrowleft\rangle$

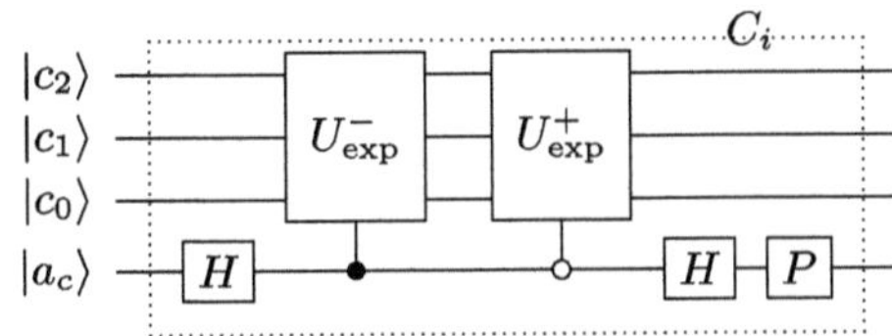

Fig. 4. Quantum circuit for the diagonal non-unitary coin operator, as proposed in [15].

The coin design results in the construction of an operator represented by a diagonal (but non-unitary) matrix $C_{|x,y\rangle}$ of dimension 8×8 for a three-qubit coin register. It can be implemented using a single ancillary qubit [15] after a first layer of hadamard gates as shown in Fig. 4. Nevertheless, the successful application of the operator requires a postselection on the ancilla qubit after each step of the walk.

$$C_{|x,y\rangle} = \begin{pmatrix} M_{A|x,y\rangle} & 0 \\ 0 & M_{S|x,y\rangle} \end{pmatrix}$$

where the local absorption and scattering matrices are given by

$$M_A = \mathrm{diag}\left(\sqrt{p_a}, \sqrt{p_a}, \sqrt{p_a}, \sqrt{p_a}\right)/4, \quad M_S = \mathrm{diag}\left(\sqrt{p_{s,\rightarrow}}, \sqrt{p_{s,\leftarrow}}, \sqrt{p_{s,\uparrow}}, \sqrt{p_{s,\downarrow}}\right)$$

We choose to implement reflective boundary conditions by modifying the local coin matrices on the edge of the geometry such that the probability of going in a prohibited direction is uniformly spread between all remaining scattering directions without altering the local absorption probability.

Shift Operator. The chosen implementation of the shift operator, in Fig. 5, is based on the generalised inverter gates implementation [16] using a series of X and Toffoli gates such that the position states amplitudes are updated according to the scattering and absorption probabilities. Since c_2 is one of the control qubits, absorbed particles are not propagated but maintained at their previous location. This model is chosen despite its failure to faithfully represent the physical distribution of particles over multiple steps of the quantum walk: absorbed particles will eventually be propagated.

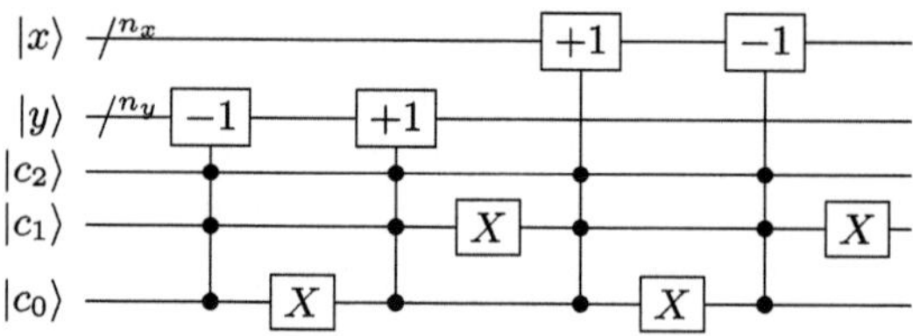

Fig. 5. Quantum circuit for the shift operator, using the toffoli-based "increment" and "decrement" operators proposed in [16].

Measured Quantum Walk. For the purpose of reproducing classical results, we overcome this limitation by constructing a circuit and measuring it for single steps of the quantum walk [17], as seen in Fig. 2. The postselection on the ancillary coin qubit gives a set of successful shots for which the circuit is well implemented while a second postselection on c_2 reduces this set to that of scattered particles. Thus, the source for the next step is no longer punctual and can be implemented from this distribution of scattered particles using the state preparation method proposed in [13] at the cost of $O(2^n)$ circuit depth.

4   Results and Discussion

The measured quantum walk (MQW) algorithm is implemented using qiskit 1.4.2 [18] and QiskitRuntimeService from qiskit-ibm-runtime 0.38.0. We use IBM's tools to simulate our circuits with the BasicSimulator backend for ideal simulations and the FakeMarrakesh backend emulating the 156-qubit ibm_marrakesh Heron r2 QPU.

Figure 6 represents the collision density obtained numerically from the different classical and quantum methods. In particular, the finite difference result is cropped once the flux is below 10^{-6}. The Monte Carlo scheme considers 500.000 particles and both the ideal and noisy quantum experiments use 20.000 initial shots at each time step. For the quantum approach, the collision density is computed as the normalized sum of distributions resulting from the circuits runs. These distributions must first be normalized such that they are represented by the same number of scattered shots.

We can observe that the collision density generated from the ideal simulations of the quantum algorithm on both 8×8 (c) and 16×16 (d) discretized domains successfully match the geometry since most collisions occur in the propagation medium, therefore showing an expected similarity with classical results on that regard. Only few particles manage to propagate through the absorbing medium due to the definition of the source, which only represents scattered particles in the measured quantum approach. In contrast, although the numerical experiment on IBM's FakeMarrakesh backend (e) encodes the punctual source distribution with reasonable errors, it fails to apply the quantum walk step as the result is a quasi-uniform superposition state over all positions. This result was expected due to the significantly large depth of the circuit describing the quantum walk

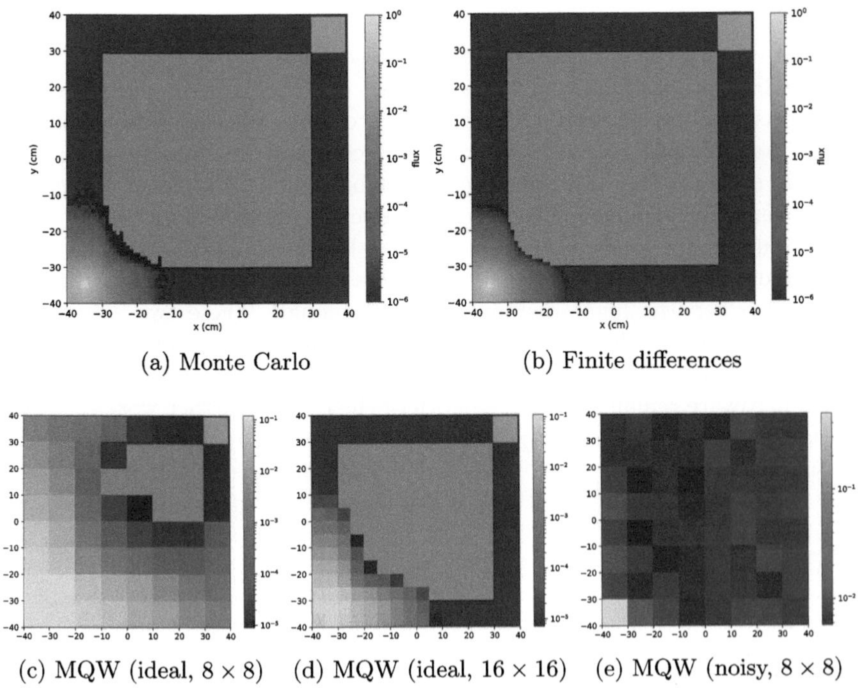

(a) Monte Carlo   (b) Finite differences

(c) MQW (ideal, 8×8)   (d) MQW (ideal, 16×16)   (e) MQW (noisy, 8×8)

Fig. 6. Comparison of the collision flux. Results are obtained via classical methods (a) and (b), and the proposed quantum approach in ideal (c,d) settings after 29 steps and noisy (e) settings after 1 step. The noisy emulation is performed on qiskit's FakeMarrakesh backend.

step (Fig. 2) after transpilation into the native gates of the machine (more than 1.1×10^6 gates). This main bottleneck mainly comes from the position-dependent coin operator, a repetition of $O(2^n)$ multi-controlled operations including $n + 1$ control qubits (Figs. 3 and 4). To a smaller extent, the depth of the circuit is also impacted by the series of toffoli-based increment and decrement operators in the shift (Fig. 5) controlled by all coin qubits, and by the implementation of the source state on the position register. Further investigation is necessary to optimize the circuits implementation to reduce its depth in view of facilitating future implementations on real hardware.

Additionally, we can observe that even in the ideal simulations, particles propagate much farther than when the discretization mesh is refined. In practice, when the mesh size becomes greater than the flight distance, scattering should also occur within a single cell. The scattering probability should therefore depend on the mesh size. In this proof-of-concept algorithm, this constant scattering probability corresponds to the probability of going from one cell to the next and does not allow scattering within a single cell (common in practice) and does not allow flights of more than one cell (rare event in practice): if the coin state is $|\circlearrowleft\rangle$ then particles are supposed to be absorbed and they are no longer considered in the next step. By overcoming this bottleneck, the results are expected to be competitive with the classical results presented here.

As a first proof of concept algorithm, we can validate the encoding of the problem on a given geometry. Nevertheless, to make this algorithm more complete, a possible perspective is to adapt the coin and shift operators such that the flight distance does not remain constant but could be sampled according to its exponential distribution. One research direction consists in increasing the size of the coin register to allow multi-cell flights. Furthermore, the coin matrix definition must also be adapted to allow scattered particles to remain within their initial cell when the flight distance is small enough. These two achievements would enable a complementary quantitative comparison to classical methods and expected competitive results. Finally, while the current bottleneck for runs on real hardware remains the large depth of the circuit, further works are necessary to identify an efficient way to adapt the measured quantum approach to a fully quantum algorithm that would not require intermediate measurements in order to leverage quantum features of the algorithm and potentially recover an expected quadratic speedup.

Acknowledgments. As part of the MetriQs-France program, this work is supported by France 2030 under the French National Research Agency grant number ANR-22-QMET-0002. We acknowledge the use of IBM Qiskit services for this work. The views expressed are those of the authors and do not reflect the official policy or position of IBM or the IBM Quantum team.

Disclosure of Interests. The authors have no competing interests to declare that are relevant to the content of this article.

References

1. Brun, E., et al.: Tripoli-4®, cea, edf and areva reference monte carlo code. In: SNA+ MC 2013-Joint International Conference on Supercomputing in Nuclear Applications+ Monte Carlo, p. 06023. EDP Sciences (2014)
2. Romano, P.K., Forget, B.: The openmc monte carlo particle transport code. Ann. Nuclear Energy **51**, 274–281 (2013)
3. Kulesza, J.A., et al.: Mcnp® code version 6.3. 0 theory & user manual. Technical report, Los Alamos National Laboratory (LANL), Los Alamos, NM (United States) (2022)
4. Tramm, J., et al.: Performance portable monte carlo particle transport on intel, nvidia, and amd gpus. arXiv preprint arXiv:2403.12345 (2024)
5. Rebentrost, P., Gupt, B., Bromley, T.R.: Quantum computational finance: monte carlo pricing of financial derivatives. Phys. Rev. A **98**(2), 022321 (2018)
6. Kaneko, K., Miyamoto, K., Takeda, N., Yoshino, K.: Quantum speedup of monte carlo integration with respect to the number of dimensions and its application to finance. Quantum Inf. Process. **20**(5), 185 (2021)
7. Montanaro, A.: Quantum speedup of monte carlo methods. Proc. Royal Soc. A: Math. Phys. Eng. Sci. **471**(2181), 20150301 (2015)

8. Yu, K., Lim, H., Rao, P.: Practical numerical integration on nisq devices. In: Quantum Information Science, Sensing, and Computation XII, vol. 11391, p. 1139106. International Society for Optics and Photonics (2020)
9. Coste-Delclaux, M., Diop, C., Nicolas, A., Bonin, B.: La neutronique (2013)
10. Hébert, A.: Applied Reactor Physics. Presses inter Polytechnique (2009)
11. Kempe, J.: Quantum random walks: an introductory overview. Contemp. Phys. **44**(4), 307–327 (2003)
12. Szegedy, M.: Quantum speed-up of markov chain based algorithms. In: 45th Annual IEEE Symposium on Foundations of Computer Science, pp. 32–41 (2004)
13. Mottonen, M., Vartiainen, J.J., Bergholm, V., Salomaa, M.M.: Transformation of quantum states using uniformly controlled rotations. arXiv preprint quant-ph/0407010 (2004)
14. Nzongani, U., Zylberman, J., Doncecchi, C.-E., Pérez, A., Debbasch, F., Arnault, P.: Quantum circuits for discrete-time quantum walks with position-dependent coin operator. Quantum Inf. Process. **22**(7), 270 (2023)
15. Zylberman, J., Nzongani, U., Simonetto, A., Debbasch, F.: Efficient quantum circuits for non-unitary and unitary diagonal operators with space-time-accuracy trade-offs. arXiv preprint arXiv:2404.02819 (2024)
16. Georgopoulos, K., Emary, C., Zuliani, P.: Comparison of quantum-walk implementations on noisy intermediate-scale quantum computers. Phys. Rev. A **103**, 022408 (2021)
17. Didi, A., Barkai, E.: Measurement-induced quantum walks. Phys. Rev. E **105**(5), 054108 (2022)
18. Javadi-Abhari, A., et al.: Quantum computing with qiskit (2024)

Pattern Matching Using Sliding Windows on Quantum Computers

Y. Ferhi[1]([⊠]) [iD], A. Geng[1] [iD], M. Kiefer-Emmanouilidis[2,3] [iD], and A. Moghiseh[1] [iD]

[1] Fraunhofer Institute for Industrial Mathematics ITWM,
67663 Kaiserslautern, Germany
yassine.ferhi@institutoptique.fr
[2] German Research Center for Artificial Intelligence DFKI,
67663 Kaiserslautern, Germany
[3] Department of Computer Science and Research Initiative QC-AI,
RPTU Kaiserslautern-Landau, 67663 Kaiserslautern, Germany

Abstract. Quantum hardware is progressing rapidly, but many high-performance algorithms are not yet feasible due to data embedding overhead and the limitations of NISQ devices. Quantum pattern matching algorithms, which on average outperform classical methods [26], rely on binary quantum embeddings and are not practical until we have larger qubit counts and fault-tolerant quantum computers. Although recent progress has reduced the overhead of amplitude encoding, there is still a lack of intermediate pattern matching algorithms that are efficient in terms of qubit usage. In this work, we introduce a sliding window template matching approach and explore suitable quantum primitives, including the standard and destructive swap tests, as well as the classical shadow method and the usage of a quantum artificial neuron, all of which have been performed, and evaluate their performance on NISQ devices. We integrate the swap test into a hybrid quantum support vector machine (QSVM) with a sliding window approach, improving noise resilience in quantum image template matching. We also illustrate, that our proposed methods are hardware-efficient, easy to implement, and suitable for near-term quantum devices.

Keywords: Quantum Pattern Matching · Quantum Computing · Quantum Image Processing · IBM Quantum Experience

1 Introduction

Detecting structured patterns in noisy images is a central problem in image processing and computer vision, with applications, for example, in astrophysics [22], and biomedical imaging [15]. Classical approaches – notably template-based convolution – are reliable, but their performance degrades rapidly at low signal-to-noise ratios [17]. Quantum computing algorithms, in principle, could speed up certain calculations [26,31]. Moreover, encoding data as quantum states

© The Author(s), under exclusive license to Springer Nature Switzerland AG 2026
F. Barbaresco and F. Gerin (Eds.): QUEST-IS 2025, CCIS 2744, pp. 196–205, 2026.
https://doi.org/10.1007/978-3-032-13855-2_18

[1,13,20], can exploit interference and entanglement to estimate similarities via overlap measurements [23,24,28]. These methods are becoming increasingly feasible in practice and applicable in their current form [26]. The applications span from automating defect detection in micro-fabricated components [16], real-time recognition of linear infrastructures or anomalies in drone and satellite captures [3], crack detection and segmentation in concrete surfaces [14,32], detecting vascular patterns, tumors, or lesions on MRI and CT scans [8] to identifying molecular binding motifs in microscopy imagery [5].

In this work, we use the *swap test* – both in its standard and destructive forms [7,10] – as a direct pattern matching primitive. Traditionally, swap tests estimate fidelity between unknown quantum states [9], or are used in quantum communication [4]. Here, each image patch is prepared as a quantum state, and its overlap with a template state is evaluated, where a high overlap probability indicates a pattern's presence.

We use a classical convolution baseline, applying a sliding window whose size matches that of the padded template. Additionally, we also incorporate the classical shadow framework to estimate fidelity without full tomography [33]. We compare these methods to quantum artificial neuron (QAN), which can be used as a phase-based quantum similarity test [34]. Finally, we integrate the swap tests into a hybrid Quantum Support Vector Machine (QSVM) sliding-window algorithm, using quantum overlaps to build the kernel matrix for a classical SVM classifier.

2 Quantum Overlap Tests

The presented pattern matching algorithm is based on the similarity provided by the Uhlman fidelity, commonly used in quantum information [6,27] $F(\rho, \sigma) = \left(\mathrm{Tr}\sqrt{\sqrt{\rho}\,\sigma\,\sqrt{\rho}}\right)^2$, which simplifies to

$$F(|\psi\rangle, |\phi\rangle) = |\langle\psi\,|\phi\rangle|^2,$$

for pure states $|\psi\rangle$ and $|\phi\rangle$. Classical fidelity estimation requires complete knowledge of the density matrices. In contrast, quantum techniques such as the swap test allow fidelity to be approximated with an error tolerance ϵ, requiring $\mathcal{O}(1/\epsilon^2)$ repetitions (shots) to achieve the desired accuracy.

This work evaluates quantum primitives for fidelity estimation in clean and noisy template matching tasks. We assess their performance across various image types, analyzing error rates and sensitivity to features such as brightness and shape. As base, we use four varying methods – standard and destructive swap test, QAN, classical shadow, and QSVM.

2.1 Standard and Destructive Swap Test

We explore quantum overlap estimation, beginning with the standard [27] and destructive swap test to evaluate the fidelity [29], where the destructive version

---

Algorithm 1. Fidelity Estimation with Classical Shadows

Input: Quantum state initializer `psi_init`, amplitudes `amps_p`, `amps_q`, number of qubits n, number of shots s

Output: Estimated fidelity between $|\psi_p\rangle$ and $|\psi_q\rangle$

1: `fidelities` $\leftarrow$ empty list
2: **for** $i = 1$ to s **do**
3:    $b \leftarrow$ randomly sampled Pauli basis of length n
4:    Construct circuit qc_ϕ on n qubits and n classical bits
5:    Apply `psi_init(amps_p)` on qc_ϕ
6:    Apply basis change corresponding to b on qc_ϕ
7:    Measure all qubits of qc_ϕ in computational basis
8:    Simulate qc_ϕ with 1 shot, extract bitstring r
9:    Compute projector P_r^b using `snapshot_projector(b, r)`
10:    Construct statevector $|\psi_q\rangle$ using `psi_init(amps_q)`
11:    Compute fidelity estimate $f_i \leftarrow \langle\psi_q|P_r^b|\psi_q\rangle$
12:    Append $\mathrm{Re}(f_i)$ to `fidelities`
13: **end for**
14: median of `fidelities`

---

uses fewer qubits than the standard one, and the same precision by the equal number of shots. We use the amplitude encoding method QPIE [35] to encode the classical pixel information of both the image and the patch into quantum states. The number of required qubits is $2 \cdot \log_2(p)$, or $2 \cdot \log_2(p) + 1$ if an ancilla qubit is used, where p denotes the number of pixels in the pattern padded to the next power of two. We use Qiskit's [21] Initialize function for that.

2.2 Classical Shadow Method

With the classical shadow [18,19] framework, as given in Algorithm 1, we can efficiently estimate fidelity using only a small number of randomized measurements. This approach is useful for near-term quantum devices, where full state tomography is infeasible.

The classical shadow procedure involves applying a randomly chosen unitary U (from the Clifford group [27]) to the quantum state, then measuring in the computational basis, and finally recording the outcome b and the unitary U used. By repeating this process N times, we obtain a set of measurement results $\{(U_i, b_i)\}_{i=1}^N$. From this data, we can reconstruct an estimator $\hat{\rho}$ that approximates the original state ρ.

Suppose we have two unknown pure quantum states, ρ and σ. By generating classical shadows $\{\hat{\rho}_i\}$ for ρ and $\{\hat{\sigma}_j\}$ for σ, we can estimate the fidelity using

$$\hat{F} = \frac{1}{MN} \sum_{i=1}^{M} \sum_{j=1}^{N} \mathrm{Tr}(\hat{\rho}_i \hat{\sigma}_j),$$

which approximates $\mathrm{Tr}(\rho\sigma)$ [6].

2.3   Approach Using a Quantum Artificial Neuron (QAN)

The algorithm proceeds in three principal stages [11,12,34]. First, an N-qubit register is initialized in the state $|0\rangle^{\otimes N}$, upon which a unitary transformation U_i is applied such that

$$U_i|0\rangle^{\otimes N} = |\psi_i\rangle. \tag{1}$$

In other words, the initial state $|0\rangle^{\otimes N}$ is transformed into the encoded input vector $|\psi_i\rangle$. This state preparation can be achieved either by employing an $m \times m$ unitary matrix whose first column corresponds to the input vector $\tilde{i}$, or by retrieving the state $|\psi_i\rangle$ directly from quantum memory.

In the second stage, a unitary operator U_w is constructed to satisfy

$$U_w|\psi_w\rangle = |1\rangle^{\otimes N} = |m-1\rangle, \tag{2}$$

thereby mapping the reference state $|\psi_w\rangle$, which encodes the weight vector $\tilde{w}$, to the computational basis state $|m-1\rangle$.

Applying both unitary operators U_w and U_i to the initialized state yields

$$U_w U_i |0\rangle^{\otimes N} = \sum_{j=0}^{m-1} c_j |j\rangle \equiv |\varphi_{i,w}\rangle, \tag{3}$$

where $c_j \in \mathbb{C}$. Using Eq. (2), the scalar product between the input and weight states can then be expressed as

$$\langle\psi_w|\psi_i\rangle = \langle\psi_w|U_w^\dagger U_w|\psi_i\rangle = \langle m-1|\varphi_{i,w}\rangle = c_{m-1}. \tag{4}$$

This establishes a direct correspondence between the scalar product of the input and weight vectors and the coefficient c_{m-1} of the final state $|\varphi_{i,w}\rangle$, up to a normalization factor:

$$\tilde{w} \cdot \tilde{i} = m\langle\psi_w|\psi_i\rangle.$$

Thus, the desired overlap required for pattern matching is encoded in the amplitude c_{m-1}. For additional details on the construction of the unitary operators, we refer to [11,12,34].

2.4   Quantum Support Vector Machine for Cross Detection

A Quantum Support Vector Machine (QSVM) [30] is a hybrid quantum–classical algorithm that leverages a quantum computer to compute a kernel matrix K in a potentially exponential–dimensional Hilbert space, which is then used by a classical SVM [25] algorithm for classification or regression tasks. In our application, we try to make the algorithms more robust to noise. Three training sets were prepared: 200 noiseless perfect crosses, 300 noisy crosses, and 500 random negative samples. For each pair (x_i, x_j) of image patches x_i and x_j, we compute the quantum kernel entry $K_{ij} = |\langle\psi(x_i)|\psi(x_j)\rangle|^2$ using a swap test (standard or destructive).

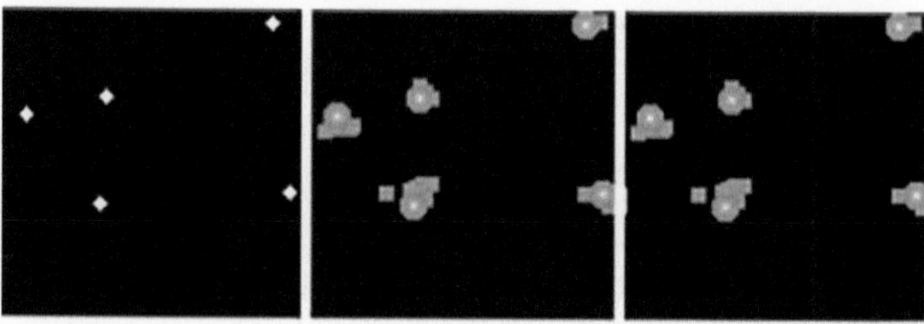

Fig. 1. 100×100 image to be analyzed (imperceptible noise) (left), standard (middle), and destructive swap test result (right).

3   Experimental Results

In the following, we present the results of our experiments, beginning with the introduction of a practical exponential threshold method for the swap tests, followed by a detailed analysis of each method on selected examples. All experiments were conducted on an Intel® Xeon® Gold 6442Y CPU (24 cores, 48 threads) operating at a base frequency of 2.6 GHz. The system included 500 GB of system memory.

Both the standard and the destructive swap test perform comparably, as illustrated, for example, in Fig. 1, for detecting a 5×5 cross pattern with gray value 1 embedded in a 100×100 image where we added noise by changing the values of random pixels values in the range $[0, 1)$.

However, noise can interfere with the swap test, leading to artificially high overlap values in noisy regions that visually resemble the target pattern. As a result, the reconstructed image does not reliably indicate the true position of the pattern. To mitigate this issue, we introduce a probabilistic threshold on the measured swap test outcomes. The threshold is defined as

$$\tau(r, p) = \exp\big(p(r - 1)\big), \tag{5}$$

where r denotes the raw swap test result normalized to values in the range $[0, 1]$ and p is a scaling parameter. This exponential form sharpens the separation between true matches and overlaps by strongly suppressing values below $r = 1$, while still retaining sensitivity near the upper range of the similarity measure. Based on experimental observations for patterns larger than 3×3, swap tests between simple patterns (such as crosses, triangles, or rhombi) and random noisy images of the same size consistently produced similarity values between 0.6 and 0.75. Thus, the threshold effectively filters out these cases, while preserving matches for identical or closely related patterns. This corresponds to a practical parameter choice of $p = 100$.

Figure 2 shows the results of the four methods described in Sect. 2. Since the standard and destructive swap tests yield similar outcomes, we report only the results of the standard swap test. Figure 2a illustrates the performance on a noise-free 100×100 image containing five 3×3 white crosses. As shown,

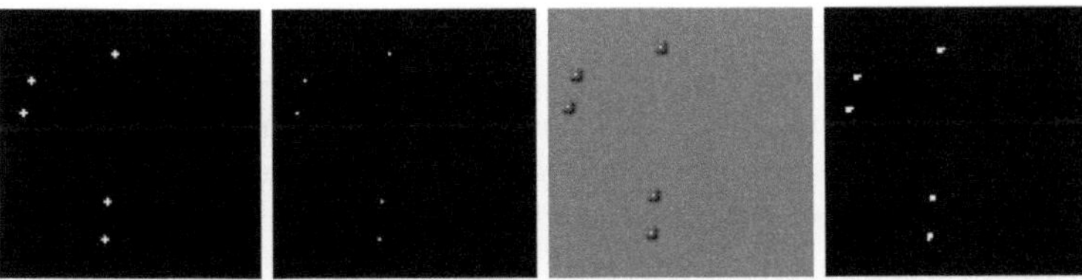

(a) Non-noisy 100×100 images with crosses. In order: Analyzed image, standard swap test, QAN's algorithm, and classical shadow (200 shots).

(b) Noisy 100×100 images with crosses. In order: Analyzed image, QSVM, swap test with exponential threshold, QAN's algorithm.

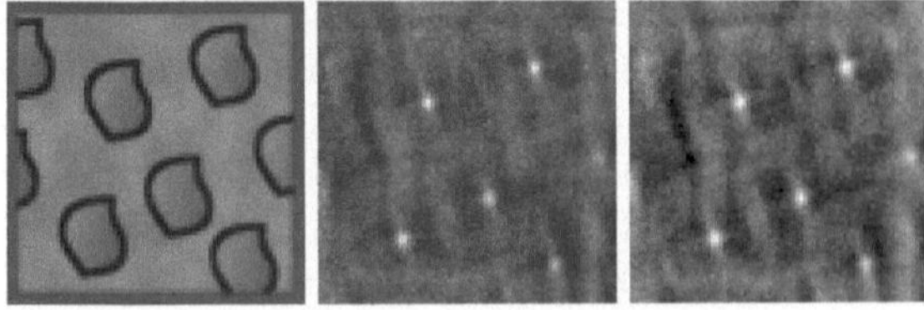

(c) 74×74 sample image. In order: analyzed image, swap test, QAN's algorithm. For a more complex pattern than a simple cross, despite the noise, the swap test needs no threshold nor QSVM to detect the pattern without false positives.

(d) In order: swap test and QAN's algorithm for non-noisy image, as in (a), then for the noisy image, as in (b) (with threshold for the swap test). The results on real hardware are not deviating much from the simulated ones.

Fig. 2. Our test results with 1024 shots, if not mentioned differently. The expected result in all samples here is a bright point in the center of each pattern. For QAN's algorithm, no threshold method is applied. (a)-(c) are simulated on Qiskit's `qasm_simulator`, (d) on IBM's real backend `ibm_aachen`.

all methods successfully detected the crosses. In the classical shadow method, however, the result exhibits an uncertainty region approximately equal in size to the pattern itself. Because generating quantum circuits for each iteration is computationally intensive, simulations quickly encounter memory limitations. Therefore, we reduced the number of iterations in the classical shadow method to 200. Nevertheless, processing the entire 100×100-pixel noisy image, as done in the other methods (see Fig. 2b), remained impractical within a reasonable computational time-frame. This observation suggests that the classical shadow method could become significantly more efficient if parallelized, for instance by executing all randomized circuits at once for each patch.

Figure 2b highlights the differences between the methods under noisy conditions. The QSVM demonstrates clearer detections and fewer artifacts than both swap test variants. The single false positive detected corresponds to a noise-generated cross of lower brightness. The swap test with exponential threshold, in contrast, produces six false positives: one caused by the low-intensity cross and five additional detections unrelated to the target pattern. These false positives arise due to normalization effects in amplitude encoding, where patches dominated by bright noise mimic the template. Future refinements, such as tuning the parameter p in Eq. (5) for different pattern classes or adopting alternative noise models, could further mitigate these false detections.

The QAN algorithm yields no false positives and demonstrates strong noise resilience, both in simulations using Qiskit's `qasm_simulator` and on IBM's real backends (e.g., `ibm_aachen`; see Fig. 2d). The hardware results closely match the simulated ones, and the patterns can be successfully identified in the image.

Finally, increasing the pattern size and complexity naturally reduces sensitivity to noise, as random fluctuations are increasingly unlikely to replicate larger templates (e.g., a 23×23 pattern) compared to smaller ones (e.g., a 3×3 cross). In these cases, the true positions of the larger patterns remain clearly identifiable. Elevated probabilities may still appear near image boundaries, where truncated patches resemble the target; however, these values remain consistently lower than those of correct matches. As demonstrated in Fig. 2c, both the swap test and the QAN algorithm successfully identify the larger patterns under such conditions.

4 Conclusion and Discussion

The choice of the optimal quantum primitive for pattern matching depends strongly on the application context. The phase-encoding approach, as employed in QAN's algorithm, is sensitive to brightness disparities between the pattern and the image; consequently, it does not detect randomly generated low-intensity crosses. This method offers rapid execution and robustness to noise, reliably identifying only the desired pattern. In contrast, swap tests can be used in two ways: with an exponential threshold to detect visually similar simple patterns (at the cost of false positives, e.g., six detections among the 10 000 patches in Fig. 2b), or without a threshold for sufficiently complex patterns. Increasing the number of shots further improves swap test accuracy by reducing the rate of false positives. The single false positive found by the QSVM, however, cannot be avoided, as the corresponding cross is genuinely present in the data. Unlike QAN, both swap tests and the QSVM identify shapes resembling the target pattern, including partial or incomplete matches. Because amplitude encoding represents the total brightness of each extracted patch, swap tests are inherently insensitive to illumination variations while requiring minimal quantum hardware resources. Importantly, their practicality is supported by experiments on real hardware (e.g., `ibm_aachen`), which produced results consistent with those obtained in simulation.

Training a QSVM kernel for classification enables finer control over which patterns are accepted or rejected as matches. While the kernel training time remains moderate (on the order of a few hours), it is not negligible. Moreover, excessively large kernels can lead to overfitting. Comparable performance can often be achieved with smaller training sets, for example, 300 samples. Although no universal rule exists to determine the optimal kernel size and avoid over- or underfitting, established approaches such as the Nyström method [2] can be employed to reduce dataset size and computational cost.

SVMs are particularly appealing when the same kernel can be reused across multiple images—for instance, when detecting the same pattern in a sequence of frames. In other contexts, where noise resilience is paramount, QAN's method or swap tests with exponential thresholding may provide sufficient accuracy. By contrast, fidelity estimation with classical shadows proves impractical for this task: although only a few measurements are required, each necessitates generating a separate circuit, resulting in significantly lower efficiency compared to other approaches.

In summary, we have demonstrated how hardware-efficient quantum algorithms for pattern matching can be effectively applied on current NISQ devices. Each primitive presents distinct trade-offs, and its suitability depends on the complexity of the target pattern, the noise characteristics of the data, and the computational resources available.

Acknowledgment. This work was supported in part by the Quantum Initiative Rheinland-Pfalz (QUIP). It was also partially funded by the Federal Ministry for Economic Affairs and Climate Action through the EniQmA project (funding number 01MQ22007A).

References

1. Araujo, I.F., Blank, C., Araújo, I.C.S., da Silva, A.J.: Low-rank quantum state preparation. IEEE Trans. Comput. Aided Des. Integr. Circuits Syst. **43**(1), 161–170 (2024). https://doi.org/10.1109/TCAD.2023.3297972
2. Bach, F.R., Jordan, M.I.: Predictive low-rank decomposition for kernel methods. In: Proceedings of the 22nd international conference on Machine learning - ICML 2005, pp. 33–40. ACM Press, Bonn, Germany (2005). https://doi.org/10.1145/1102351.1102356
3. Bozcan, I., Kayacan, E.: UAV-AdNet: unsupervised anomaly detection using deep neural networks for aerial surveillance. In: 2020 IEEE/RSJ International Conference on Intelligent Robots and Systems (IROS), pp. 1158–1164. IEEE (2020). https://doi.org/10.1109/IROS45743.2020.9341790
4. Buhrman, H., Cleve, R., Watrous, J., De Wolf, R.: Quantum fingerprinting. Phys. Rev. Lett. **87**(16), 167902 (2001). https://doi.org/10.1103/PhysRevLett.87.167902

5. Caicedo, J.C., et al.: Data-analysis strategies for image-based cell profiling. Nat. Methods **14**(9), 849–863 (2017). https://doi.org/10.1038/nmeth.4397

6. Chen, R., Song, Z., Zhao, X., Wang, X.: Variational quantum algorithms for trace distance and fidelity estimation. Quantum Sci. Technol. **7**(1), 015019 (2021). https://doi.org/10.1088/2058-9565/ac38ba

7. Das, S., Zhang, J., Martina, S., Suter, D., Caruso, F.: Quantum pattern recognition on real quantum processing units. Quantum Mach. Intell. **5**(1), 16 (2023). https://doi.org/10.1007/s42484-022-00093-x

8. ElâĂŘDeredy, W.: Pattern recognition approaches in biomedical and clinical magnetic resonance spectroscopy: a review. NMR Biomed. **10**(3), 99–124 (1997). https://doi.org/10.1002/(SICI)1099-1492(199705)10:3<99::AID-NBM461>3.0.CO;2-

9. Fanizza, M., Rosati, M., Skotiniotis, M., Calsamiglia, J., Giovannetti, V.: Beyond the swap test: optimal estimation of quantum state overlap. Phys. Rev. Lett. **124**(6), 060503 (2020). https://doi.org/10.1103/PhysRevLett.124.060503

10. Garcia-Escartin, J.C., Chamorro-Posada, P.: swap test and hong-ou-mandel effect are equivalent. Phys. Rev. A **87**(5), 052330 (2013). https://doi.org/10.1103/PhysRevA.87.052330

11. Geng, A.: Application of hybrid quantum machine learning for image processing in the NISQ era. PhD Thesis, Rheinland-Pfälzische Technische Universität Kaiserslautern-Landau (2024). https://doi.org/10.26204/KLUEDO/8207

12. Geng, A., Moghiseh, A., Redenbach, C., Schladitz, K.: A hybrid quantum image edge detector for the NISQ era. Quantum Mach. Intell. **4**(2), 15 (2022). https://doi.org/10.1007/s42484-022-00071-3

13. Geng, A., Moghiseh, A., Redenbach, C., Schladitz, K.: Improved FRQI on superconducting processors and its restrictions in the NISQ era. Quantum Inf. Process. **22**(2) (2023). https://doi.org/10.1007/s11128-023-03838-0

14. Geng, A., Moghiseh, A., Redenbach, C., Schladitz, K.: Hybrid quantum transfer learning for crack image classification on NISQ hardware. In: AIP Conference Proceedings, vol. 3182, p. 130001. AIP Publishing LLC (2025). https://doi.org/10.1063/5.0246500

15. Glumov, N.I., Kolomiyetz, E.I., Sergeyev, V.V.: Detection of objects on the image using a sliding window mode. Optics Laser Technol. **27**(4), 241–249 (1995). https://doi.org/10.1016/0030-3992(95)93752-D

16. Guijo, D., et al.: Quantum artificial vision for defect detection in manufacturing (2024). https://doi.org/10.48550/arXiv.2208.04988

17. Hashemi, N.S., Aghdam, R.B., Ghiasi, A.S.B., Fatemi, P.: Template matching advances and applications in image analysis (2016). https://doi.org/10.48550/arXiv.1610.07231

18. Huang, H.Y.: Learning quantum states from their classical shadows. Nat. Rev. Phys. **4**(2), 81–81 (2022). https://doi.org/10.1038/s42254-021-00411-5

19. Huang, H.Y., Kueng, R., Preskill, J.: Predicting many properties of a quantum system from very few measurements. Nat. Phys. **16**(10), 1050–1057 (2020). https://doi.org/10.1038/s41567-020-0932-7

20. Iten, R., Colbeck, R., Kukuljan, I., Home, J., Christandl, M.: Quantum circuits for isometries. Phys. Rev. A **93**(3), 032318 (2016). https://doi.org/10.1103/PhysRevA.93.032318

21. Javadi-Abhari, A., et al.: Quantum computing with Qiskit (2024). https://doi.org/10.48550/ARXIV.2405.08810

22. Juvela, M.: Template matching method for the analysis of interstellar cloud structure. Astron. Astrophys. **593**, A58 (2016). https://doi.org/10.1051/0004-6361/201628727
23. Kankeu, I., Fritsch, S.G., Schönhoff, G., Mounzer, E., Lukowicz, P., Kiefer-Emmanouilidis, M.: Quantum-inspired embeddings projection and similarity metrics for representation learning (2025). https://doi.org/10.48550/arXiv.2501.04591
24. Liu, W., Li, Y.Z., Yin, H.W., Wang, Z.R., Wu, J.: Quantum multi-state Swap Test: an algorithm for estimating overlaps of arbitrary number quantum states. EPJ Quantum Technol. **11**(1), 46 (2024). https://doi.org/10.1140/epjqt/s40507-024-00259-5
25. Mammone, A., Turchi, M., Cristianini, N.: Support vector machines. WIREs Comput. Stat. **1**(3), 283–289 (2009). https://doi.org/10.1002/wics.49
26. Montanaro, A.: Quantum pattern matching fast on average. Algorithmica **77**(1), 16–39 (2015). https://doi.org/10.1007/s00453-015-0060-4
27. Nielsen, M.A., Chuang, I.L.: Quantum Computation and Quantum Information. Cambridge university press (2010)
28. Nishimura, H.: A survey: SWAP test and its applications to quantum complexity theory. In: Algorithmic Foundations for Social Advancement: Recent Progress on Theory and Practice, pp. 243–261. Springer, Singapore (2025). https://doi.org/10.1007/978-981-96-0668-9_16
29. Niu, S., Suau, A., Staffelbach, G., Todri-Sanial, A.: A hardware-aware heuristic for the qubit mapping problem in the NISQ era. IEEE Trans. Quantum Eng. **1**, 1–14 (2020). https://doi.org/10.1109/TQE.2020.3026544
30. Rebentrost, P., Mohseni, M., Lloyd, S.: Quantum support vector machine for big data classification. Phys. Rev. Lett. **113**(13), 130503 (2014). https://doi.org/10.1103/PhysRevLett.113.130503
31. Sasaki, M., Carlini, A., Jozsa, R.: Quantum template matching. Phys. Rev. A **64**(2), 022317 (2001). https://doi.org/10.1103/PhysRevA.64.022317
32. Srinivasan, A., Geng, A., Macaluso, A., Kiefer-Emmanouilidis, M., Moghiseh, A.: A comparative study of q-seg, quantum-inspired techniques, and u-net for crack image segmentation **92**(7), 321–331 (2025). https://doi.org/10.1515/teme-2025-0017
33. Struchalin, G., Zagorovskii, Y.A., Kovlakov, E., Straupe, S., Kulik, S.: Experimental estimation of quantum state properties from classical shadows. PRX Quantum **2**(1), 010307 (2021). https://doi.org/10.1103/PRXQuantum.2.010307
34. Tacchino, F., Macchiavello, C., Gerace, D., Bajoni, D.: An artificial neuron implemented on an actual quantum processor. npj Quantum Inf. **5**(1), 1–8 (2019). https://doi.org/10.1038/s41534-019-0140-4
35. Yao, Xi-Wei et al.: Quantum image processing and its application to edge detection: theory and experiment (2017). https://doi.org/10.1103/PhysRevX.7.031041

Session: 9 Quantum Algorithms, Computing; Simulation – Simulation and Quantum Chemistry A

Solving the 3D Heat Equation with VQA via Remeshing-Based Warm Starts

Samuel Donachie[1], Ulysse Remond[1,2], Arthur Mathorel[1],
and Kyryl Kazymyrenko[1]

[1] EDF R&D, 7 Bd Gaspard Monge, 91120 Palaiseau, France
kyrylo.kazymyrenko@edf.fr
[2] LUX, Sorbonne University, 4 Pl. Jussieu, 75005 Paris, France

Abstract. Quantum computing holds great promise for solving classically intractable problems such as linear systems and partial differential equations (PDEs). While fully fault-tolerant quantum computers remain out of reach, current noisy intermediate-scale quantum (NISQ) devices enable the exploration of hybrid quantum-classical algorithms. Among these, Variational Quantum Algorithms (VQAs) have emerged as a leading candidate for near-term applications. In this work, we investigate the use of VQAs to solve PDEs arising in stationary heat transfer. These problems are discretized via the finite element method (FEM), yielding linear systems of the form $Ku = f$, where K is the stiffness matrix. We define a cost function that encodes the thermal energy of the system, and optimize it using various ansatz families. To improve trainability and bypass barren plateaus, we introduce a remeshing strategy which gradually increases resolution by reusing optimized parameters from coarser discretizations. Our results demonstrate convergence of scalar quantities with mesh refinement. This work provides a practical methodology for applying VQAs to PDEs, offering insight into the capabilities and limitations of current quantum hardware.

Keywords: Variational Quantum Algorithm · Mesh Refinement · Quantum Computer Benchmark

1 Introduction

In the current state of quantum computing, fully fault-tolerant quantum computers (FTQCs) remain out of reach [3]. Instead, we operate in the so-called NISQ (Noisy Intermediate-Scale Quantum) era, characterized by quantum devices that are both limited in qubit count and prone to noise. Despite these constraints, NISQ devices can still be leveraged for meaningful tasks [8].

As the timeline toward FTQC remains uncertain, both academic and industrial communities are actively exploring use cases that can provide value using today's imperfect quantum hardware. Surpassing classical solutions typically requires a large number of qubits—something that current hardware is not yet capable of. While classical solvers are powerful, they often struggle with scalability and can require extensive computational resources. This challenge, called the curse of dimensionality, refers to the growth in computational complexity as the number of variables and/or degrees of freedom increases. The typical size of challenging industrial simulations reaches 10^6 degrees of freedom. Quantum computing offers a potential way to overcome this limitation: with n qubits, one can represent 2^n states, enabling the compact encoding of high-dimensional vectors and matrices. With 50 qubits, we expect to show one day quantum supremacy on some of those complex problems [14]. Among the many problems of interest, solving linear equations plays a central role in scientific computing, engineering, and industry. In particular, many physical systems are modeled by Partial Differential Equations (PDEs), which are typically discretized into large linear systems. Efficiently solving these is therefore currently critical for simulations in fields as diverse as fluid dynamics [20], structural mechanics [2], and electromagnetism [9].

One particularly active area is optimization, where quantum computing may already offer practical benefits through the use of Variational Quantum Algorithms (VQAs) [5]. These hybrid quantum-classical algorithms are well-suited for NISQ devices and have shown promise in addressing linear algebra problems [4,6,10,11,17]. In this work, we investigate the use of VQAs for solving PDE-derived linear systems for a 3D object, with a focus on designing ansätze that are both expressive and hardware-efficient. To this end, we also introduce a cascading remeshing strategy, aimed at improving the convergence and stability of VQA-based linear equation solvers. The goal of this strategy is to use a coarse quantum representation of a solution, requiring less qubits, as a warm start in a more refined mesh. This procedure creates tailored warm starts for PDE-solving VQAs.

Finally, recognizing that current quantum hardware may not yet support the full execution of such VQAs at scale, we propose to use the ideal final state obtained from our simulated VQA as a benchmark. Since this state corresponds to the solution of a well-defined physical problem (in our case, the 3D heat equation $\Delta u = f$), it provides a concrete, structured target for quantum hardware. We suggest that reproducing such physically meaningful states could serve as a valuable and application-driven benchmark for future quantum devices.

2   Problem and Method

2.1   Problem Definition

Our objective here is to solve the heat equation in 3D : $\Delta u = f$ with u being the temperature field and f the heat source term. One common approach to solve Partial Differential Equations is to use the Finite Element Method (FEM), which discretizes the PDE into a linear system [17]. For instance, in the case of the heat equation, the system can be written as:

$$Ku = f \tag{1}$$

Here, K is the stiffness matrix arising from the FEM discretization [2]. It encodes the relationships between the degrees of freedom in the system and reflects both the geometry of the domain and the material or physical properties involved. In the quantum setting, the solution vector $\boldsymbol{u}$ and the source term (charge vector) $\boldsymbol{f}$ are encoded as quantum states $|u\rangle$ and $|f\rangle$, enabling a compact representation and allowing the use of variational algorithms to approximate the solution. Putting this equation into its variational form leads us to define the energy E_c of a given trial state $|x\rangle$:

$$E_c(|x\rangle) = \frac{1}{2}\langle x|K|x\rangle - \frac{\langle x|f\rangle + \langle f|x\rangle}{2} \tag{2}$$

We use the energy E_c as a cost function, which is minimized by the solution vector $|u\rangle$. However, we're working with quantum states, which are normalized, whereas the true solution vector $\boldsymbol{u}$ is not necessarily normalized. We therefore introduce a parameterized quantum state $|\psi(\theta)\rangle$, a norm $r \in \mathbb{R}$, such as a trial state is expressed as $|x\rangle = r|\psi(\theta)\rangle$. The equation we need to optimize then becomes:

$$|u\rangle = argmin_{r,\theta}\left(\frac{1}{2}r^2\langle\psi(\theta)|K|\psi(\theta)\rangle - r\frac{\langle\psi(\theta)|f\rangle + \langle f|\psi(\theta)\rangle}{2}\right) = \Re(r_{opt}|\psi(\theta_{opt})\rangle) \tag{3}$$

Analytically minimizing this variational equation with respect to r, prior to θ, we obtain:

$$r_{opt} = \frac{\langle\psi(\theta)|f\rangle + \langle f|\psi(\theta)\rangle}{2\langle\psi(\theta)|K|\psi(\theta)\rangle} \tag{4}$$

Leading us to the final expression of the cost function we use in our VQA [17]:

$$C(\psi(\theta)) = -\frac{1}{8}\frac{(\langle\psi(\theta)|f\rangle + \langle f|\psi(\theta)\rangle)^2}{\langle\psi(\theta)|K|\psi(\theta)\rangle} \tag{5}$$

By finding the state that minimizes this equation, we obtain the solution to our problem. Variational Quantum Algorithms (VQAs) leverage a hybrid quantum-classical loop to perform this minimization: a classical optimizer proposes parameters θ for a parameterized quantum circuit (ansatz), which is executed on a quantum processor. The resulting state is measured, and the outcome of the cost function C seen in (5) is used by the classical optimizer to update the parameters. Performance depends on each component of the loop: the choice of ansatz, the classical optimizer, and the measurement strategy. As with classical optimizers and preconditioning, different tradeoffs between the complexities of the quantum computations regarding K, $|f\rangle$ and $|u\rangle$ are possible. As a comparatively low number of parameters can be stored to represent large states, QPUs provide a sizeable advantage with regard to the efficient generation and evaluation of high-dimensional states. It is also possible to interact with K and $|f\rangle$ using measurements, scaling polylogarithmically with system size [11,16]. It is worth noting that a high fidelity with the target state $|\langle\psi(\theta|u\rangle|^2$ does not necessarily imply convergence in energy Sect. 3.2.

2.2   The Choice of the Ansatz

In a Variational Quantum Algorithm (VQA), an ansatz is a parameterized quantum circuit designed to generate trial quantum states. It serves as a flexible model that is potentially optimized to approximate the solution to a given problem. The expressiveness and structure of the ansatz critically affect the performance and accuracy of the algorithm [18]. Firstly, since one of the main goals of this project is to ensure compatibility with today's NISQ devices, the ansatz must be designed with hardware constraints in mind. This means using quantum gates with high fidelity and limiting the overall circuit depth. Typically, ansätze consist of alternating layers of rotation gates and layers of entangling gates. It is well known that entangling gates—especially those involving two or more qubits—are significantly more error-prone than single-qubit gates, with their implementation difficulty increasing rapidly with the number of qubits involved [1,13].

We employ the NLocal class from Qiskit (Qiskit version: 1.4.2) [15] to construct our ansatz. The circuit is constructed by alternating layers of parameterized rotation and entangling gates, with a final layer of rotation gates. In each entangling layer, a linear chain of CNOTs is applied, where every qubit except the last controls an X gate on the subsequent qubit. This pattern is repeated throughout the circuit, while the final layer contains only rotation gates. When two rotation gates are used, each is applied in a separate layer. Using this Qiskit class, we constructed and compared a limited selection of ansätze, each chosen based on specific criteria relevant to our problem. Since we are addressing a heat diffusion problem, it seems that the target quantum state should be real-valued. All numerical results reported in the next section are obtained from noiseless state-vector simulations where the optimizer provides parameters to our simulated PQC, producing the ideal resulting state. The cost function (5) is then computed from this state, and the optimizer adjusts the parameters accordingly, repeating the process until it found the optimal set. We used the NEWUOA [19] gradient-free optimization algorithm throughout this article. We studied an inhomogeneous problem to showcase that our model can handle arbitrary boundary conditions. Specifically, we imposed Dirichlet conditions, by fixing the temperature on two faces of the 3D object, and Neumann conditions consisting in a uniform heat flux on an entire third face and an additional flux over half of a fourth face. Then, using Qiskit's NLocal, we benchmark the following ansätze—on 6 qubits with 36 parameters—each relevant to a real-valued heat-diffusion solution (Table 1):

Let us note that we could have used fewer parameters for Ansatz 1 and 4 (namely 6 and 12, respectively). Given the small number of parameters, this did not affect the results presented below. However, if we had chosen those ansätze, we would have used 6/12 parameters in the later stages.

Table 1. List of Ansätze candidates we considered

	Name	Rotation Layer	Entangling Layer	Comments
1	Ry-only	RY	—	This baseline illustrates the limitations of an ansatz without entanglement
2	Ry-Cnot	RY	CNOT	The simplest real-valued ansatz: RY yields purely real amplitudes, and the we chose CNOT as the 2 qubit entangling gate
3	Ry-Toffoli	RY	CCNOT	Replaces the CNOT with a more complex three-qubit gate to test whether deeper entanglement outweighs its higher error rate
4	Ry-1Cnot	RY	CNOT	Rotation layers: $RY(\theta)$; exactly one CNOT layer inserted just before the final rotation layer Provides sparse entanglement while keeping depth minimal
5	RxRy-Cnot	RX then RY	CNOT	Adding RX enlarges the reachable Hilbert space potentially giving the optimizer better control through a more expressible ansatz [12], it introduces complex amplitudes
6	RxRy-Toffoli	RX then RY	CCNOT	Same expressiveness as above but with a higher-order entangling gate

3   Results

3.1   Cold Start

To compare these ansätze on equal footing, we tuned each circuit so that the total number of trainable parameters was approximately the same, instead of fixing the number of layers. We then ran the optimization 100 times on 6 simulated noiseless qubits, each time starting from random initial parameters θ generated using a pseudo-random number generator (PCG64, from NumPy version: 1.23.5). It is worth noting that changing the number of layers or trainable parameters can sometimes have a drastic impact on the performance of an ansatz. We also note that we let the optimizer run until it could no longer find any improvement, using an absolute tolerance of 10^{-12} and a relative tolerance of 10^{-9}, or until it was forcefully stopped after 2 h. We can interpret these thresholds as resource limits. This second condition was not triggered until we reached 12 qubits. In order to compare them, we used the accuracy of the energy as a metric of convergence. Energy accuracy is defined as $\frac{C(|\psi(\theta)\rangle)}{C(|u\rangle)} \times 100\%$ reflecting how closely the optimization converges to the true solution. This metric is designated as "Energy" in the Tables 2.

Table 2. Ansätze scores on 6 qubits (averaged over 100 2 h-optimization runs, each with 36 parameters).

	Ansatz	Mean Energy (%)	NEWUOA Iterations	Max Energy (%)	Min Energy (%)
1	Ry	88.00 % ± 0.32	452 ± 82	91.17 %	87.97 %
2	Ry-Cnot	86.7 % ± 6.7	2645 ± 1331	96.89 %	68.70 %
3	Ry-Toffoli	86.31 % ± 6.52	2430 ± 1203	96.70 %	71.92 %
4	Ry-1Cnot	86.1 % ± 11.0	1104 ± 525	93.38 %	50.75 %
5	Rx-Ry-Cnot	93.0 % ± 4.2	3887 ± 1875	97.14 %	54.33 %
6	Rx-Ry-Toffoli	93.0 % ± 1.7	5793 ± 2720	97.17 %	88.48 %

Compared to the 6 qubits-discretized target energy all ansatz achieve 85% and above mean energy accuracy. We observe that multi-qubit entangling gates beyond pairs do not provide a noticeable advantage at this scale Rather than requiring every run to succeed perfectly, we only need most runs to perform well and at least one to converge very well. Based on this, we can distinguish three performance tiers among the ansätze. On average, using 2 rotation gates seems to help but considering only the maximum accuracy, it doesn't have such an impact. These performance insights can help guide the choice of quantum hardware when deploying the VQA in practice. For the rest of the article, thanks to the scores we obtained, we chose to work with the ansätze 2 and 5. We won't be using Toffoli gates since we can see no clear advantage at this scale, moreover, they perform usually worse than the other existing gates on current hardware. We could also take the optimization time into account by considering the number of iterations required to reach the final result. Adding more gates tends to increase this number, and since the overall waiting time scales with the size of the system, this becomes an important factor to consider. Therefore, we need to be cautious when adding complexity to the circuit.

When examining these results, one might notice that some runs perform quite poorly, with minimum accuracies sometimes dropping to around 50%. This is due to local minima and/or barren plateaus, where the optimizer gets stuck in regions of vanishing gradients. This accuracy can still be considered high and textbook strategies like a restart could help enhance this score. However, in this case and for our sake, it might not be a major concern, as some runs still manage to perform very well. The problem arises as the system becomes bigger, requiring more qubits: barren plateaus become more prevalent and harder to avoid. Such random cold start-based runs with 9 to 15 qubits can be observed in Table 3. For 9 qubits we had 108 parameters, for 12 qubits we had 252 parameters and for 15 qubits, we had 330 parameters. We will keep the same number of parameters for the rest of the article.

Table 3. Ansätze scores with cold starts (averaged over 100 2h-optimization runs).

Qubits	Par	Ansatz	Mean Energy (%)	NEWUOA Iterations	Max Energy (%)	Min Energy (%)
9	108	Rx-Ry-Cnot	83.3 ± 5.5	8506 ± 7313	92.9	78.5
9	108	Ry-Cnot	23.6 ± 2.4	4106 ± 1308	29.4	17.6
12	252	Rx-Ry-Cnot	38.0 ± 32	11716 ± 4160	66.9	0.70
12	252	Ry-Cnot	4.0 ± 0.7	4625 ± 2965	5.3	2.5
15	330	Rx-Ry-Cnot	0.21 ± 0.44	5352 ± 714	2.71	0.02
15	330	Ry-Cnot	0.53 ± 0.03	4867 ± 796	0.57	0.25

The results show lower convergence when working with a more complex system of 9 qubits especially with the second ansatz Ry-Cnot. Using 12 qubits or more, the optimizer always fails to find a way to optimize the parameters and/or it stops after the 2-h limit we imposed. We can notice a great difference between the two ansatz for 12 qubits. The Ry-Cnot ansatz showcases very low accuracy as it is twice as deep, and only outputs real states making it less expressible. The largest systems' low accuracy can also be explained by the optimizer getting stuck on a barren plateau and/or in local minima. In such cases, several classical strategies can accelerate convergence like dual-objective optimization, restarting the optimization process, or a warm start.

3.2 Uniform Start

In a warm start, the initial state is already close to the target, which helps the optimizer find better parameters faster. Since we optimize parameters and not quantum states directly, we need to identify both a state near the objective and a parameterization within our ansatz that produces it. Our first strategy is the **Uniform start**. The parameters of the first layer set a Hadamard-like transformation, by setting the parameters of R_y gates of the first layer to $\frac{\pi}{2}$ and the R_x gates to 0, while the remaining parameters are initialized randomly from $[-0.01, 0.01]$:

$$R_y(\theta) = \begin{bmatrix} \cos\left(\frac{\theta}{2}\right) & -\sin\left(\frac{\theta}{2}\right) \\ \sin\left(\frac{\theta}{2}\right) & \cos\left(\frac{\theta}{2}\right) \end{bmatrix}, \quad R_x(\theta) = \begin{bmatrix} \cos\left(\frac{\theta}{2}\right) & -i\sin\left(\frac{\theta}{2}\right) \\ -i\sin\left(\frac{\theta}{2}\right) & \cos\left(\frac{\theta}{2}\right) \end{bmatrix}, \quad R_y\left(\frac{\pi}{2}\right)|0\rangle = d|+\rangle \quad (6)$$

This prepares a near to equal superposition $\approx |+\rangle^{\otimes n}$, a more structured start than full randomness. In physical terms, it is like setting all components of the system close to the same temperature, rather than assigning random temperatures to each one. This uniform starting condition can help the optimizer avoid getting lost in barren plateaus by giving it a more meaningful direction to begin with.

Table 4. Ansätze scores with Uniform starts (averaged over 100 2 h-optimization runs).

Qubits	Par	Ansatz	Mean Energy (%)	NEWUOA Iterations	Max Energy (%)	Min Energy (%)
9	108	Rx-Ry-Cnot	92.4 ± 0.77	5489 ± 3162	94.4	85.9
9	108	Ry-Cnot	92.3 ± 6.2	4727 ± 2381	94.2	33.4
12	252	Rx-Ry-Cnot	71.6 ± 23	4665 ± 2288	91	0.32
12	252	Ry-Cnot	83.7 ± 20	7409 ± 2583	92.2	11.1
15	330	Rx-Ry-Cnot	46.7 ± 2.90	5585 ± 594	51	35
15	330	Ry-Cnot	54.71 ± 5.02	4707 ± 240	63	44

The results improve with regard to the cold starts. Nevertheless, as the system becomes more complex, even this form of structured initialization begins to lose effectiveness. It still works better than a cold start; for 12 and 15 qubits, most of the runs that were doing well stopped because of the time running out and not because the optimizer couldn't find better parameters. In the 15-qubit case using the Ry–Cnot ansatz, the mean fidelity reaches 95% even though energy convergence is poor, highlighting that fidelity can be misleading as a performance metric. Moreover, convergences using fidelity disregard the physics of the object, leading to discontinuous states, and therefore cannot be used as a cost function [16]. It will be useful later on when evaluating the performances of a QPU's output. Usual warm starts are inherently problem-dependent, since they leverage specific knowledge about the target state or solution landscape. For instance, in combinatorial optimization, a warm start might be obtained by using a classical heuristic solution mapped into the quantum circuit's parameter space [7]. This start only induces a uniform state, meaning that it merely leverages broad information about the solution's smoothness. That said, it remains valuable in scenarios like ours, where we are primarily concerned with properties such as the temperature of an object—typically a uniform value across the object—rather than a more specialized configuration.

In that sense, Hadamard initialization can be better described as a structured cold start: it provides a more uniform and reproducible starting point than random parameters, but it still lacks the informed guidance that characterizes a true warm start. To overcome this limitation, we introduce the Cascade protocol.

3.3   The Cascade Protocol

To choose our ansatz, we compared the performance of various circuits, and all of them performed well on smaller systems (6 qubits). However, as we increase the number of qubits, the Hilbert space grows exponentially, and the optimizer struggles more. The quantum states become more complex, and even the number of parameters becomes harder to manage. In this regime, a warm start is currently one of the only viable strategies to get a VQA to converge effectively. Inspired by recent results on quantum solutions in structural mechanics [16] obtained by our research group, we propose a novel strategy to improve the scalability of variational quantum algorithms (VQAs) by extending optimized solutions obtained on smaller, tractable systems as structured initializations for larger problems. The central idea is to use the solution of a lower-dimensional instance as a coarse-grained approximation, which can then be embedded and refined within a higher-dimensional Hilbert space. To illustrate this, consider a 6-qubit system whose 64 computational basis states can be naturally interpreted as a $4 \times 4 \times 4$ 3D grid:

$$|u\rangle_{6qb} = \sum_{\bar{x},\bar{y},\bar{z}} |z_{coarse} z_{fine} y_{coarse} y_{fine} x_{coarse} x_{fine}\rangle, \text{ where } \bar{w} = w_{coarse} w_{fine} \in \{0,1\} \times \{0,1\} \tag{7}$$

This smaller problem can typically be addressed efficiently either using classical techniques and quantum state preparation, or using a simpler VQA. To extend the solution to a more complex instance—such as a 9-qubit system representing an $8 \times 8 \times 8$ grid—we construct an initialization that embeds the optimized 6-qubit state into the larger state space. Concretely, each basis state of the smaller system becomes embedded in a local $2 \times 2 \times 2$ neighborhood of the larger configuration, effectively increasing the resolution of the space. This hierarchical strategy provides a structured and informed starting point, potentially enhancing convergence and solution quality in the variational optimization. This Cascade protocol creates a natural way to scale the problem up while preserving useful structure from the lower-dimensional solution. To visualize a bit better how it works, here is a remeshing of a 2D grid (Fig. 1).

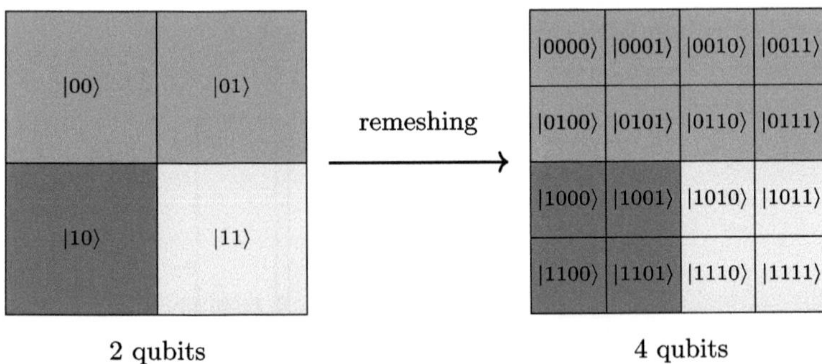

Fig. 1. Remeshing from a 2-qubit 2D-grid into a 4-qubit 2D-grid. Each 2-qubit state expands into a structured patch of finer resolution made of $2^2 = 4$ states.

Since we want to keep the same element geometry (a cube) we need to add 8 new states. In order to obtain those 8 new finite elements per old element, we need to introduce 3 additional qubits, since $2^3 = 8$. Suppose that an optimized solution has already been obtained for the 6-qubit instance of the problem. To take advantage of this solution when scaling up, we begin by initializing the full 9 qubit system in the state

$$|u\rangle_{6qb} \otimes |000\rangle \tag{8}$$

Next, we apply Hadamard gates to each of the 3 newly added qubits. This creates a uniform superposition over all their basis states, yielding:

$$|u\rangle_{6qb}|+\rangle|+\rangle|+\rangle = |u\rangle_{6qb} \otimes \frac{1}{\sqrt{8}} \sum_{i=0}^{7} |i\rangle \qquad (9)$$

However, unlike in one-dimensional systems—where an additional qubit can simply be appended to the end of the register—in three-dimensional settings, the mapping between basis states and spatial positions is inherently more complex. In particular, the tensor product structure does not directly preserve spatial locality across axes. To ensure that newly added qubits are correctly positioned to encode higher-resolution features along each spatial dimension, it is necessary to introduce a sequence of SWAP gates. These gates effectively rearrange the qubits, allowing the additional degrees of freedom to control fine-grained resolution along the x, y, and z axes of the enlarged spatial grid. Concretely, we can think of the 6-qubit state as encoding coordinates with pairs of qubits representing coarse and fine positions (7) as $\bar{w} = w_{coarse}w_{fine} \in \{0,1\} \times \{0,1\}$. The procedure can be illustrated as $|\bar{z}\rangle|\bar{y}\rangle|\bar{x}\rangle \overset{+3}{\mapsto} |\bar{z}\rangle|\bar{y}\rangle|\bar{x}\rangle|000\rangle \overset{H}{\mapsto} |\bar{z}\rangle|\bar{y}\rangle|\bar{x}\rangle|+\rangle|+\rangle|+\rangle \overset{swap}{\mapsto} |\bar{z}\rangle|+\rangle|\bar{y}\rangle|+\rangle|\bar{x}\rangle|+\rangle$.

By carefully swapping, we ensure the new qubits replace the fine control qubits where higher resolution is needed. Nevertheless, this step is unnecessary as you could construct your circuit to leave unused wires in the right spots. Using this technique, we should get a warm start by remeshing the solution of a previous optimization with a lower resolution (Fig. 2).

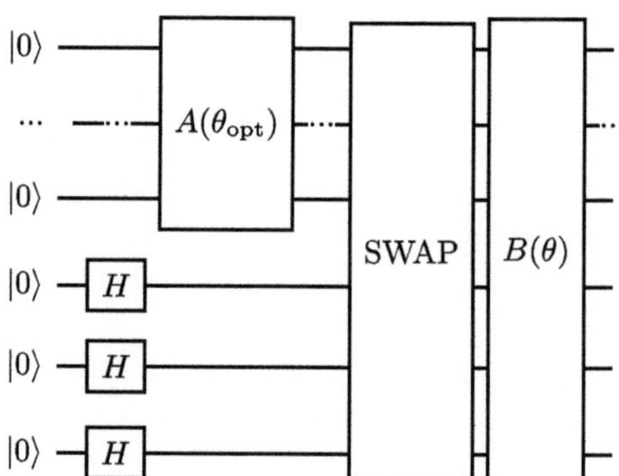

Fig. 2. Circuit of the remeshing cascade strategy using previous results $A(\theta_{\mathrm{opt}})$ and superposition expansion via Hadamard gates and swaps.

We define the second ansatz $B(\theta)$ to be the composition of two NLocal ansätze: $Ansatz(\theta) \times Ansatz(0)^{\dagger}$, such that $B(0) = \mathbb{I}$. If it is possible to implement an identity gate $\mathbb{I}$ easily with a specific ansatz by simply setting the correct first parameters, this last part isn't needed. However, composing an ansatz in this way simplifies the process, as we only need to initialize all the parameters θ to 0 to get $\mathbb{I}$, but obviously complexifies greatly the circuit. We can now compare the results obtained using the Cascade strategy to the Uniform start and the cold start from 9 to 15 qubits on our 3D object.

Using this strategy, we ran again the optimizations with the two different ansätze. Let's note that the parameters chosen for the first part of the Cascade protocol are the one obtained through a previous optimization on a coarser system. As shown in the previous table, those parameters made us obtain the maximum accuracy recorded for each type of ansatz and were at least 91% accurate in energy.

The Cascade protocol not only improves convergence but also enhances its consistency, as show by the significantly lower standard deviation of the energy accuracy. Notably, for the 15-qubit case, the protocol yields higher energy accuracy, successfully achieving our objective of constructing an effective warm-start strategy. Specifically, we observe a 15–20% (Tables 4, 5) improvement compared to the Uniform start, which enables us to meaningfully visualize one of the layers of the target 3D volume Appendix B.1. As expected, at 9 and 12 qubits the overhead of applying an additional layers of gates—including the cost of possible swapping—renders the strategy less relevant: the improvement at 9 qubits is negligible, while at 12 qubits the process becomes prohibitively slow relative to the modest gains. At 15 qubits, however, the benefits dominate, with faster convergence and higher accuracy.

Table 5. Ansätze scores with the Cascade protocol (averaged over 100 2h-optimization runs).

Qubits	Par	Ansatz	Mean Energy (%)	NEWUOA Iterations	Max Energy (%)	Min Energy (%)
6 to 9	108	Rx-Ry-Cnot	90.0 ± 0.28	4456 ± 1907	90.2	89.6
6 to 9	108	Ry-Cnot	91.5 ± 0.897	10513 ± 955	92.0	84.6
9 to 12	252	Rx-Ry-Cnot	87.4 ± 2.04	5209 ± 870	88.3	66.3
9 to 12	252	Ry-Cnot	87.1 ± 2.37	7563 ± 2156	90.2	80.9
12 to 15	330	Rx-Ry-Cnot	72.1 ± 1.45	4377 ± 379	74.0	62
12 to 15	330	Ry-Cnot	62.5 ± 4.84	3725 ± 892	70.3	44

4   Conclusion

In this work, we explored the use of Variational Quantum Algorithms (VQAs) for solving partial differential equations, focusing on the 3D heat equation discretized via FEM. We applied a noise-free simulation of a VQA to minimize a physically meaningful cost function related to thermal energy. After demonstrating the growing difficulty of such an optimization as system size increases, we introduced the Cascade protocol, a remeshing-based warm-start strategy that reuses optimized parameters from coarser discretizations to initialize larger systems, allowing us to converge with a 15 qubits system. Through extensive simulation, we benchmarked various ansätze and initialization strategies. Basic circuits like Ry-Cnot were gradually outperformed by more expressive configurations such as Rx-Ry-Cnot as system size increased. Moreover, since it's using fewer two-qubit gates, the latter are also more noise resistant. The Cascade protocol significantly improved convergence stability and accuracy at larger scales, surpassing both cold and Uniform start baselines. This strategy offers a scalable, physically motivated approach to training VQAs for PDEs, mitigating barren plateaus and improving overall performance. Moreover, this approach opens the possibility of investigating other initialization schemes, for instance selecting tailored parameter sets at the start of optimization, to evaluate their influence on VQA performance.

Even if current quantum hardware remains limited, we propose through this article a path towards a realistic and structured benchmark for evaluating quantum processing units. By comparing noise-free simulations, noisy models, and real QPU outputs, we highlight both the potential and the current limitations of those devices. In particular, fidelity and energy-based accuracy metrics suggest that small systems can already be meaningfully benchmarked using our approach as seen in Appendix B.2. Moving forward, we envision further improvements through problem-specific warm starts, optimization algorithms, new observables and hardware-aware ansatz design. As quantum hardware matures, the methodology introduced here may serve as a foundation for solving real-world PDEs and also assessing quantum advantage in scientific computing.

Acknowledgments. We would like to thank Joseph Mikael from EDF R&D for his many fruitful discussions. We are also particularly grateful to Heidi Nelson-Quillin (IonQ) for her valuable technical assistance and the whole IonQ team for their support.

Disclosure of Interests. The authors have no competing interests to declare that are relevant to the content of this article.

A   Using Our Results as a Benchmark

Building on the results obtained through simulation, we introduce a benchmark strategy for quantum hardware based on physically meaningful quantum states. These states arise from the solution of a 3D heat equation and offer a concrete, interpretable target for evaluating quantum processors. While executing the full optimization process directly on quantum hardware remains challenging, we propose a more accessible benchmark: given a set of optimized parameters θ, how accurately can a quantum device reproduce the corresponding state $|\psi(\theta)\rangle$? This approach isolates the hardware's ability to generate complex quantum states without requiring full in-loop optimization. To assess this, we compare the output of the IonQ Aria 1 noise model—designed to emulate the behavior of the Aria 1 QPU—to both the ideal solution and the noiseless simulation. Specifically, we evaluate the accuracy with respect to the ideal solution's energy, the accuracy with respect to the optimized simulation's energy, and the fidelity between the quantum output and each of these reference states. As a baseline, we consider the effect of stochastic sampling noise. Using the best parameters previously obtained, we simulate 100 000 measurement shots to produce a reference stochastic state, which allows us to quantify the expected variance in metrics due to finite sampling alone (Table 6).

Table 6. Simulation results with stochastic noise

Qubits	Ansatz	Energy		Fidelity	
		stoch/solution	stoch/simulation	stoch/solution	stoch/simulation
6	Rx-Ry-Cnot	93.3	99.9	96.9	99.8
6	Ry-Cnot	97.8	99.9	98.4	99.9
9	Rx-Ry-Cnot	89.3	97.2	97.8	99.8
9	Ry-Cnot	91.2	97.0	98.8	99.9
12	Rx-Ry-Cnot	40.2	46.2	97.1	98.7
12	Ry-Cnot	42.8	47.1	97.7	98.8

A clear trend emerges: as the number of qubits increases, the accuracy decreases. Since the simulations are noise-free, this decline can be attributed solely to stochastic sampling noise. The exponential growth of the Hilbert space with the number of qubits leads to a corresponding increase in the number of possible measurement outcomes, requiring more shots to obtain statistically reliable estimates. For example, in the 12 qubit case, the Hilbert space is made of 4096 basis states, and 100,000 shots begin to fall short of delivering accurate statistics across the entire state space. This effect becomes even more pronounced in the 15-qubit case, where the degradation in accuracy due to insufficient sampling is both more severe and more evident (Table 7).

Table 7. Noisy simulation results

Qubits	Ansatz	Energy		Fidelity	
		noisy/solution	noisy/simulation	noisy/solution	noisy/simulation
6	Rx-Ry-Cnot	92.4	99.0	96.8	99.6
6	Ry-Cnot	93.5	95.7	96.3	97.2
9	Rx-Ry-Cnot	80.2	87.4	96.8	98.1
9	Ry-Cnot	64.4	68.5	92.4	93.1
12	Rx-Ry-Cnot	20.8	23.9	89.1	89.6
12	Ry-Cnot	7.65	8.41	64.3	64.6

We observe a clear difference between the two ansatz. To match the same number of total parameters, the Rx-Ry-CNOT ansatz uses half as many entangling layers as the Ry-CNOT ansatz. Since two-qubit gates have lower success rates, adding more entangling layers reduces performance, as seen above. Other factors, such as gate types and qubit count, also influence results. Given hardware costs and simulation limits, we restricted our comparison to 6-qubit circuits with the Ry-CNOT ansatz, running 10,000 shots to obtain the results shown in Appendix B.2.

B   Imaging Our Results

B.1   The Cascade Results

As explained above, we used the results obtained through the best optimization thanks to the Cascade warm-start, and applied matplotlib's Lánczos interpolation on the raw statevectors (Fig. 3).

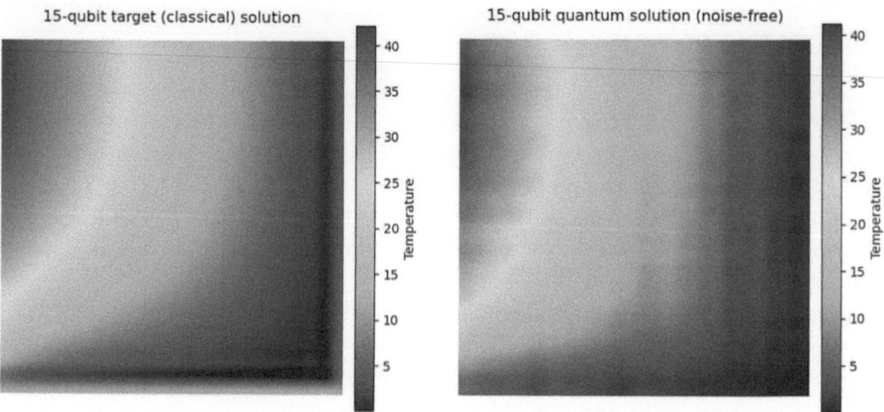

Fig. 3. Comparison between the target and the quantum solution for a 15-qubit system.

To visualize this 2D slice of the 3D volume, we used the quantum state obtained from the best optimization run (15 qubits, 330 parameters, Rx-Ry-Cnot Ansatz, with 74% energy accuracy). We can see how closely the two states match, by noticing the preservation of the general shape and gradients (Fig. 4).

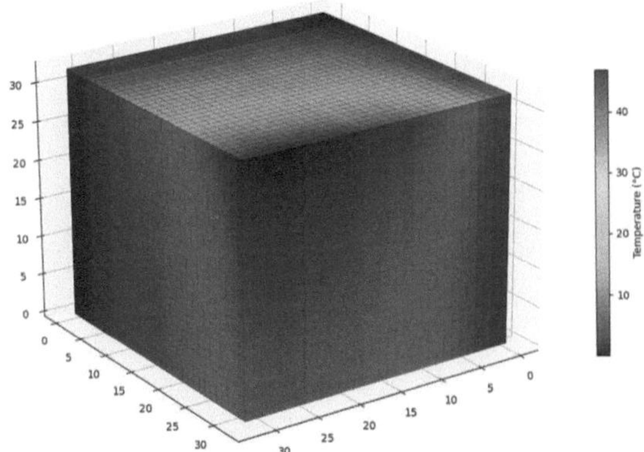

Fig. 4. 3D Heatmap of the 15 qubits noise-free simulation using the Cascade warm-start

B.2   Imaging the Noisy Simulation and QPU's Output

We ran our PQC on IonQ's QPU Aria 1, allowing us to run it against their noise model, the stochastic noise and the noise-free simulation. We retrieved the following results (Table 8):

Table 8. Performance of the Aria 1 noise model and QPU from IonQ for 6 qubits

Metric	Ideal	Stochastic	Noisy (Aria 1) simulation	Aria 1 (QPU)
Energy (state/solution)	96.8	97.1	93.5	92.2
Energy (state/simulation)	100	99.3	95.7	97.8
Fidelity (state/solution)	98.6	98.2	96.3	95.3
Fidelity (state/simulation)	100	99.8	97.2	96.3

These performances show us that all results are close to one another, and both the QPU output and the noisy simulation remain close to the ideal simulated state for such small qubit counts. Naturally, some minor differences and reduced accuracy can still be observed, likely due to noise—whether stochastic or arising from other sources. We can visualize those differences by imaging a slice of our 3D object (made with 6 qubits) as shown previously for 15 qubits (Fig. 5).

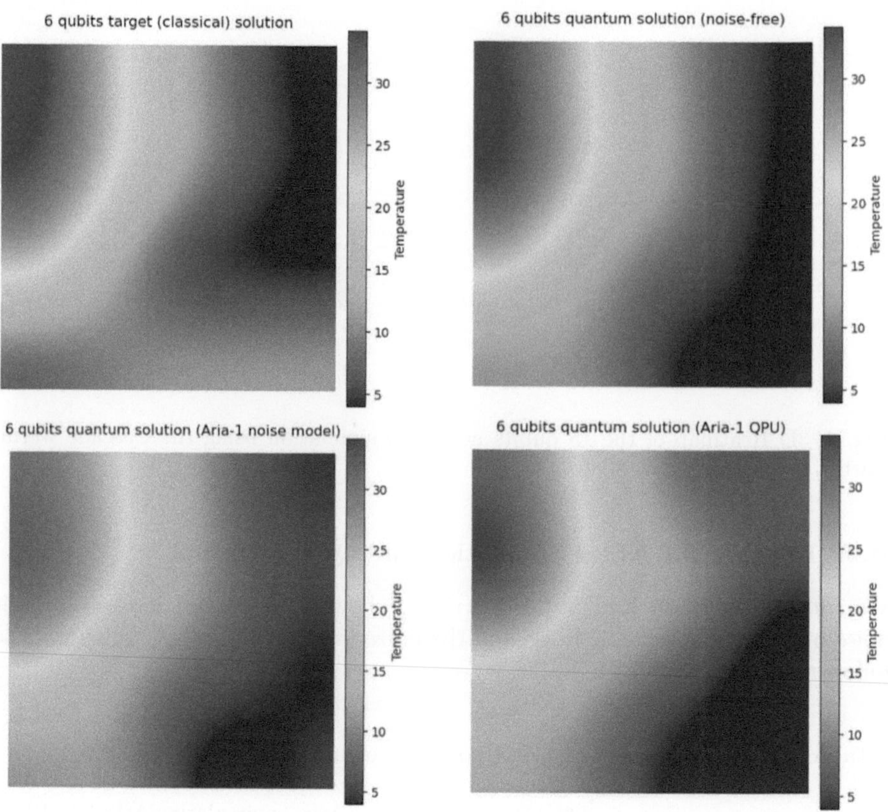

Fig. 5. Comparison between the classical target and quantum solutions for a 6-qubit system using Lánczos interpolation.

The noise-free simulation closely reproduces the overall structure, suggesting that the chosen ansatz captures the essential features of the problem. When introducing the Aria 1 noise model (third), we observe a shift in temperature as observed with the color distribution and the color bar, indicating the degradation caused by noise. The result from the real QPU (rightmost) exhibits similar features to the noisy simulation but with slightly higher variation, that is likely due to hardware imperfections or shot noise. Despite these limitations, the general shape and gradients are preserved, validating the approach for small systems under realistic condition.

References

1. Abdurakhimov, L., et al.: Technology and performance benchmarks of IQM's 20-qubit quantum computer (2024). https://arxiv.org/abs/2408.12433
2. Bathe, K.: Finite Element Procedures. Prentice Hall (2006). https://books.google.fr/books?id=rWvefGICfO8C

3. Battistel, F., et al.: Real-time decoding for fault-tolerant quantum computing: progress, challenges and outlook. Nano Futures **7**(3), 032003 (2023). https://doi.org/10.1088/2399-1984/aceba6

4. Cao, Y., Papageorgiou, A., Petras, I., Traub, J., Kais, S.: Quantum algorithm and circuit design solving the Poisson equation (2012). https://doi.org/10.1088/1367-2630/15/1/013021

5. Cerezo, M., et al.: Variational quantum algorithms. Nat. Rev. Phys. **3**(9), 625–644 (2021). https://doi.org/10.1038/s42254-021-00348-9

6. ColibriTD: Solving partial differential equations on IBM quantum processors with a variational quantum algorithm. Technical report (2025). https://docsend.com/view/9xsbyyi7pxfkgt7q

7. Egger, D.J., Mareček, J., Woerner, S.: Warm-starting quantum optimization. Quantum **5**, 479 (2021). https://doi.org/10.22331/q-2021-06-17-479

8. Emeriau, P.E.: AI image processing: Quandela's quantum leap in day-to-night scene translation (2024). https://www.quandela.com/resources/blog/ai-image-processing-quandelas-quantum-leap-in-day-to-night-scene-translation. December 2024

9. Hoole, S.R.H. (ed.): Finite Elements. Electromagnetics and Design. Elsevier, Amsterdam (1995)

10. Jaffali, H., et al.: H-DES: a quantum-classical hybrid differential equation solver (2024). https://arxiv.org/abs/2410.01130

11. Liu, H., et al.: Variational quantum algorithm for Poisson equation (2020). https://doi.org/10.1103/PhysRevA.104.022418

12. Liu, Y., Baba, K., Kaneko, K., Takeda, N., Koyama, J., Kimura, K.: Analysis of parameterized quantum circuits: on the connection between expressibility and types of quantum gates (2024). https://arxiv.org/abs/2408.01036

13. Mai, X., Zhang, L., Yu, Q., Zhang, J., Lu, Y.: Scalable entangling gates on ion qubits via structured light addressing (2025). https://arxiv.org/abs/2506.19535

14. Pollinger, T., Craen, A.V., Offenhäuser, P., Pflüger, D.: Realizing joint extreme-scale simulations on multiple supercomputers—two superfacility case studies, pp. 1–17 (2024). https://doi.org/10.1109/SC41406.2024.00104

15. Qiskit Development Team: Qiskit: An Open-source Framework for Quantum Computing. Zenodo (2019). https://doi.org/10.5281/zenodo.2562111

16. Rémond, U., Emeriau, P.E., Lysaght, L., Ruel, J., Mikael, J., Kazymyrenko, K.: VQA for structural mechanical problems (2025). Private communication

17. Sato, Y., Kondo, R., Koide, S., Takamatsu, H., Imoto, N.: Variational quantum algorithm based on the minimum potential energy for solving the Poisson equation (2021). https://doi.org/10.1103/PhysRevA.104.052409

18. Sim, S., Johnson, P.D., Aspuru-Guzik, A.: Expressibility and entangling capability of parameterized quantum circuits for hybrid quantum-classical algorithms. Adv. Quantum Technol. **2**(12) (2019). https://doi.org/10.1002/qute.201900070

19. Zaslavski, A., Powell, M.: The NEWUOA software for unconstrained optimization without derivatives, pp. 255–297 (2006). https://doi.org/10.1007/0-387-30065-1_16

20. Zienkiewicz, O., Taylor, R., Nithiarasu, P.: The finite element method for fluid dynamics. In: The Finite Element Method for Fluid Dynamics (Seventh Edition), 7th edn., p. i. Butterworth-Heinemann, Oxford (2014). https://doi.org/10.1016/B978-1-85617-635-4.00016-9

K-ADAPT-VQE: Optimizing Molecular Ground State Searches by Chunking Operators

Tatiana A. Bespalova[1,2]([✉]) [iD], Oumaya Ladhari[1,2] [iD], and Guido Masella[1] [iD]

[1] QPerfect, 23 Rue du Loess, 67000 Strasbourg, France
`tatiana.a.bespalova@gmail.com`
[2] Université de Strasbourg and CNRS, ISIS (UMR 7006), 67000 Strasbourg, France

Abstract. Classical simulation of molecular systems is fundamentally limited by exponential scaling, motivating the development of quantum algorithms such as the Variational Quantum Eigensolver (VQE). While the Adaptive VQE (ADAPT-VQE) improves upon VQE by dynamically constructing ansätze from an operator pool, its one-operator-at-a-time strategy can lead to high computational overhead. In this work, we introduce K-ADAPT-VQE, a generalization of ADAPT-VQE that selects operators in chunks of K per iteration, thereby accelerating convergence. Using the MIMIQ quantum simulator, we benchmark the method on BeH_2, LiH, and N_2, demonstrating that K-ADAPT-VQE consistently reduces the number of iterations and quantum function calls required to achieve chemical accuracy, with improvements of up to an order of magnitude relative to ADAPT-VQE. These findings indicate that chunked operator selection offers a scalable path for applying adaptive ansätze construction to increasingly complex molecular systems.

Keywords: Quantum Chemistry · Quantum Computing · VQA

1 Introduction

Quantum chemistry is one of the most computationally demanding fields in science due to the exponential growth of the Hilbert space with system size. This exponential scaling renders many accurate quantum chemistry calculations intractable for classical computers, particularly for systems exhibiting strong electron correlations. Since molecular systems are fundamentally governed by quantum mechanics, quantum computing presents a natural and highly promising approach for their study [2].

Although fault-tolerant quantum computers are not yet available, researchers are actively exploring algorithms that could demonstrate a quantum advantage using today's Noisy Intermediate-Scale Quantum (NISQ) devices [11]. Variational Quantum Algorithms (VQAs), such as the Variational Quantum Eigensolver (VQE) [10], have become prominent in this context. VQEs employ a

F. Barbaresco and F. Gerin (Eds.): QUEST-IS 2025, CCIS 2744, pp. 226–235, 2026.
https://doi.org/10.1007/978-3-032-13855-2_20

hybrid quantum-classical approach: a parametrized quantum circuit, known as the ansatz, prepares a trial wavefunction on a quantum processor. The energy of this state is then estimated on a quantum processor, while a classical optimizer updates the circuit parameters to minimize the energy.

However, conventional VQE methods can suffer from several limitations, including inefficient parameter optimization and the presence of numerous local minima in the energy landscape, especially for larger systems. To address these issues, the Adaptive Derivative-Assembled Pseudo-Trotter ansatz Variational Quantum Eigensolver (ADAPT-VQE) [3] was introduced. ADAPT-VQE dynamically builds a compact, problem-specific ansatz by iteratively adding operators from a predefined pool, selecting them based on their contribution to lowering the energy gradient. Extensions of this approach involve adding multiple operators per iteration to further improve efficiency. As an example, TETRIS-ADAPT-VQE [1] selects multiple operators with disjoint qubit supports to build denser circuits that are significantly shallower, a key consideration for hardware with limited coherence times.

Interestingly, this philosophy of adaptively building a compact and efficient wavefunction is not new to computational chemistry. In classical electronic structure theory, Selected Configuration Interaction (SCI) methods were developed to address the challenge of Hilbert space explosion in Full Configuration Interaction (FCI) calculations. Rather than considering all possible electron configurations, SCI algorithms iteratively select only the most energetically important determinants. Different selection criteria have led to several SCI variants, such as Configuration Interaction using a Perturbative Selection made Iteratively (CIPSI) [5], Heat-Bath Configuration Interaction (HCI) [4], and Adaptive Sampling CI [15]. Like ADAPT-VQE, these methods share a common adaptive, iterative philosophy: the wavefunction is built progressively, guided by quantitative measures of importance, such as energy contributions or Hamiltonian couplings.

In this work, we propose several improvements to the ADAPT-VQE algorithm aimed at enhancing its efficiency without compromising accuracy. Our modifications include an optimized subroutine that reduces the operator pool by pruning unnecessary operators and an improved selection process to further minimize the number of addresses to the quantum function. We call our selection process K-ADAPT-VQE. The key difference from the standard procedure is the simultaneous addition of K operators to the ansatz at each iteration. This "chunking" strategy significantly reduces the total number of quantum function evaluations required to achieve chemical accuracy.

Our benchmark focuses on gradient-free optimizers, such as COBYLA and CMA-ES, which are particularly interesting for NISQ-era hardware where gradient evaluations are costly. We assess the performance of our enhanced ADAPT-VQE implementation across a diverse set of small molecules at various bond lengths, demonstrating consistent improvements in convergence, accuracy, and overall computational resource requirements.

2   Method

This section outlines the methodology employed in our K-ADAPT-VQE approach. We begin by describing the preparation of the molecular Hamiltonian and the construction of the operator pool. We then detail the operator selection mechanism, highlighting the K-ADAPT-VQE criterion, and conclude with the full algorithmic procedure.

2.1   Hamiltonian Preparation

The starting point of the K-ADAPT-VQE algorithm is the formulation of the chemical problem, given by the electronic structure Hamiltonian for a molecular system, typically expressed in the second quantization formalism. The general form of this Hamiltonian is:

$$\hat{H} = \sum_{pq} h_{pq}\hat{a}_p^\dagger\hat{a}_q + \frac{1}{2}\sum_{pqrs} h_{pqrs}\hat{a}_p^\dagger\hat{a}_q^\dagger\hat{a}_r\hat{a}_s, \tag{1}$$

where $\hat{a}_p^\dagger$ and $\hat{a}_q$ are fermionic creation and annihilation operators acting on spin-orbitals, and the coefficients h_{pq} and h_{pqrs} are the one- and two-electron integrals arising from the molecular orbital basis [7]. These interactions account for the kinetic energy of the electrons and the Coulomb interactions between them.

To obtain these integrals, we first use PySCF [14], an open-source quantum chemistry package, to perform initial Hartree-Fock (HF) calculations and define the molecular orbitals. The resulting integrals are then processed using Open-Fermion [8], a library tailored for quantum simulations of chemistry, to construct the second-quantized Hamiltonian.

We then map this fermionic Hamiltonian into a qubit Hamiltonian suitable for quantum computation using the Jordan-Wigner (JW) transformation, a standard method that preserves the antisymmetric nature of fermionic wavefunctions. We choose the JW mapping for its conceptual and implementation simplicity, although we note that for other physically (chemically) inspired fermion-to-qubit mappings [9] the ADAPT-VQE pipeline would remain unchanged aside from the mapping step.

For small to moderate bond lengths, in the molecular systems studied in this work (e.g. BeH_2, LiH, N_2) a good approximation to the ground-state can be obtained from restricted HF methods. Conversely, for systems near-degeneracies or stretched bond lengths unrestricted HF may offer better approximations.

2.2   Operator Pool

The ADAPT-VQE framework relies on an operator pool from which the VQE ansatz is dynamically constructed. In many classical quantum chemistry approaches, coupled-cluster (CC) theory is used to capture electron correlation beyond mean-field theory. In quantum computing, this leads to the Unitary

Coupled-Cluster (UCC) ansatz [13], which is compatible with VQE frameworks. In this work, we follow the UCCSD approach, which includes single and double excitations. This can, in principle, be extended to include triple and higher excitations, although the computational cost increases significantly.

The operator pool for the K-ADAPT-VQE used in this work is constructed from UCCSD excitation operators, including all symmetry-allowed double fermionic excitations, which are then mapped to qubit operators and used iteratively in the adaptive ansatz construction.

The UCC ansatz introduces a unitary form of the CC operator by symmetrizing it $|\Psi_{\mathrm{UCC}}\rangle = e^{\hat{T}-\hat{T}^\dagger}|\Psi_{\mathrm{HF}}\rangle$, where the CC operator $\hat{T} = \sum_k \hat{T}_k$ is composed of excitation operators, and $\hat{T} - \hat{T}^\dagger$ is anti-Hermitian, ensuring that the exponential is unitary and preserves normalization. Here, $\hat{T}$ is truncated to include only single ($\hat{T}_1 = \sum_{i,j} \theta_{ij}\hat{a}_i^\dagger\hat{a}_j$) and double ($\hat{T}_2 = \sum_{i,j,k,l} \theta_{ij}^{kl}\hat{a}_i^\dagger\hat{a}_j^\dagger\hat{a}_k\hat{a}_l$) excitations. The exponential is then expanded using the Lie-Trotter formula into a product of operators.

In the JW transformation, these operators have a simple form. Let us consider, for example, double excitation for spin-orbitals p, q, r, and s. For compactness, we define the anti-Hermitian two-body generator $\hat{G}_{pqrs} \equiv \hat{a}_p^\dagger\hat{a}_q^\dagger\hat{a}_r\hat{a}_s - \hat{a}_r^\dagger\hat{a}_s^\dagger\hat{a}_p\hat{a}_q$. Under Jordan–Wigner, $\hat{G}^{pqrs}$ maps to a sum of eight Pauli strings with appropriate Z-chains:

$$\hat{G}_{pqrs} = \hat{a}_p^\dagger\hat{a}_q^\dagger\hat{a}_r\hat{a}_s - \hat{a}_r^\dagger\hat{a}_s^\dagger\hat{a}_p\hat{a}_q =$$
$$= 2i(X_pX_qX_rY_s + X_pX_qY_rX_s - X_pY_qX_rX_s - Y_pX_qX_rX_s-$$
$$-Y_pY_qY_rX_s - Y_pY_qX_rY_s + Y_pX_qY_rY_s + X_pY_qY_rY_s) \bigotimes_{k=q+1}^{p-1} \hat{Z}_k \bigotimes_{l=s+1}^{r-1} \hat{Z}_l. \tag{2}$$

We note that everything here commutes, so the expansion of $e^{\theta\hat{G}_{pqrs}} = e^{\sum_\alpha \pm 2i\theta\hat{P}_\alpha}$ to a product of exponents $\prod_\alpha e^{\pm 2i\theta\hat{P}_\alpha}$ is exact. Further in the paper we will be referring to an arbitrary sum of Pauli strings as $\hat{P}_{\mathrm{sum}} := \sum_\alpha c_\alpha\hat{P}_\alpha$. We also note that the double excitations do not act on any qubits between q and r.

To enhance computational efficiency, and focus the ansatz construction, when starting from the HF state we exclude from the operator pool: all single excitations (as HF is already optimized for those), all operators that do not conserve the total z-projection of spin (as the state we are searching for here is in the symmetry sector of zero spin projection), and we only allow hopping from two occupied in HF orbitals to two unoccupied as the influence of the rest is small on HF. A similar methodology for operator pool construction is used in [12], although in our case we include $\hat{a}_{\uparrow i}^\dagger\hat{a}_{\uparrow j}^\dagger\hat{a}_{\uparrow k}\hat{a}_{\uparrow l}$ and $\hat{a}_{\downarrow i}^\dagger\hat{a}_{\downarrow j}^\dagger\hat{a}_{\downarrow k}\hat{a}_{\downarrow l}$ operators in the pool, but exclude the operators representing electron jumps into already occupied spin-orbitals or from already vacant spin-orbitals.

2.3   Selecting the Operators

A core component of ADAPT-VQE is the iterative selection of operators based on the largest contribution to the energy gradient. The energy gradient resulting from adding an operator $\hat{P}_{\mathrm{sum}}$ to the ansatz $(|\psi\rangle)$ has a convenient form: $\frac{\partial E}{\partial \theta}\big|_{\theta=0} = \frac{\partial}{\partial \theta}\langle\Psi|e^{-i\theta\hat{P}_{\mathrm{sum}}}He^{i\theta\hat{P}_{\mathrm{sum}}}|\Psi\rangle\big|_{\theta=0} = -i\langle\Psi|e^{-i\theta\hat{P}_{\mathrm{sum}}}[\hat{P}_{\mathrm{sum}}, H]e^{i\theta\hat{P}_{\mathrm{sum}}}|\Psi\rangle\big|_{\theta=0} = -i\langle\Psi|[\hat{P}_{\mathrm{sum}}, H]|\Psi\rangle$.

To optimize computational cost we precompute the analytical form of the commutators and, at each step of the adaptive procedure, the expectation values of all operator pool commutators. We note that the expectation values for all of the needed Pauli strings could also be precomputed at each step and then combined into the commutator expectations.

The main addition of the K-ADAPT-VQE to the ADAPT-VQE procedure is in the way operators from the pool are chosen and added to the ansatz. Instead of adding only the operator with the largest energy gradient, we choose the K operators with the highest absolute values of the commutator expectation and add them to the ansatz from the highest gradient to the lowest with the initial value of the parameters equal to zeros. The choice of $K = 5$ was guided by preliminary tests aimed at balancing efficiency and accuracy. For very small values of K (e.g., $K = 1$, corresponding to the original ADAPT-VQE), convergence is reliable but often requires many iterations and quantum function evaluations. Conversely, very large values of K risk overshooting by introducing too many operators simultaneously, which can increase ansatz redundancy and degrade optimization stability. In practice, $K = 5$ provided a consistent reduction in the number of VQE iterations and quantum function calls across all benchmarked systems, while maintaining chemical accuracy. Although this value worked robustly for the molecules considered here, the optimal choice of K may depend on the system size, electronic structure, and operator pool, and a systematic exploration of this dependence remains an interesting avenue for future work.

2.4   K-ADAPT-VQE Algorithm

1. **Prepare the system:** First, we compute the molecular integrals using a classical electronic structure package (e.g. PySCF), then we construct the second-quantized fermionic Hamiltonian which is mapped to qubit operators using one of the fermion-to-qubit transformations (e.g. JW)
2. **Initialize the ansatz:** The first step is to set the initial quantum state $|\Psi_0\rangle$ to the HF state. We then start with an empty ansatz.
3. **Construct the operator pool:** The operator pool includes all symmetry-allowed spin-conserving double excitation operators that correspond to electron hopping from two HF-occupied orbitals to two virtual orbitals. We proceed by mapping all allowed fermionic operators to qubit operators.
4. **Repeat until convergence:**
   a. Compute the energy gradient for each operator $\hat{P}_i$ in the pool.
   b. Select the top K operators with the largest absolute gradient magnitudes.

   c. Append these K operators to the ansatz (in order of decreasing gradient).

   d. Optimize the full ansatz parameters including previously added ones using a classical optimizer (e.g. COBYLA), allowing a maximum number of VQE iterations per ADAPT step (e.g. 200)

   e. Check convergence: if all gradients are below a threshold ϵ, or if the energy difference between the last computed and the newly computed energy is below a threshold ϵ', stop. Otherwise, return to step 4a.

5. Output: We get the final optimized ansatz, parameter set, as well as the ground state energy estimate.

3   Results

We implemented the K-ADAPT algorithm using the MIMIQ simulator [6] to compute the ground state energies for several molecules, with systems up to 20 qubits. This section presents a comparative analysis of our K-operator chunking strategy (specifically with $K = 5$, termed 5-ADAPT) against the standard one-operator-at-a-time ADAPT-VQE (termed 1-ADAPT). All calculations presented utilize the STO-3G basis set and the COBYLA optimizer with a tolerance of 10^{-3} Hartree for the VQE subroutine. The gradient threshold ϵ was set to 10^{-5} and the energy difference threshold ϵ' was set to 10^{-15}. The HF and Full Configuration Interaction (FCI) energies, obtained from PySCF, serve as benchmarks.

Figure 1 showcases the performance benefits of the K-ADAPT strategy across various molecules and bond lengths.

For the linear BeH_2 molecule Fig. 1(a) shows the computed ground state energies as a function of the Be-H bond length. For both the 5- and 1-ADAPT procedures, the total number of VQE iterations is 1000 (200 iterations for 5-ADAPT chunks, 40 iterations for each 1-ADAPT operator), and the resulting ansatz consists of 25 operators. The 5-ADAPT approach consistently yields energies closer to FCI, achieving chemical accuracy (error $< 10^{-3}$ Ha) for most of the presented bond lengths. In contrast, the 1-ADAPT procedure generally results in energies with errors approximately an order of magnitude larger. The improved convergence of the K-ADAPT method is further detailed in Fig. 1(b), which shows the energy error relative to FCI as a function of the total number of VQE iterations for BeH_2 at the fixed bond length of 1.3 Å.

To reach chemical accuracy, the 5-ADAPT strategy required significantly fewer calls to the quantum function. Assuming three evaluations per COBYLA iteration, the 5-ADAPT approach (5 chunks of 5 operators, 200 VQE iterations per chunk for the first 800 iterations, then adapting) reached chemical accuracy with approximately 3300 total quantum function evaluations ($3 \times 800 + 180 \times \frac{25}{5}$). The 1-ADAPT strategy (25 operators added one-by-one, 40 VQE iterations per operator) required roughly 14100 evaluations ($3 \times 3200 + 180 \times 25$). This signifies a computation cost reduction of ~ 4.3 times.

We observe that, in the ADAPT-VQE procedure, the parameters associated with the most recently added operators undergo the most significant changes during each VQE optimization. However, allowing previously added parameters

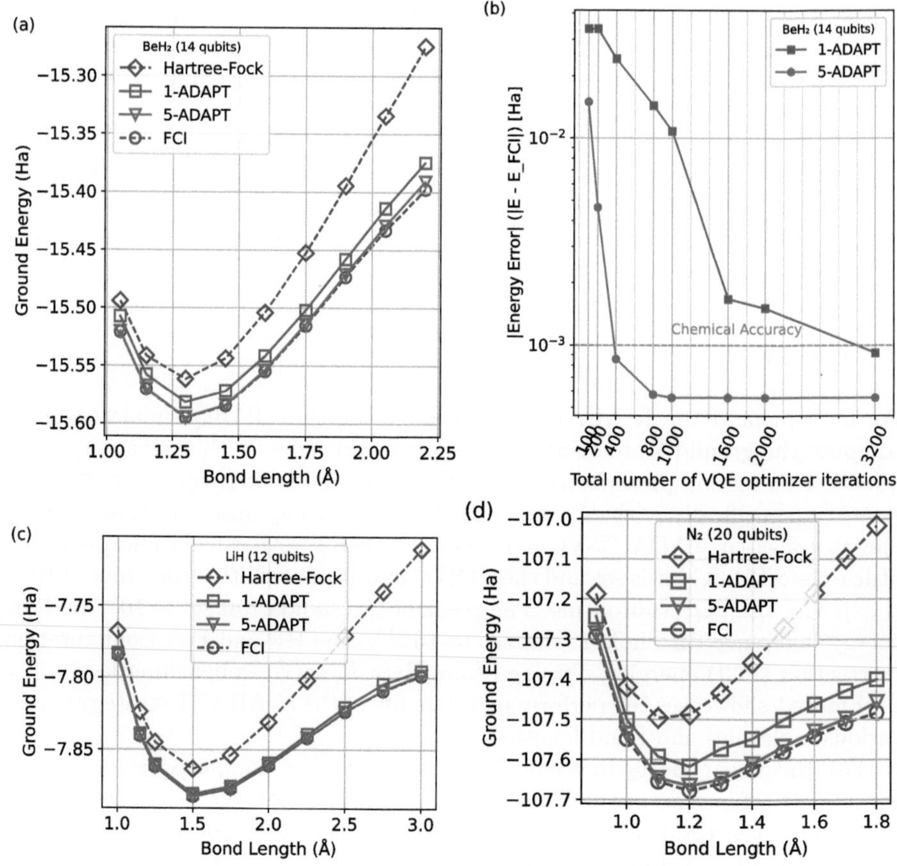

Fig. 1. Performance of 5-ADAPT and 1-ADAPT. All calculations employ the STO-3G basis, the HF and FCI energies serve as benchmarks. In figures a, c and d, the ground state energy (in Hartree) was computed as a function of the bond length for 3 different molecules using ADAPT-VQE with COBYLA optimizer (tolerance 1e-3) and a maximum of 200 VQE iterations per 5-ADAPT step, 40 VQE iterations per 1-ADAPT step. (a) BeH_2 molecule (14 qubits): with 25 operators added. (b) Convergence behavior at a fixed bond length for BeH_2 of 1.3 Å, showing the energy error (difference from FCI) versus the number of VQE iterations. (c) LiH molecule (12 qubits): with 25 operators added. (d) N_2 molecule (20 qubits): with 50 operators added.

to be fine-tuned in subsequent steps is crucial in finding the global energy minima and contributes to the overall efficiency of the optimization.

Figure 1(c, d) further demonstrates the advantages of the 5-ADAPT approach for other molecules. Fig. 1(c) shows results for the LiH molecule (12 qubits), where 25 operators were added in total. Fig. 1(c) shows results for N_2 (20 qubits), with 50 operators added. In both cases, the 5-ADAPT method demonstrates robust performance across different bond lengths.

In Fig. 2 we show the resulting ansatz structure for the BeH_2 molecule at a bond length of 1.3 Å, generated by the 5-ADAPT algorithm. We observe that operators within the same chunk frequently act on intersecting sets of qubits. This is a consequence of K-ADAPT which selects the top K operators based purely on gradient magnitude, without imposing a disjoint support criterion, distinguishing it from methods like TETRIS-ADAPT-VQE [1], which adds several operators at a time but only to non-overlapping qubits and allows mostly the reduction in depth of the circuit and not in the total number of gates needed. A direct numerical comparison of these two strategies would be a valuable direction for future work. We hypothesize that this way of focusing only on the physically relevant operators, even if overlapping, contributes to a more efficient search of the optimal parameters, without increasing the ansatz complexity with less impactful, yet parallelizable operators.

Furthermore, we note that the presence of spin-conserving excitations within the same spin type, in the fourth chunk of the ansatz, for example (Fig. 2), indicates their meaningful contribution to the ansatz, and hence supports our choice of including them in the operator pool.

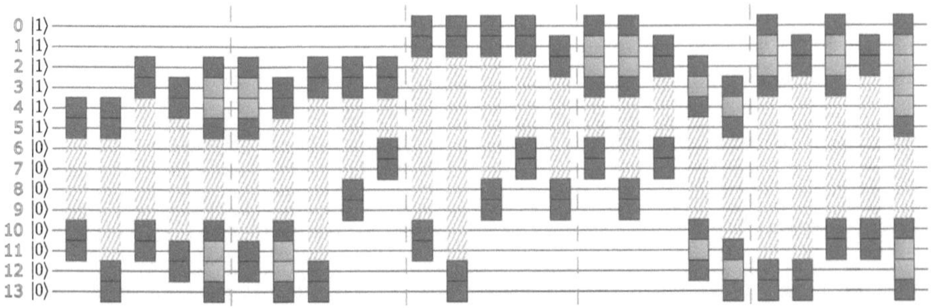

Fig. 2. The resulting structure of the ansatz for BeH_2 molecule with 1.3 Å bond length in STO-3G basis. Here each gate denotes operator $e^{\theta \hat{G}_{pqrs}}$ (see Eq. 2), where each operator is expanded into eight sequential Pauli rotations $e^{\pm 2i\theta \hat{P}}$. Here θ is the variable coefficient, it is independent for each gate but same for each Pauli within one gate, p, q, r, s correspond to the positions of dark-gray squares, in the Pauli strings $\hat{P}$ there are X and on these positions, on the positions corresponding to light-gray squares there are Z, the rest are I.

4   Conclusion

In this work, we introduced K-ADAPT-VQE, a strategy that enhances the efficiency of the ADAPT-VQE algorithm by adding operators in chunks of K (empirically set to $K = 5$ in our demonstrations). This approach significantly reduces overall computational costs, most notably by decreasing the total number of

VQE optimization iterations and, consequently, the calls to the quantum function required to reach chemical accuracy. For instance, our 5-ADAPT variant achieved a roughly 4.3-fold reduction in quantum function calls for the BeH_2 molecule compared to the standard one-operator-at-a-time ADAPT-VQE while attaining comparable accuracy.

The K-ADAPT-VQE method is effective even when operators within a single chunk act on overlapping sets of qubits, a practical scenario that distinguishes it from approaches mandating disjoint supports. This focus on adding the K most gradient-impactful operators, irrespective of their qubit intersection, prioritizes the inclusion of physically relevant correlations, thereby streamlining the ground-state search. These findings suggest that chunking operators is a beneficial strategy for molecular simulations, especially in resource-limited computational environments.

While we demonstrated that $K = 5$ provided a robust balance between accuracy and efficiency across BeH_2, LiH, and N_2, the optimal choice of K may depend on system size and operator pool characteristics, and thus represents an interesting direction for further investigation.

Acknowledgments. We thank Rosario Roberto Riso for insightful discussions on quantum chemistry. This research has received funding from the European Union's Horizon 2020 research and innovation programme under the Marie Skłodowska-Curie Grant Agreement No. 955479 (MOQS - Molecular Quantum Simulations).

Disclosure of Interests. Guido Masella is a shareholder of QPerfect.

References

1. Anastasiou, P.G., Chen, Y., Mayhall, N.J., Barnes, E., Economou, S.E.: Tetris-adapt-vqe: an adaptive algorithm that yields shallower, denser circuit ansätze. Phys. Rev. Res. **6**, 013254 (2024). https://doi.org/10.1103/PhysRevResearch.6.013254
2. Feynman, R.P.: Simulating physics with computers. Int. J. Theor. Phys. **21**(6–7), 467–488 (1982). https://doi.org/10.1007/BF02650179
3. Grimsley, H.R., Economou, S.E., Barnes, E., Mayhall, N.J.: An adaptive variational algorithm for exact molecular simulations on a quantum computer. Nature Commun. **10**(1) (2019). https://doi.org/10.1038/s41467-019-10988-2
4. Holmes, A.A., Tubman, N.M., Umrigar, C.: Heat-bath configuration interaction: an efficient selected configuration interaction algorithm inspired by heat-bath sampling. J. Chem. Theory Comput. **12**(8), 3674–3680 (2016)
5. Huron, B., Malrieu, J., Rancurel, P.: Iterative perturbation calculations of ground and excited state energies from multiconfigurational zeroth-order wavefunctions. J. Chem. Phys. **58**(12), 5745–5759 (1973)
6. Leonteva, A., Masella, G., Outteryck, M., Orioli, A.P., Whitlock, S.: Comparative benchmarking of utility-scale quantum emulators. arXiv preprint arXiv:2504.14027 (2025). https://doi.org/10.48550/arXiv.2504.14027
7. McArdle, S., Endo, S., Aspuru-Guzik, A., Benjamin, S.C., Yuan, X.: Quantum computational chemistry. Rev. Mod. Phys. **92**(1), 015003 (2020)

8. McClean, J.R., et al.: Openfermion: the electronic structure package for quantum computers. Quantum Sci. Technol. **5**(3), 034014 (2020). https://doi.org/10.1088/2058-9565/ab8ebc

9. Parella-Dilmé, T., Kottmann, K., Zambrano, L., Mortimer, L., Kottmann, J.S., Acín, A.: Reducing entanglement with physically inspired fermion-to-qubit mappings. PRX Quantum **5**, 030333 (2024). https://doi.org/10.1103/PRXQuantum.5.030333

10. Peruzzo, A., et al.: A variational eigenvalue solver on a photonic quantum processor. Nature Commun. **5**(1) (2014). https://doi.org/10.1038/ncomms5213

11. Preskill, J.: Quantum computing in the nisq era and beyond. Quantum **2**, 79 (2018). https://doi.org/10.22331/q-2018-08-06-79

12. Ramôa, M., Anastasiou, P.G., Santos, L.P., Mayhall, N.J., Barnes, E., Economou, S.E.: Reducing the resources required by adapt-vqe using coupled exchange operators and improved subroutines. npj Quantum Inf.**11**(1), 1–19 (2025). https://doi.org/10.48550/arXiv.2407.08696

13. Romero, J., Babbush, R., McClean, J.R., Hempel, C., Love, P., Aspuru-Guzik, A.: Strategies for quantum computing molecular energies using the unitary coupled cluster ansatz (2018).https://arxiv.org/abs/1701.02691

14. Sun, Q., et al.: Pyscf: the python-based simulations of chemistry framework. WIREs Comput. Molecular Sci. **8**(1), e1340 (2018). https://doi.org/10.1002/wcms.1340

15. Tubman, N.M., Lee, J., Takeshita, T.Y., Head-Gordon, M., Whaley, K.B.: A deterministic alternative to the full configuration interaction quantum monte carlo method. J. Chem. Phys. **145**(4), 044112 (2016)

Neural Network Assisted Fermionic Compression Encoding: A Lossy-QSCI Framework for Scalable Quantum Chemistry Simulations

Yu Cheng Chen[1], Ronin Wu[2(✉)] (iD), M. H. Cheng[3,4], and Min Hsiu Hsieh[1]

[1] Hon Hai Research Institute, Taipei, Taiwan
kesson.yc.chen@foxconn.com
[2] QunaSys Europe, Copenhagen, Denmark
ronin@qunasys.com
[3] Department of Physics, Blackett Laboratory, Imperial College London, London SW7 2AZ, UK
[4] Fraunhofer Institute for Industrial Mathematics, Fraunhofer-Platz 1, 67663 Kaiserslautern, Germany

Abstract. Quantum-chemistry calculations on noisy hardware are bottlenecked by qubit count and measurement overhead. We introduce **Lossy-QSCI**–a compact form of Quantum Selected CI that (i) uses a chemistry-aware lossy Random Linear Encoder (CHEMICAL-RLE) to compress an M-orbital, N-electron Hamiltonian to $O(N \log M)$ qubits, and (ii) restores observables via a lightweight neuralnetwork Fermionic Expectation Decoder (NN-FED). Applied to C_2 and LiH, Lossy-QSCI attains chemical accuracy with roughly half the qubits and determinants required by standard QSCI, pointing to a practical route for accurate quantum-chemistry on NISQ and early fault-tolerant devices.

1 Introduction

Quantum computers promise chemically exact solutions of many-electron problems that overwhelm classical methods, from small molecules to catalytic clusters [1,3]. Near-term devices, however, are constrained by few qubits, noise, and the heavy sampling overhead of hybrid algorithms such as VQE and Quantum-Selected CI (QSCI) [5,8].

VQE stalls on barren plateaus and requires millions of measurements, while standard QSCI avoids optimisation but still stores the <u>full</u> Fock space, exhausting qubits and amplifying post-processing cost [9].

We tackle both hurdles with **Lossy-QSCI**: a chemistry-aware, lossy Random Linear Encoder that compresses an M-orbital, N-electron problem to $\mathcal{O}(N \log M)$ qubits, and a neural-network Fermionic Expectation Decoder that restores observables with microsecond latency [2]. Iterative sampling plus fast decoding yields compact CI subspaces that keep measurement counts low.

F. Barbaresco and F. Gerin (Eds.): QUEST-IS 2025, CCIS 2744, pp. 236–242, 2026.
https://doi.org/10.1007/978-3-032-13855-2_21

On C_2 and LiH we reach chemical accuracy using roughly <u>half</u> the qubits and 25% fewer determinants than qubit-uncompressed QSCI, pointing to a practical route for accurate quantum chemistry on NISQ and early fault-tolerant hardware (Table 1).

Table 1. Comparison of QSCI variants.

Method	QSCI [5]	Scaled-up QSCI [10]	TE-QSCI [6]	Lossy-QSCI [This work]
Qubit Efficiency	O(M)	O(M)	O(M)	O(Nlog(M))
Bit-flip Error Mitigation	Post-Selection	Configuration Recovery	Post-Selection	Compression+Post-Selection
Chemical Prior	Customizable	LUCJ Ansatz	Time-Evolved-HF/UCCSD	Chemical-RLE

Related work. Early mappings such as JordanWigner encode the <u>entire</u> Fock space, consuming one qubit per spin-orbital [4]. Recent number-conserving schemes shrink this to $\mathcal{O}(N \log M)$ qubits, but at the cost of either an exponential measurement count (non-linear variants) or a hard decoding problem, only partially resolved by the Fermionic-Expectation-Decoder (FED) [2,11].[1] On the algorithmic side, Quantum-Selected CI (QSCI) samples a trial wavefunction, keeps the most frequent determinants, and classically diagonalises the Hamiltonian [5]. Its 77-qubit hardware demo added self-consistent recovery and a LUCJ ansatz to fight noise [10]; ADAPT-QSCI grows the circuit operator-by-operator [7]; TE-QSCI sidesteps optimisation by time-evolving a simple reference state [6]. All still store the full JW register, rely on deep or problem-specific ansätze, and require heavy sampling [9].

Our Contribution. Lossy-QSCI closes this gap: a chemistry-biased lossy Random Linear Encoder (Chemical-RLE) compresses directly to the $\mathcal{O}(N \log M)$ register, while a neural-network FED (NN-FED) decodes observables in microseconds. The resulting workflow attains chemical accuracy on C_2 and LiH using roughly half the qubits and fewer determinants than uncompressed QSCI, offering a practical path toward resource-frugal quantum chemistry on near-term and early fault-tolerant hardware.

2 Numerical Result of Lossy-QSCI

Benchmark Protocol. We evaluate Lossy-QSCI on three systems, each defined by an active space $(N_{\text{Spin Orb}}, N_e)$:

- C_2. Vary the lossy–compression rate to reveal the trade-off between qubit count and energy accuracy.
- LiH. Combine a hardwareefficient ansatz (HEA) with different QSCI basis sizes R to probe basisset convergence.

---

[1] Full derivations, scaling proofs and decoder benchmarks are given in of the Supplement.

– H_2. Compare LOSSY-QSCI and vanilla QSCI under a Pauli-X bit-flip channel to gauge noise robustness.

For every molecule we iteratively sample the R most-probable determinants from successive trial states, thereby refining a compact CI subspace that retains the dominant weight of the exact ground state.

2.1 C_2 Molecule

We begin by classically diagonalising the full 6–31G Hamiltonian for C_2 at a bond length of 2.5 Å. The resulting eigenvector is taken as the exact ground state, $|\psi_g\rangle$. Figure 1 illustrates the performance of LOSSY-QSCI as a function of the number of qubits. We compare our results with three other reference energies: (i) ED in the active space of size (22,4) (marked as "Chemical Accuracy (22, 4)"); (ii) ED in the active space of size (26, 4) (marked as "Chemical Accuracy (26, 4)"); (iii) uncompressed QSCI with R values of 65 at 26 qubits. The dashed blue and black lines respectively represent the values for Chemical Accuracy (22,4) and (26,4), respectively. The dashed green line depicts the uncompressed QSCI results for $R = 65$ at 26 qubits. The solid red line represents the LOSSY-QSCI method with $R = 50$.

LOSSY-QSCI with the number of qubits below 14 evaluates to a final energy that is higher than the (22, 4) Chemical Accuracy, but it quickly converges to similar values as the exact QSCI results with R values of 65 at 26 qubits, when the number of qubits increases to beyond 16.

The results demonstrate that LOSSY-QSCI, which maps a larger active space to a smaller qubit representation, can outperform exact diagonalization in a smaller active space with the same number of qubits. Specifically, the LOSSY-QSCI method with $R = 50$ achieves an energy closer to the exact QSCI results as the qubit count increases, surpassing the ED (22,4) benchmark. This suggests that LOSSY-QSCI offers a promising approach for achieving more accurate quantum simulations with constrained qubit resources and circuit depth, effectively balancing computational efficiency and accuracy.

2.2 *LiH* Molecule

In the study of LiH molecule at a bond length of 2.5 Å with the STO-3G basis set, the candidate configuration sets, essential for a single round CHEMICAL-RLE, is constructed from the 5 most frequently sampled configurations of the approximate ground states prepared via VQE. Figure 2 illustrates the result of the LOSSY-QSCI method encoded in five qubits. The two dashed lines – colored in black and blue – provide references for the chemical accuracy in the (6, 2) and (10, 2) active space respectively. The red line depicts the energy convergence of the LOSSY-QSCI method as we increase R, starting at slightly above the chemical accuracy of (6,2) active space for $R = 4$, converges quickly to the chemical accuracy of (10,2) active space and stabilizes as $R \geq 12$. The green

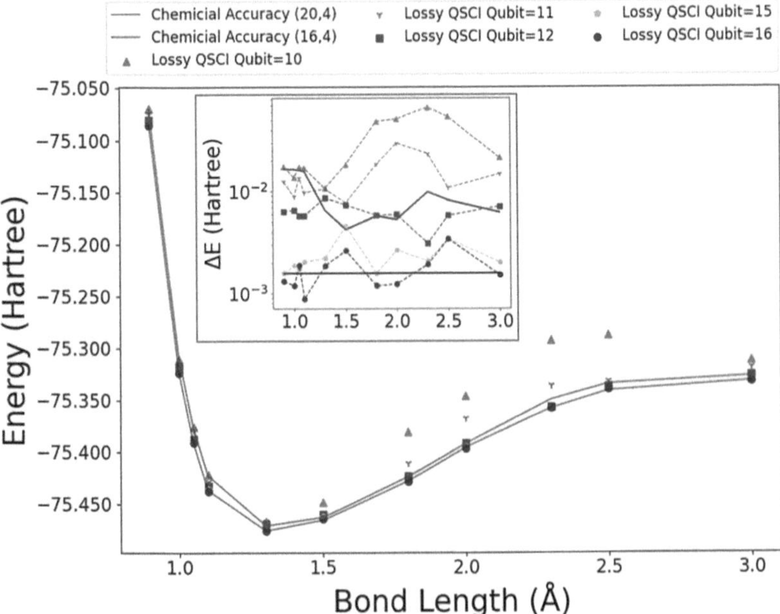

Fig. 1. The C_2 molecule Lossy-QSCI results with approximate ground state prepared by adding random noise to the exact ground state. The x-axis represents the number of qubits, while the y-axis shows the corresponding energy. The greater number of qubits will result in a less lossy rate of injectivity.

line represents the collective optimized energy of VQE on randomly compressed Hamiltonians for five qubits.

The results demonstrate that iterative sampling of CI states from different compressed subspaces can significantly enhance the performance of Lossy-QSCI. Specifically, the Lossy-QSCI method achieves the (10,2) chemical accuracy with as few as 12 basis states, outperforming all collective VQE results that lies above -7.8. Improved convergence indicates a potential of Lossy-QSCI to deliver more accurate ground state energies with fewer basis states. This approach achieves the required accuracy with fewer qubit counts compared to conventional QSCI approaches through iterative VQE, thereby improving resource efficiency in quantum simulations, and boosts the simulation power of noisy hardwares.

2.3 H_2 Molecule

We investigate the performance of noisy VQE optimization on the H_2 molecule with the 6-31G basis set at bond length 4 Å. Figure 3 illustrates the performance of the Lossy-QSCI and QSCI under the influence of bit-flip error. For the Lossy-QSCI method, the candidate configuration set was constructed by sampling the 10 most frequent configurations from 20 subspaces of ground states

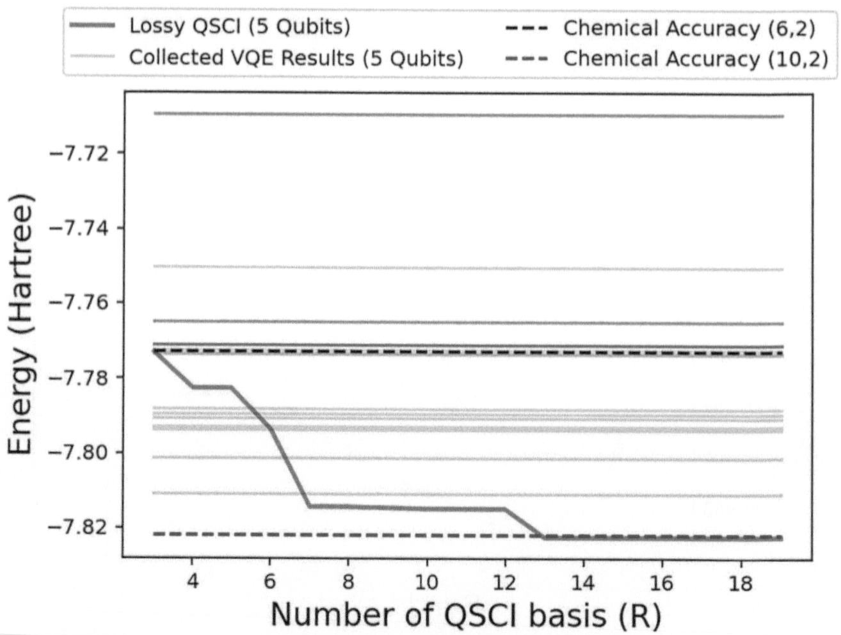

Fig. 2. The LiH molecule Lossy-QSCI Results with VQE. The x-axis represents the number of QSCI basis states (R) and the y-axis shows the computed energy in Hartree (Ha).

(due to random compression) prepared by VQE. On the other hand, the QSCI method's candidate configuration set was constructed by sampling the 10 most frequent configurations from 50 subspaces of ground states prepared by VQE. Since Lossy-QSCI with Chemical-RLE is only optimized and sampled on the number-conserving subspace, we post-process the QSCI candidate configuration set only containing the number-conserving state to improve the construction of the effective Hamiltonian.

Both results demonstrate that enforcing number-conserving constraints can provide error robustness for both QSCI and noisy QSCI and enable them to recover from noisy VQE results, as depicted by the red and blue lines. Meanwhile, the Lossy-QSCI method (blue) outperforms QSCI (red), achieving energy below chemical accuracy with as few as 12 basis states compared to the 15 basis states in QSCI. The total sampled configurations from Lossy-QSCI are 200 samples of 4-qubit configurations compared to the 500 samples of 8-qubit configurations for QSCI. This approach highlights the potential of Lossy-QSCI to deliver better searching efficiency. Robustness against bit-flip error shows the potential for making it far more suitable than QSCI in near-term devices.

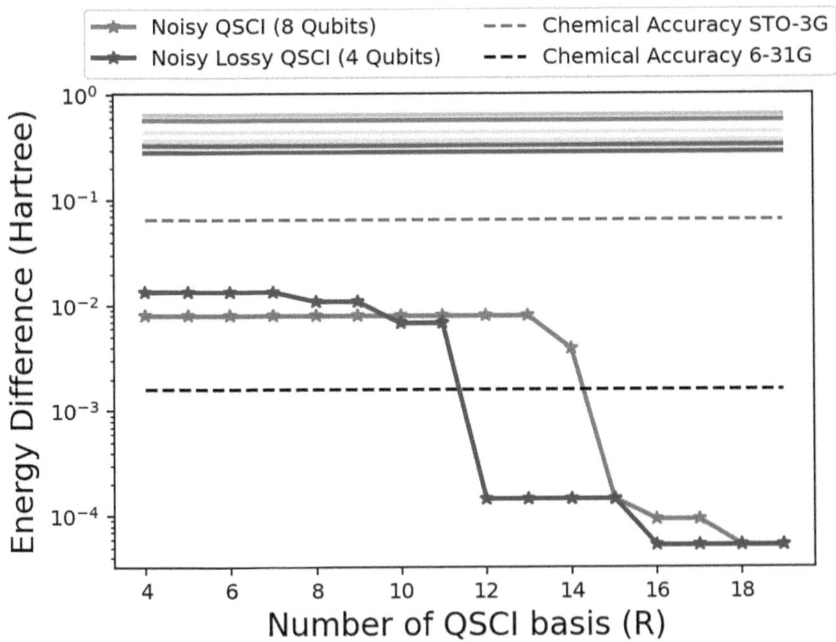

Fig. 3. This figure illustrates the H_2 molecule noisy simulation. A lossy compressed QSCI simulation not only significantly reduce the qubit cost, but also enhances the convergence w.r.t. increasing QSCI basis.

3 Conclusion

LOSSY-QSCI marries a chemistry-aware lossy Random Linear Encoder with a neural-network Fermionic Expectation Decoder, compressing the register to $\mathcal{O}(N \log M)$ qubits while retaining all chemically relevant determinants.

The trained decoder recovers full-space observables in $\sim 10^{-3}$ s per shot, eliminating the exponential lookup table of earlier FED schemes.

Benchmarks on C_2, LiH, and H_2 achieve chemical accuracy with $\approx \frac{1}{2}$ the qubits and $\sim 25\%$ fewer CI states than uncompressed QSCI, and show built-in robustness to bit-flip noise.

These resource savings yield shallower circuits and lower sampling budgets, opening a practicable route to selected-CI on NISQ hardware and early fault-tolerant platforms.

Ongoing work will chart qubitgate trade-offs for deeper encodings and extend the encoderdecoder paradigm to excited states and real-time dynamics.

References

1. Aspuru-Guzik, A., Dutoi, A.D., Love, P.J., Head-Gordon, M.: Simulated quantum computation of molecular energies. Science **309**(5741), 1704–1707 (2005). https://doi.org/10.1126/science.1113479, https://www.science.org/doi/10.1126/science.1113479
2. Cheng, M.H., et al.: Optimal Particle-Conserved Linear Encoding for Practical Fermionic Simulation (2024). https://doi.org/10.48550/arXiv.2309.09370, http://arxiv.org/abs/2309.09370, arXiv:2309.09370 [quant-ph]
3. Feynman, R.P.: Simulating physics with computers. Int. J. Theor. Phys. **21**(6-7), 467–488 (1982). https://doi.org/10.1007/BF02650179, http://link.springer.com/10.1007/BF02650179
4. Jordan, P., Wigner, E.: ber das Paulische quivalenzverbot. Zeitschrift fr Physik **47**(9-10), 631–651 (1928). https://doi.org/10.1007/BF01331938, http://link.springer.com/10.1007/BF01331938
5. Kanno, K., et al.: Quantum-Selected Configuration Interaction: classical diagonalization of Hamiltonians in subspaces selected by quantum computers (2023). https://doi.org/10.48550/arXiv.2302.11320, http://arxiv.org/abs/2302.11320, issue: arXiv:2302.11320arXiv:2302.11320 [quant-ph]
6. Mikkelsen, M., Nakagawa, Y.O.: Quantum-selected configuration interaction with time-evolved state (2025). https://doi.org/10.48550/arXiv.2412.13839, http://arxiv.org/abs/2412.13839, arXiv:2412.13839 [quant-ph]
7. Nakagawa, Y.O., Kamoshita, M., Mizukami, W., Sudo, S., Ohnishi, Y.V.: Adapt-qsci: adaptive construction of an input state for quantum-selected configuration interaction. J. Chem. Theory Comput. **20**(24), 10817–10825 (2024). https://doi.org/10.1021/acs.jctc.4c00846, pMID: 39642269
8. Peruzzo, A., et al.: A variational eigenvalue solver on a photonic quantum processor. Nat. Commun. **5**(1), 4213 (2014). https://doi.org/10.1038/ncomms5213, https://www.nature.com/articles/ncomms5213
9. Reinholdt, P., Ziems, K.M., Kjellgren, E.R., Coriani, S., Sauer, S.P.A., Kongsted, J.: Fundamental Limitations in Sample-Based Quantum Diagonalization Methods (2025). https://doi.org/10.48550/arXiv.2501.07231, http://arxiv.org/abs/2501.07231, arXiv:2501.07231 [physics]
10. Robledo-Moreno, J., et al.: Chemistry Beyond Exact Solutions on a Quantum-Centric Supercomputer (2024). https://doi.org/10.48550/arXiv.2405.05068, http://arxiv.org/abs/2405.05068, arXiv:2405.05068 [quant-ph]
11. Shee, Y., Tsai, P.K., Hong, C.L., Cheng, H.C., Goan, H.S.: Qubit-efficient encoding scheme for quantum simulations of electronic structure. Phys. Rev. Res. **4**(2), 023154 (2022). https://doi.org/10.1103/PhysRevResearch.4.023154, http://arxiv.org/abs/2110.04112, arXiv:2110.04112 [cond-mat, physics:physics, physics:quant-ph]

Towards a Quantum Generative Graph-Based Clustering for Molecule Discovery

Julien Rauch[1]([envelope]) [iD], Damien Rontani[2,3] [iD], and phane Vialle[4] [iD]

[1] LISN UMR-9015, Quantum-Saclay, CNRS, Université Paris-Saclay, Orsay, France
`julien.rauch@universite-paris-saclay.fr`
[2] LMOPS EA-4423, CentraleSupélec, Université de Lorraine, Metz, France
[3] Chair in Photonics, CentraleSupélec, Metz, France
[4] LISN UMR-9015, CentraleSupélec, CNRS, Université Paris-Saclay, Orsay, France

Abstract. In the field of drug discovery, it is necessary to speed up the exploration and analysis of very large datasets of molecules. A possible approach is the clustering of molecules with similar structures to limit the exploration of the chemical-space. This computationally-intensive operation can be mapped to the quantum domain offering inherent parallelism and providing possible higher-quality solutions.

We propose in this article a novel quantum-based clustering pipeline based on the use of quantum version of Generative Adversarial Network (QuGAN) to perform the clustering of open-access molecular data. Our contribution capitalizes on a previously-developped hybrid quantum data clustering algorithm that incorporates a molecule-specific quantum generative model for small molecular graphs (*e.g.*: QuMolGAN [11]).

We present preliminary results based on noiseless simulations of a quantum processing unit with 4 qubits. It shows gains in clustering quality after training our quantum generative pipeline on graph-based data from a quantum mechanical dataset (QM9) of small organic molecules. This motivates further investigation, in particular the fine-tuning hyperparameter optimization of the generative model, to achieve more distinct cluster separation and increased intra-cluster homogeneity.

Keywords: Generative Approach · Graph-based Clustering · Molecule Generation · Quantum Machine Learning

1 Motivation and Related Works

1.1 Molecule Graph Clustering

Molecule clustering is a fundamental cheminformatics technique to group molecules based on their structural and/or chemical similarities. The idea is that molecules belonging to the same cluster share multiple features whereas they differ significantly from molecules in other clusters. This technique is particularly

F. Barbaresco and F. Gerin (Eds.): QUEST-IS 2025, CCIS 2744, pp. 243–251, 2026.
https://doi.org/10.1007/978-3-032-13855-2_22

interesting to accelerate the exploration of large-scale chemical compounds data sets [2,6].

Molecular Data can be represented using a graph-theoretic approach called *molecular graph*, where vertices and edges represents the atoms and the chemical bonds, respectively. Grouping molecular graph data according to maximum common subgraphs enables more efficient chemical space exploration. For example, by identifying families of similar graphs, one can dramatically speed up the search for novel molecules [17].

There are two main ways of grouping molecular graphs: directly using the graphs or using their fingerprints [19]. A molecule's fingerprint encodes its structural features in the form of a vector [8], which enables us to apply classical clustering algorithms (on these vectors). Fingerprints are widely used in machine learning, but we cannot use them to deduce the corresponding molecules in return. They therefore become ineffective in the context of a clustering algorithm based on a generative model that aims at learning how to directly generate candidate molecules.

Graph clustering refers to the process of classifying the nodes of a graph and is widely used in social networks, for example [14]. In our case, we perform graph-level clustering. This means that we group graphs together [10].

1.2 Quantum Generative Clustering

Many quantum clustering algorithms already exist, such as q-means [13], quantum spectral clustering [12], and others. However, these algorithms assume the availability of *Fault-Tolerant Quantum Computers* (FTQC), which are not yet a reality. To take advantage of current quantum hardware and get initial feedback, we use hybrid clustering algorithms in which certain subroutines only are implemented as quantum algorithms. The most suitable approaches for today's *Noisy Intermediate-Scale Quantum* (NISQ) devices are *Variational Quantum Algorithms* (VQAs), as they can be adapted to specific hardware constraints. For instance, the *Variational Quantum Eigensolver* (VQE) [5,22] can be employed as a quantum subroutine.

In a previous work [16], we developed a novel clustering algorithm where quantum subroutines are discrete *Quantum Circuit Born Machine* (QCBM), trained and controlled by an *Expectation-Maximization* (EM) algorithm [3,4]. A discrete QCBM is a variational, trainable quantum circuit, which output probability distribution matches that of our data [18]. In the context of data clustering, we use a mixture (*i.e.* a set) of QCBMs, where each of them is trained to reproduce the distribution of the data located in one given cluster. See Fig. 1 and Sect. 2 for a description of our ML hybrid pipeline tailored for Molecule clustering.

1.3 Quantum Generative Molecule Models

MolGAN is a generative model designed to produce molecular graphs [7], based on the framework of generative adversarial networks (GANs). The generator is

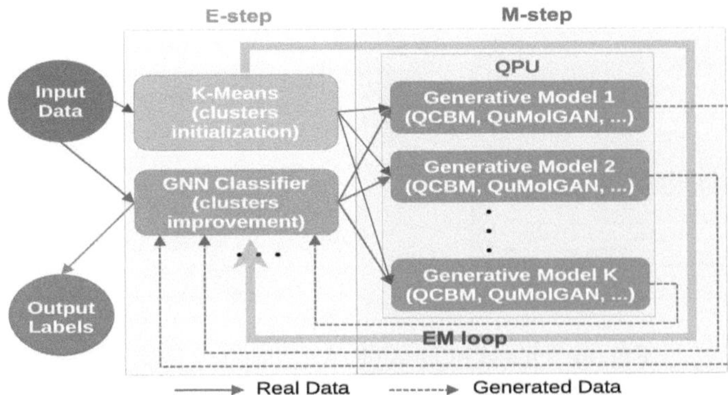

Fig. 1. Hybrid ML pipeline for data clustering

a classical neural network that outputs two matrices. The first is the adjacency matrix of the molecular graph, with dimensions $V \times V \times L$, where V is the maximum number of atoms in a molecule, and L is the number of chemical bond types (represented using one-hot encoding). The second matrix encodes the types of atoms in the molecule, with dimensions $V \times M$, where M is the number of distinct atom types (also using one-hot encoding).

In the literature, several quantum variants of MolGAN have been proposed. These approaches adapt the GAN generator to a quantum or hybrid quantum-classical architecture, such as Hybrid Quantum MolGAN (HQ-MolGAN) [9], HQ-Cycle-MolGAN [1], and MolGAN-QC [11]. The latter also introduces three different versions: MolGAN-CQ with a quantum discriminator, MolGAN-QC with a quantum generator and QuMolGAN, which exploits quantum noise. The Gaussian noise usually used as input to the GAN generator is thus replaced by noise generated by a quantum circuit.

QuMolGAN has shown some quantum advantage in the generation of small molecules. Although it is more challenging to generate molecules with QuMol-GAN compared to the classical MolGAN, the molecules it produces generally shows higher *Quantitative Estimate of Drug-likeness* (QED) scores. This mean that they possess better drug properties [11].

2   Development of Our Hybrid Molecule Clustering

2.1   Hybrid Clustering Algorithm

We propose to adapt our generative clustering pipeline for molecules clustering and generation:

- For the first speed clustering we use the k-means clustering applicated on fingerprints of molecules.
- For the classifiers we took a Graph convolutional network (GNN). This ensures that molecule graphs sharing similar structures are located in the same cluster.

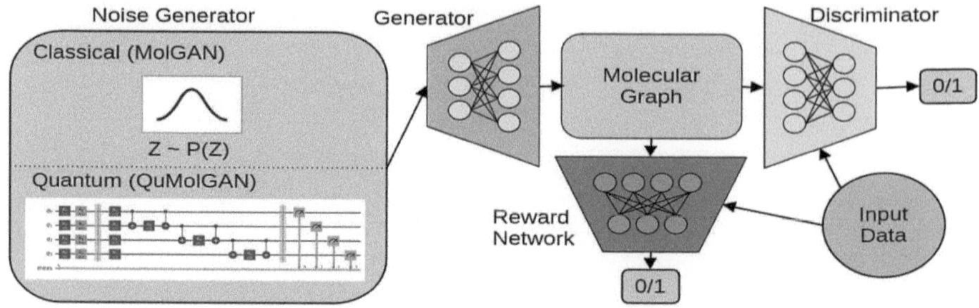

Fig. 2. Schema of QuMolGAN (inspired from [7])

- For the generative model we took a Quantum MolGAN (QuMolGAN) exploiting quantum noise.

Our ML hybrid algorithm (see Fig. 1) starts by splitting the data in clusters with a naive and fast clustering algorithm (*e.g.*: k-means), then distribute the data to our mixture of generative models. During a first M step each generative model learns its cluster attributed by the fast clustering algorithm. During a first E-step, new data clusters are generated by our generative models and we train a GNN classifier to match these new clusters while adjusting the input data clustering. Then, we enter a EM loop: The GNN classifier redistributes our data to the mixture of generative models (M-step). If data samples have a high probability to be in two clusters, then the GNN classifier can duplicate and send them to two distinct generative models. This mechanism guarantees an efficient exploration of the data space during the mixture evolution. The generative models generate new groups of data (E-step) and the classifier learns these new groups, adjusts the grouping of the input data one last time and returns the most probable label for each input data. Note that the GNN performs the calculation of distance, graph similarity in our algorithm.

As mentioned earlier, we use QuMolGAN as our generative model. The quantum circuit is located in the noise generator (see Fig. 2). In the original MolGAN, noise is sampled from a normal distribution and fed into the generator. In QuMolGAN, however, this noise is generated by a quantum parametric circuit. In our case, we use the same circuit architecture as the one proposed in [11]. In the MolGAN implementation we are using, the noise generator outputs a vector of dimension 4. To produce a comparable output, our quantum generator must also return a 4-dimensional vector. This requires a circuit of 4 qubits, where each component of the vector corresponds to the expectation value of the Pauli-Z observable on each qubit.

The reward network learns to assign a differentiable score to each molecule generated in order to drive the reinforcement learning process and enable Mol-Gan to generate molecules with interesting chemical properties. The reward network is assisted in this task by the RDKit library, which provides it with non-differentiable scores (as QED score) for reference purposes.

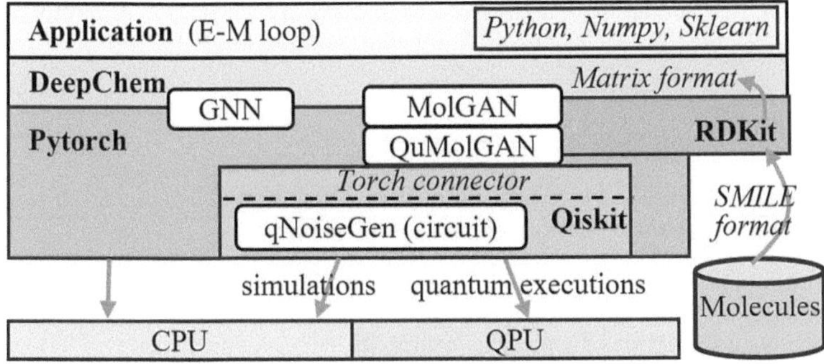

Fig. 3. Software stack for our hybrid application development

2.2 Hybrid Software Stack

Figure 3 introduces the software stack of our python-based hybrid application. It includes several libraries (all with at less a Python interface): DeepChem [20] and RDKit[1] are chemical libraries, Pytorch is used to easily implement and control ML models, and Qiskit allows to implement quantum circuit and to run quantum simulations or to use real quantum hardware. The implementation of a consistent software stack was a lengthy process, requiring the identification of a set of compatible versions of all the libraries used. Finally, we managed to obtain a consistent stack based on Python 3.11.11, and from there, developing our application was easy.

We were therefore able to easily reuse our variational algorithm and its implementation in Python and Qiskit [16] by integrating MolGAN or qMolGAN objects, a GNN object, the qNoiseGen quantum circuit, and calls to RDKit. The latter library was mainly used to perform transformations of molecules from SMILE format to matrix format and to evaluate the chemical properties of the generated molecules.

To date, all of our executions and benchmarks have been performed using noise-free simulations. However, our software stack is ready to leverage real QPUs with Qiskit.

3 First Experiments

3.1 Quantum Mechanical Dataset QM9

The QM9 Dataset is composed of 133,885 small organic molecules [21]. It has been used for machine learning tasks and the identification of structure-property relationships, making it adequate to benchmark our algorithm.

In this first phase of development of our work, we restrict ourselfs to a subset of 3,000 molecules from the initial data set mainly due to computational cost

---

[1] RDKit: Open-source cheminformatics. https://www.rdkit.org.

Table 1. Average scores over 5 runs of our hybrid algorithm for different clustering metrics on molecules of QM9 dataset (underlined scores indicate improved metrics).

Metrics	3 clusters before training	3 clusters after training	5 clusters before training	5 clusters after training	15 clusters before training	15 clusters after training
Silhouette	−0.0465	−0.0409	−0.1571	−0.0008	−0.1890	−0.0696
Davies-Bouldin Index	6.0249	5.5015	4.2534	5.5445	3.786	4.3931
Calinski-Harabasz Index	22.794	33.017	21.128	47.4092	16.1305	39.4427

of the perfect simulation of noiseless quantum processing units (QPU) on multi-core GPUs (*e.g.* The simulation of one run with 200 iterations of the EM steps of our pipeline may take up to two days).

3.2   Clustering Metrics

To evaluate the quality of our clustering we use several well-established *clustering metrics* [15]. We did not use the Normalized Mutual Information (NMI) or Adjusted Rand Index (ARI), since we do not have access to a base-line of properly clusterized molecular-graph data.

- The *Silhouette Score*: This metric assesses how similar an object is to its own cluster compared to other clusters. The score ranges from -1 to 1, where a value close to 1 indicates that the object is well-matched to its own cluster and poorly matched to neighboring clusters. A value close to -1 indicates the opposite.

- The *Davies-Bouldin Index* (DBI): This index evaluates the average similarity between each cluster and the one most similar to it, based on the ratio between intra-cluster distances and inter-cluster distances. It ranges in $[0; +\infty[$, and lower values indicate better clustering.

- The *Calinski-Harabasz Index*: Also known as the variance ratio criterion, this indicator assesses the ratio between inter-cluster dispersion and intra-cluster dispersion. It ranges in $[0; +\infty[$, and higher values indicate better-defined clusters.

However, it is difficult to apply these metrics directly to molecular graphs, as this involves defining a distance between two graphs. We therefore prefer to apply these metrics to the molecular fingerprint vectors established by RDKit, by calculating simple distances between two vectors. In our implementation all these metrics and distances are calculated by scikit-learn.

3.3   Preliminary Results

With the QM9 dataset, the exact number of clusters is unknown. As a result, we start with three clusters and gradually increase it to five. Our first results are introduced in Table 1. They show some improvements in clustering metrics between the start of the algorithm and the 200−th iteration of the EM learning loop. Unfortunately, these first results are still quite low despite measurable effect associated with the evolution of mixture models during the learning phase. Thee underlying reasons for this limited performance level needs to be explored but could stem from sub-optimal selection of hyperparameters or improper choice in the number of clusters.

4   Conclusion and Next Steps

We have developed an hybrid quantum-classical algorithm and software architecture. They are executable and aim at generating candidate molecule structures directly (once trained) with similar properties to other existing molecules belonging to the same cluster rather than just generating fingerprints or features that would not allow the structures of the molecules to be retrieved.

Following this development phase and initial simulation experiments, we plan to continue this ongoing work in two main stages:

Short term developments: hybrid algorithm improvement

- Identify adapted graph-based clustering metrics to drive graph-based learning algorithm (instead of fingerprint-based metrics).
- Conduct an experimentation campaign to optimize the hyper-parameters (QuMolGAN hyper-parameters, GNN hyper-parameters, number of clusters, ...)
- Replace the fingerprint-based fast clustering method during algorithm initialization with a graph-based method.
- Compare our quantum-generative approach to classical graph-based clustering techniques.

Medium term work: experiments on real quantum architectures

- Switch to noisy quantum simulations and evaluate robustness of our hybrid algorithms.
- Test our hybrid algorithm on real (noisy) quantum computers.
- Use generated molecules scores as global quality metric to compare classic and hybrid approaches.
- Use larger molecule dataset (not limited to 3000 molecules).

Acknowledgments. This work was supported by the French government under the France 2030 program (QuanTEdu-France) under reference ANR-22-CMAS-0001. The authors acknowledge also the financial support of the Chair in Photonics and the Région Grand-Est, France.

References

1. Anoshin, M., et al.: Hybrid quantum cycle generative adversarial network for small molecule generation. IEEE Trans. Quantum Eng. **5**, 1–14 (2024). https://doi.org/10.1109/tqe.2024.3414264
2. Backman, T.W.H., Cao, Y., Girke, T.: ChemMine tools: an online service for analyzing and clustering small molecules. Nucleic Acids Res. **39**(2), W486–W491 (2011). https://doi.org/10.1093/nar/gkr320
3. Benedetti, M., Garcia-Pintos, D., Perdomo, O., Leyton-Ortega, V., Nam, Y., Perdomo-Ortiz, A.: A generative modeling approach for benchmarking and training shallow quantum circuits. NPJ Quantum Inf. **5**(1), 45 (2019). https://doi.org/10.1038/s41534-019-0157-8
4. Benedetti, M., Lloyd, E., Sack, S., Fiorentini, M.: Parameterized quantum circuits as machine learning models. Quantum Sci. Technol. **4**, 043001 (2019). https://doi.org/10.1088/2058-9565/ab4eb5
5. Bermejo, P., Orus, R.: Variational quantum and quantum-inspired clustering (2024). https://arxiv.org/abs/2206.09893
6. Butina, D.: Unsupervised data base clustering based on daylight's fingerprint and Tanimoto similarity: a fast and automated way to cluster small and large data sets. J. Chem. Inf. Comput. Sci. **39**(4), 747–750 (1999)
7. Cao, N.D., Kipf, T.: MolGAN: an implicit generative model for small molecular graphs (2022). https://arxiv.org/abs/1805.11973
8. Capecchi, A., Probst, D., Reymond, J.L.: One molecular fingerprint to rule them all: drugs, biomolecules, and the metabolome. J. Cheminformatics **12**(1) (2020). https://doi.org/10.1186/s13321-020-00445-4
9. Jain, P., Ganguly, S.: Hybrid quantum generative adversarial networks for molecular simulation and drug discovery (2022). https://arxiv.org/abs/2212.07826
10. Ju, W., et al.: GLCC: a general framework for graph-level clustering (2023). https://arxiv.org/abs/2210.11879
11. Kao, P.Y., et al.: Exploring the advantages of quantum generative adversarial networks in generative chemistry. J. Chem. Inf. Model. **63**(11), 3307–3318 (2023). https://doi.org/10.1021/acs.jcim.3c00562. pMID: 37171372
12. Kerenidis, I., Landman, J.: Quantum spectral clustering. Phys. Rev. A **103**(4) (2021). https://doi.org/10.1103/physreva.103.042415
13. Kerenidis, I., Landman, J., Luongo, A., Prakash, A.: q-means: a quantum algorithm for unsupervised machine learning (2018). https://arxiv.org/abs/1812.03584
14. Liu, R., Feng, S., Shi, R., Guo, W.: Weighted graph clustering for community detection of large social networks. Procedia Comput. Sci. **31**, 85–94 (2014). https://doi.org/10.1016/j.procs.2014.05.248, https://www.sciencedirect.com/science/article/pii/S1877050914004256, 2nd International Conference on Information Technology and Quantitative Management, ITQM 2014
15. Pedregosa, F., et al.: Scikit-learn: machine learning in python. J. Mach. Learn. Res. **12**, 2825–2830 (2011)
16. Rauch, J., Rontani, D., Vialle, S.: Data clustering on hybrid classical-quantum NISQ architecture with generative-based variational and parallel algorithms. J. Syst. Archit., 103431 (2025). https://doi.org/10.1016/j.sysarc.2025.103431
17. Raymond, J.W., Blankley, C., Willett, P.: Comparison of chemical clustering methods using graph- and fingerprint-based similarity measures. J. Mol. Graph. Model. **21**(5), 421–433 (2003). https://doi.org/10.1016/S1093-3263(02)00188-2

18. Riofrio, C.A., et al.: A characterization of quantum generative models. ACM Trans. Quantum Comput. **5**(2), 1–34 (2024). https://doi.org/10.1145/3655027
19. Seeland, M., Johannes, A.K., Kramer, S.: Structural clustering of millions of molecular graphs. In: Proceedings of the 29th Annual ACM Symposium on Applied Computing, SAC '14, pp. 121–128. Association for Computing Machinery, New York (2014). https://doi.org/10.1145/2554850.2555063
20. Shreyas, V., Siguenza, J., Bania, K., Ramsundar, B.: Open-source molecular processing pipeline for generating molecules (2024). https://arxiv.org/abs/2408.06261
21. Valdés, J.J., Tchagang, A.B.: Understanding the structure of QM7b and QM9 quantum mechanical datasets using unsupervised learning (2023). https://arxiv.org/abs/2309.15130
22. Yung, C., Usman, M.: Clustering by contour coreset and variational quantum eigensolver. Adv. Quantum Technol. **7**(8) (2024). https://doi.org/10.1002/qute.202300450

Session: 10 Quantum Algorithms, Computing; Simulation – Simulation and Quantum Chemistry B

Quantum Algorithm for Anisotropic Diffusion and Convection Equations with Vector Norm Scaling

Julien Zylberman[1(✉)], Thibault Fredon[2], Nuno F. Loureiro[2], and Fabrice Debbasch[1]

[1] Sorbonne Université, Observatoire de Paris, Université PSL, CNRS, LUX, 75005 Paris, France
`julien.zylberman@gmail.com`
[2] Plasma Science and Fusion Center, Massachusetts Institute of Technology, Cambridge, MA 02139, USA

Abstract. In this work, we tackle the resolution of partial differential equations (PDEs) on digital quantum computers. Two fundamental PDEs are addressed: the anisotropic diffusion equation and the anisotropic convection equation. We present a quantum numerical scheme consisting of three steps: quantum state preparation, evolution with diagonal operators, and measurement of observables of interest. The evolution step relies on a high-order centered finite difference and a product formula approximation, also known as Trotterization. We provide novel vector-norm analysis to bound the different sources of error. We prove that the number of time-steps required in the evolution can be reduced by a factor $\Theta(16^n)$ for the diffusion equation, and $\Theta(4^n)$ for the convection equation, where n is the number of qubits per dimension, an exponential reduction compared to the previously established operator-norm analysis.

Keywords: quantum algorithms · partial differential equations · quantum computing · quantum numerical scheme · convection equation · diffusion equation

1 Introduction

The development of quantum numerical schemes for solving linear and non-linear partial differential equations is an active field of research, with open question about the computational efficiency, the reachable accuracies and the potential speedups when compared to classical methods. While the simulation of quantum systems by quantum computers has been extensively studied [1–6], the simulation of classical (non-quantum) systems requires the development of novel quantum numerical schemes. Those need to take into account non-quantum evolution, i.e., non-unitary or non-linear evolution, that are not directly implementable on quantum computers.

In the following, we address the problem of solving anisotropic convection and anisotropic diffusion equations on quantum computers. First, we introduce a quantum numerical scheme based on high-order central finite difference and product formula approximation, also known as Trotter-splitting. Then, we present a numerical analysis based on vector norm instead of generic operator norm analysis[1]. We show that the vector norm analysis predicts a reduction of the number of time-steps by an exponential factor $\theta(4^n)$ for the anisotropic convection equation, and by a factor $\theta(16^n)$ for the anisotropic diffusion equation, where n is the number of qubits per dimension.

2   Problem and Approach

The first considered PDE is the d-dimensional anisotropic diffusion equation:

$$\partial_t \phi = \boldsymbol{\nabla} \cdot (K(\boldsymbol{x}, t) \boldsymbol{\nabla} \phi), \tag{1}$$

where ϕ is the unknown, $\boldsymbol{\nabla} = (\partial_{x_1}, \dots, \partial_{x_d})$ and $K(\boldsymbol{x}, t) = \mathrm{diag}(\kappa_1(\boldsymbol{x}, t), \dots, \kappa_d(\boldsymbol{x}, t))$ is a space- and time-dependent thermal conductivity. The second PDE of interest is the d-dimensional anisotropic convection equation:

$$\partial_t f + \boldsymbol{c}(\boldsymbol{x}, t).\boldsymbol{\nabla} f = 0, \tag{2}$$

where $\boldsymbol{c}(\boldsymbol{x}, t) = (c_1(\boldsymbol{x}, t), \dots, c_d(\boldsymbol{x}, t))$ is a space- and time-dependent velocity field. Both equations are defined on the d-dimensional torus $\mathbb{T}^d$, i.e., the box $[0, 1]^d$ with periodic boundary conditions, with a time $t \in [0, T]$ for some $T > 0$, and with initial conditions $\phi(\boldsymbol{x}, t = 0) = \phi_0(\boldsymbol{x})$ and $f(\boldsymbol{x}, t = 0) = f_0(\boldsymbol{x})$ respectively. Additionally, we assume that the c_j and the κ_j functions are $2p+1$ differentiable, for some integer p, and do not depend on the j-th variables: $\forall j \in \{1, \dots, d\}$, $\partial_{x_j} c_j = 0$, $\partial_{x_j} \kappa_j = 0$[2]. This last assumption is typical for different examples, such as the collisionless Boltzmann equation with a Lorentz force, the Liouville equation associated with non-linear Hamiltonian systems, or for diffusion processes in some functionally graded materials.

Real-space Encoding. In order to numerically solve these equations on quantum computers, we defined the real space encoding of the solution function into a $n = n_1 + \dots + n_d$ qubit state as: $|f\rangle_t = \sum_{\boldsymbol{x}} f(\boldsymbol{x}, t)|\boldsymbol{x}\rangle \in \mathcal{H}$, where $\mathcal{H} = \mathcal{H}_{x_1} \otimes \dots \otimes \mathcal{H}_{x_d}$ is the Hilbert space associated with the n qubits and each $\mathcal{H}_{x_j}$ is a 2^{n_j}-dimensional Hilbert space. The ket vector $|\boldsymbol{x}\rangle$ is defined as the tensor product of the ket vector of each space $|\boldsymbol{x}\rangle = |x_1\rangle \otimes \dots \otimes |x_d\rangle$. Each ket vector $|x_j\rangle$ is defined as $|x_j\rangle = |q_0\rangle \otimes \dots \otimes |q_{n_j-1}\rangle$, with $x_j = \sum_{k=0}^{n_j-1} q_k/2^{k+1} \in \{0, 1/2^{n_j}, \dots, (2^{n_j} - 1)/2^{n_j}\}$ being the dyadic expansion of x and $\forall k, q_k \in \{0, 1\}$.

---

[1] In the literature, quantum algorithms for simulating Hamiltonian systems, or solving differential equations, often have a complexity given in terms of the spectral norm of the associated operator, such as the Hamiltonian or the operator $\hat{A}$ when solving a linear ordinary differential equation $df/dt = \hat{A}f$.

[2] This assumption is necessary to derive an efficient quantum circuit made only of quantum Fourier transforms and diagonal operators for the evolution step.

Quantum State Preparation. The quantum numerical scheme is based on three steps. The first one is a quantum state preparation routine that encodes the initial condition into an n-qubit state: $\phi_0 \rightarrow |\phi_0\rangle = \frac{1}{\mathcal{N}} \sum_x \phi_0(\boldsymbol{x})|\boldsymbol{x}\rangle$, $f_0 \rightarrow |f_0\rangle = \frac{1}{\mathcal{N}'} \sum_x f_0(\boldsymbol{x})|\boldsymbol{x}\rangle$, where $\mathcal{N}, \mathcal{N}'$ are normalization constants. Many quantum state preparation routines have been published in the literature: exact protocols generically require a number of quantum gates proportionate to the number of components of the target quantum states. In other words, optimal circuit size[3] for preparing arbitrary n-qubit state is $O(2^n)$ [7–9]. Fortunately, efficient methods[4] have been developed for qubit states depending on differentiable functions, which leverage the structure of the state by using Walsh, Fourier or polynomial series [10–13] (Fig. 1).

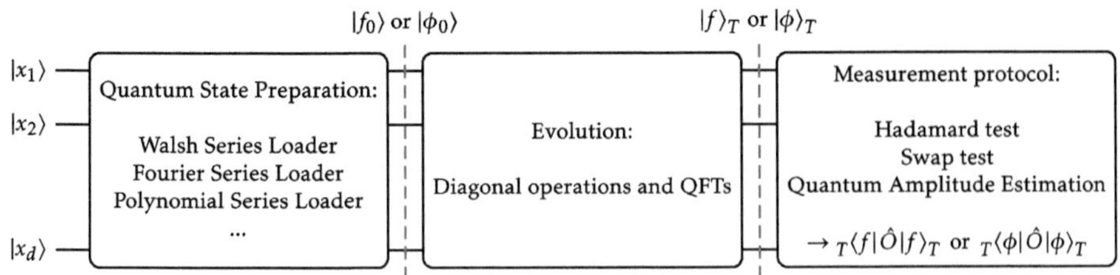

Fig. 1. Quantum numerical scheme associated with the resolution of the d-dimensional convection or diffusion equations with position- and time- dependent coefficients.

Evolution. The second step reproduces the PDE evolution using primitive quantum gates, evolving the initial qubit state $|\phi_0\rangle$ (resp. $|f_0\rangle$) toward a qubit state that encodes the PDE's solution at a target time $T > 0$: $|\phi_0\rangle \rightarrow |\phi\rangle_T = \frac{1}{\mathcal{N}''} \sum_x \phi(\boldsymbol{x},T)|\boldsymbol{x}\rangle$ (resp. $|f_0\rangle \rightarrow |f\rangle_T = \frac{1}{\mathcal{N}''} \sum_x f(\boldsymbol{x},T)|\boldsymbol{x}\rangle$). The evolution step is based on a high-order centered finite difference $\partial_{x_i}\phi(\boldsymbol{x},t) \rightarrow (1/\Delta x_i) \sum_{k=-p}^{p} a_k \phi(\boldsymbol{x} + k\boldsymbol{\Delta x_i},t) + O(\Delta x_i^{2p})$, where $a_0 = 0$ and, for $k \in \{-p, -p+1, ...p-1, p\}\backslash\{0\}$, $a_k = (-1)^{k+1}(p!)^2/(k(p-k)!(p+k)!)$[5]. By defining the discrete derivative operator associated with the j−th axis as $\hat{D}_j = \hat{I}_1 \otimes \cdots \otimes \hat{I}_{j-1} \otimes -i\frac{\sum_{k=-p}^{p} a_k(\hat{S}_j)^k}{\Delta x_j} \otimes \hat{I}_{j+1} \otimes \cdots \otimes \hat{I}_d$, where $\hat{S}_j = \sum_{x_j} |(x_j + \Delta x_j)\mathrm{mod}1\rangle\langle x_j|$ is the shift operator over the j-th axis, one can recast the PDE problems as ordinary differential equation (ODE) problems:

$$\frac{d|\tilde{f}\rangle_t}{dt} = -i\Big(\sum_{j=1}^{d} \hat{c}_j(t)\hat{D}_j\Big)|\tilde{f}\rangle_t, \tag{3}$$

---

[3] The size of a quantum circuit is the number of primitive quantum gates.

[4] A quantum circuit is said efficient if its circuit complexity expressed in terms of size, depth and ancilla qubits scale polynomially with n and $1/\epsilon$.

[5] The a_k coefficients are the finite difference coefficients, solution of the system of equations $\sum_{k=-p}^{p} a_k k^j = \delta_{j,1}$, where $j \in \{0, \ldots, 2p\}$.

$$\frac{d|\tilde{\phi}\rangle_t}{dt} = -\left(\sum_{j=1}^{d} \hat{\kappa}_j(t)\hat{D}_j^2\right)|\tilde{\phi}\rangle_t, \tag{4}$$

with the initial conditions $|\tilde{\phi}\rangle_{t=0} = |\phi_0\rangle$, $|\tilde{f}\rangle_{t=0} = |f_0\rangle$ and where $\hat{c}_j(t) = c_j(\hat{x}, t)$, $\hat{\kappa}_j(t) = \kappa_j(\hat{x}, t)$ with $\hat{x} = (\hat{x}_1, ...\hat{x}_d)$, $\hat{x}_j = \hat{I}_1 \otimes \cdots \otimes \hat{I}_{j-1} \otimes \sum_{x_j} x_j |x_j\rangle\langle x_j| \otimes \hat{I}_{j+1} \otimes \cdots \otimes \hat{I}_d$. To be explicit, we distinguish $|\tilde{\phi}\rangle_t$ and $|\tilde{f}\rangle_t$, the solutions of the ODE problems, to the real space encoding $|\phi\rangle_t$ and $|f\rangle_t$ of the solution ϕ and f of the diffusion equation Eq. 1 and convection equation Eq. 2. Furthermore, the "hat" notation only refers to operators.

The solutions of the ODEs are given by time-ordered evolution operators as: $|\tilde{f}\rangle_t = \mathcal{T}e^{-i\int_0^t \sum_{j=1}^{d} \hat{c}_j(s)\hat{D}_j ds}|f_0\rangle$ and $|\tilde{\phi}\rangle_t = \mathcal{T}e^{-\int_0^t \sum_{j=1}^{d} \hat{\kappa}_j(s)\hat{D}_j^2 ds}|\phi_0\rangle$, where $\mathcal{T}$ is the time-ordering operator. In general, it is not possible to implement directly time-ordered exponentials on computers, whether classical or quantum. To overcome this issue, we consider first-order product formula to decompose the time-ordered evolution operators into efficiently implementable unitary operators. The standard product formula denoted by the index s and the generalized one denoted by g are defined as [14]:

$$\hat{U}_s(t+h, t) = \overleftarrow{\prod_{j=1}^{d}} \exp(-ih\hat{c}_j(t+h)\hat{D}_j), \hat{U}_g(t+h, t) = \overleftarrow{\prod_{j=1}^{d}} \exp\left(-i\int_t^{t+h} \hat{c}_j(s)ds\hat{D}_j\right),$$

$$\hat{V}_s(t+h, t) = \overleftarrow{\prod_{j=1}^{d}} \exp(-h\hat{\kappa}_j(t+h)\hat{D}_j^2), \hat{V}_g(t+h, t) = \overleftarrow{\prod_{j=1}^{d}} \exp\left(-\int_t^{t+h} \hat{\kappa}_j(s)ds\hat{D}_j^2\right),$$

$$\tag{5}$$

where $\overleftarrow{\prod_{j=1}^{d}} A_j = A_d...A_1$ is a product with decreasing indexes for non-commuting operators. The product formula approximation is as accurate as h is small. Therefore, we discretize the time span $[0, T]$ into L equal segments and we define $|\tilde{\phi}_\alpha\rangle_t$ and $|\tilde{f}_\alpha\rangle_t$, with $\alpha = s, g$, the qubit states given by L trotterized-time-steps as:

$$|\tilde{\phi}_\alpha\rangle_T = \prod_{l=1}^{L} \hat{V}_\alpha\left(\frac{lT}{L}, \frac{(l-1)T}{L}\right)|\phi_0\rangle,$$

$$|\tilde{f}_\alpha\rangle_T = \prod_{l=1}^{L} \hat{U}_\alpha\left(\frac{lT}{L}, \frac{(l-1)T}{L}\right)|f_0\rangle. \tag{6}$$

The implementation of the operators $\hat{V}_\alpha$ and $\hat{U}_\alpha$ is performed using quantum Fourier transforms and diagonal operators. The quantum Fourier transform (QFT) diagonalizes the discrete derivative operators $\hat{D}_j$ as $\hat{D}_j = \widehat{QFT}_j^{-1}\left(\sum_{x_j} d(x_j)|x_j\rangle\langle x_j|\right)\widehat{QFT}_j$, where $\widehat{QFT}_j$ is the QFT acting on the j-th register and $d(x_j) = \frac{2}{\Delta x_j}\sum_{q=0}^{p} a_q \sin(2\pi q x_j)$ is one of the eigenvalues of the operator $\hat{D}_j$. Consequently, the operators $\hat{D}_j^2$ and each of the exponential defined in Eq. 5 are also diagonalized by the QFT. The associated quantum circuit is well established for the QFT [15]. The diagonal operators depend on differentiable functions and can be efficiently implemented using (sparse) Walsh or Fourier series approximations (see references [11,16] for the detailed quantum circuit constructions).

Measurement Protocol. The last step of the numerical scheme is the extraction of information from the final qubit state $|\phi\rangle_T$ (resp. $|f\rangle_T$) using measurement protocols such as quantum amplitude estimation, the Hadamard test or the Swap test [15,17–21]. These protocols enable the measurement of averaged values of observables $|\phi\rangle_T \to_T \langle\phi|\hat{O}|\phi\rangle_T$, where $\hat{O}$ is designed to correspond to relevant information, such as the value of the solution at a given position, the average position of the solution, its standard deviation, or higher-order moments[6].

3 Error Analysis

Space Discretization Error. The quantum numerical scheme introduced two approximations: the space-discretization error and the product formula error. The first one is quantified by the difference between $|f\rangle_t$, the normalized vector encoding the solution of the convection equation Eq. (2) at time t, and $|\tilde{f}\rangle_t$, the solution of the ODE Eq. (3). One can bound their difference by writing the following equation for $|f\rangle_t$: $\partial_t|f\rangle_t = -i\sum_{j=1}^{d}\hat{c}_j(t)\hat{D}_j|f\rangle_t + |r(t)\rangle$ with $|r(t)\rangle = \sum_{j=1}^{d}\hat{c}_j(t)(i\hat{D}_j|f\rangle_t - |\partial_{x_j}f\rangle_t)$. The solution of this equation is given by variation of parameter formula as $|f\rangle_t = \hat{U}(t,0)|f_0\rangle + \int_0^t \hat{U}(t,s)|r(s)\rangle ds$, where $\hat{U}$ is the time-ordered evolution operator associated with the ODE Eq. (3). The variation of parameter formula makes it possible to bound the difference as $|||f\rangle_t - |\tilde{f}\rangle_t||_{2,N} \leq t\max_{s\in[0,t]}|||r(s)\rangle||_{2,N}$, where $||.||_{2,N}$ is simply the vector 2-norm of a vector of N-components. Then, using Taylor inequality, one can show $i\hat{D}_j|f\rangle = |\partial_{xj}f\rangle + O((\Delta x_j)^{2p})$. Finally, the difference is bounded as $|||\tilde{f}\rangle_T - |f\rangle_T||_{2,N} \leq tK\sum_{j=1}^{d}||c_j||_\infty(\Delta x_j)^{2p}$, where K is a constant depending on p and the maximum value over $j \in \{0,...,d\}$ and time of the $(2p+1)$ derivative of f_0 with respect to axis j. Since $\Delta x = 1/2^n$, one can choose a number of qubits associated with the j-th axis as $n_j = \lceil\frac{1}{2p}\log_2(TK||c_j||_\infty/\epsilon)\rceil$ to ensure an ϵ approximation level.

One can proceed analogously to bound the discretization error associated with the diffusion equation. Since the diffusion equation has a non-unitary evolution[7], one needs to consider the difference between the normalized vectors as:

---

[6] Remark that, for the diffusion equation, the overall computational cost to estimate an observable of interest is dependent on the norm of the solution: either the evolution includes some projective measurement to ensure that $|||\tilde{\phi}_\alpha\rangle_T||_{2,N} = 1$ with a probability of success proportional to $|||\tilde{\phi}_\alpha\rangle_T||_{2,N}$, either one perform the measurement protocol with $|||\tilde{\phi}_\alpha\rangle_T||_{2,N} \leq 1$ and the cost factor $|||\tilde{\phi}_\alpha\rangle_T||_{2,N}$ is passed on the probabilistic cost of the measurement protocol.

[7] The evolution operator $\hat{U}$ associated with the discretized convection equation Eq. (3) is unitary thanks to the central finite difference and the assumption that $\forall j, \partial_{x_j}c_j = 0$, while the evolution operator associated with the discretized diffusion equation Eq.

$\|\frac{|\tilde\phi\rangle_t}{\||\tilde\phi\rangle_t\|_{2,N}} - \frac{|\phi\rangle_t}{\||\phi\rangle_t\|_{2,N}}\|_{2,N} \le TK \sum_{j=1}^d \|\kappa_j\|_\infty (\Delta x_j)^{2p}$, where K is independent of N and depends on the the $2p+1$ derivatives of ϕ.

Product Formula Approximation. A direct expansion of the Trotter formula gives $\|\big(e^{i(\hat H_1+\hat H_2)\Delta t} - e^{i\hat H_1\Delta t}e^{i\hat H_2\Delta t}\big)|\psi\rangle\|_{2,N} = \frac{\Delta t^2}{2}\|[\hat H_2,\hat H_1]|\psi\rangle\|_{2,N} + O(\Delta t^3)$. In the literature on quantum algorithms, most of the scaling uses the operator norm inequality by bounding the commutator as $\|[\hat H_2,\hat H_1]|\psi\rangle\|_{2,N} \le \|[\hat H_2,\hat H_1]\|_2 \le 2\|\hat H_1\|_2\|\hat H_2\|_2$[8]. However, in the case of the convection or the diffusion equation, the summand $\hat H_j$ depends on the discrete derivative operator such as $\|\hat H_j\|_2 = \|c_j\hat D_j\|_2 = O(1/\Delta x_j) = O(2^{n_j})$ or $\|\hat H_j\|_2 = \|\kappa_j\hat D_j^2\|_2 = O(1/\Delta x_j^2) = O(4^{n_j})$, resulting in exponential-with-n error bounds. In the following, we overcome these exponential scaling by considering the action of the operator $\hat D_j$ on the qubit state encoding the solution $|f\rangle_t$ (or $|\phi\rangle_t$), deriving a novel bound in vector norm with terms of the form $\|[\hat H_2,\hat H_1]|f\rangle_t\|_{2,N} = O(1)$. Precisely, one can prove that for the convection equation and $\alpha = s, g$[9] :

$$\||\tilde f_\alpha\rangle_T - |\tilde f\rangle_T\|_{2,N} \le a_\alpha \frac{T^2}{L} + r_\alpha, \tag{7}$$

where r_α contains higher order terms that are asymptotically negligible in the large $L \to +\infty$ limit and

$$a_g = \frac{1}{2}\sum_{j=1}^{d-1}\sum_{m=j+1}^d \left(\|c_j\|_\infty\|\partial_{x_j}c_m\|_\infty \max_{t\in[0,T]}\frac{\|\partial_{x_m}f_t\|_{2,N}}{\|f_0\|_{2,N}} + \|c_m\|_\infty\|\partial_{x_m}c_j\|_\infty \max_{t\in[0,T]}\frac{\|\partial_{x_j}f_t\|_{2,N}}{\|f_0\|_{2,N}}\right),$$

$$a_s = a_g + \frac{1}{2}\sum_{j=1}^d \|\partial_t c_j\|_\infty \max_{t\in[0,T]}\frac{\|\partial_{x_j}f_t\|_{2,N}}{\|f_0\|_{2,N}} \le a_g + \frac{d}{2}\max_{1\le j\le d,\, t\in[0,T]}\|\partial_t c_j\|_\infty \frac{\|\partial_{x_j}f_t\|_{2,N}}{\|f_0\|_{2,N}}.$$

Remark that $\|\partial_{x_m}f_t\|_{2,N}/\|f_0\|_{2,N}$ converges toward $\|\partial_{x_m}f_t\|_{L^2}/\|f_0\|_{L^2} = \|\partial_{x_m}f_0\|_{L^2}/\|f_0\|_{L^2}$ in the large N limits, which depends only on f_0[10].

Similarly, for the diffusion equation, one can show:

$$\left\|\frac{|\tilde\phi_\alpha\rangle_T}{\||\tilde\phi_\alpha\rangle_T\|_{2,N}} - \frac{|\tilde\phi\rangle_T}{\||\tilde\phi\rangle_T\|_{2,N}}\right\|_{2,N} \le a'_\alpha \frac{T^2}{L} + r'_\alpha, \tag{8}$$

---

(4) is non-unitary, but has a spectral norm $\|\hat V_\alpha\| = 1$, since 0 is an eigenvalue of the discrete derivative operators: $\forall j, d_j(0) = 0$, which expresses the conservation over time of the average value of the solution function.

[8] The first reference introducing vector norm analysis for quantum algorithm is [14], which address the problem of solving the Schrödinger equation with a bounded potential, i.e., in the case where $\hat H_1$ is asymptotically unbounded and $\hat H_2$ is bounded. Our results generalize their approach since all the summands are composed of discrete derivative operators that are asymptotically unbounded for the convection and diffusion equations.

[9] Proof for the convection equation can be found in [22], and its generalization to the diffusion equation is straightforward under the assumptions of this work.

[10] The L^2 norm of any derivative of f is constant over time, equal to its initial value at time $t = 0$.

where r'_α contains higher order terms that are asymptotically negligible and

$$a'_g = \frac{1}{2} \sum_{j=1}^{d-1} \sum_{m=j+1}^{d} \left(\|\kappa_j\|_\infty \|\partial^2_{x_j} \kappa_m\|_\infty \max_{t \in [0,T]} \frac{\|\partial^2_{x_m} \phi_t\|_{2,N}}{\|\tilde{\phi}_T\|_{2,N}} + \|\kappa_m\|_\infty \|\partial^2_{x_m} \kappa_j\|_\infty \max_{t \in [0,T]} \frac{\|\partial^2_{x_j} \phi_t\|_{2,N}}{\|\tilde{\phi}_T\|_{2,N}} \right),$$

$$a'_s = a'_g + \frac{1}{2} \sum_{j=1}^{d} \|\partial_t \kappa_j\|_\infty \max_{t \in [0,T]} \frac{\|\partial^2_{x_j} \phi_t\|_{2,N}}{\|\tilde{\phi}_T\|_{2,N}} \leq a_g + \frac{d}{2} \max_{1 \leq j \leq d, t \in [0,T]} \|\partial_t \kappa_j\|_\infty \frac{\|\partial^2_{x_j} \phi_t\|_{2,N}}{\|\tilde{\phi}_T\|_{2,N}},$$

where $\|\tilde{\phi}_T\|_{2,N} = \|\widehat{QFT}\tilde{\phi}_T\|_{2,N} \geq |\langle 0|^{\otimes n}\widehat{QFT}|\phi\rangle_T| = |\sum_X \tilde{\phi}(X,T)|/\sqrt{N} = |\sum_X \phi_0(X)|/\sqrt{N}$, where the average value of $\tilde{\phi}$ is conserved over time and equal to the average value of ϕ_0. Then, by assuming that the average value of ϕ_0 is non-zero, one deduce that $1/\|\tilde{\phi}_T\|_{2,N} \leq \sqrt{N}/(|\sum_X \phi_0(X)|) \xrightarrow{N \to +\infty} 1/(\sqrt{N}\langle\phi_0\rangle_{\mathbb{T}^d})$ with $\langle\phi_0\rangle_{\mathbb{T}^d} = \int_{\mathbb{T}^d} \phi_0(x)dx$ and $\|\partial^2_{x_m} \phi_t\|_{2,N}/\|\tilde{\phi}_T\|_{2,N} \leq \sqrt{N}\|\partial^2_{x_m} \phi_t\|_{2,N}/|\sum_X \phi_0(X)| \xrightarrow{N \to +\infty} \|\partial_{x_m} \phi_t\|_{L^2}/\langle\phi_0\rangle_{\mathbb{T}^d}$.

In both case, the product formula approximation scales as $O(T^2/L)$, with a prefactor independent of the discretization step. Consequently, the number of time steps L to reach an arbitrary precision $\epsilon > 0$ should scale as $L = O(T^2/\epsilon)$. It represents an exponential-with-n reduction compared to operator-norm analysis that predicts a number of time-step scaling with the operator norms $\|\hat{D}_j^2\|_2 = \Theta(4^{n_j})$ for the convection equation, and $\|\hat{D}_j^4\|_2 = \Theta(16^{n_j})$ for the diffusion equation.

These scalings are validated through simulations of a two-dimensional diffusion equation with space- and time-dependent thermal conductivity, as shown in Fig. 2. In Fig. 2a, the vector-norm error and operator-norm bounds are plotted against the number of qubits per axis, confirming that the vector-norm error is asymptotically independent of n. Figure 2b then shows the same quantities as a function of the number of time steps, exhibiting the expected linear decrease with L. Notably, while the operator-norm bounds suggest that achieving an error of 10^{-3} would require approximately $L \simeq 200000$ time-steps, the simulations demonstrate that the same accuracy is effectively attained with only $L \simeq 64$ time-steps, proving the relevance of vector-norm scalings. Additionally, simulations of anisotropic convection equations that illustrate the vector-norm scalings can be found in reference [22].

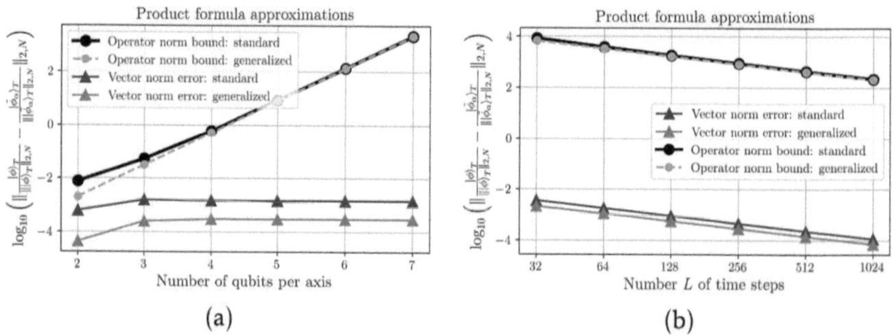

Fig. 2. Product formula errors for the generalized product formula $\alpha = g$ and the standard product formula $\alpha = s$ as a function of the number of qubits per axis with $L = 256$ (left) and, as a function of the number of time steps L with $n_1 = n_2 = 5$ (right). The simulations are performed for the diffusion equation of dimension $d = 2$ with $\kappa_1(x_2, t) = \cos(\omega t - kx_2) + 1, \kappa_2(x_1, t) = \cos(\omega t - kx_1) + 1$ and parameters $\omega = 128\pi, k = 2\pi$. The numerical solution $|\tilde{\phi}_\alpha\rangle_T$ is compared to the solution of the diffusion equation computed via a Runge-Kunta classical solver of order 5. Time is $T = 0.01$ for the left plot and $T = 0.1$ for the right plot. The initial condition is $f_0(x_1, x_2) \propto e^{-((x_1 - \mu_1)^2 + (x_2 - \mu_2)^2)/(2\sigma^2)}$ with $\mu_1 = 0.5, \mu_2 = 0.5$ and $\sigma = 1/(10\sqrt{2})$.

4   Conclusion

In this work, we introduced quantum numerical schemes for an anisotropic convection equation and for an anisotropic diffusion equation. The numerical schemes are composed of three steps: a quantum state preparation to load the initial condition into a qubit state; an evolution step made of diagonal operators and quantum Fourier transforms; and a measurement protocol to estimate quantities of interest from the numerical solution. We presented an error analysis of the space discretization and product formula approximations. In particular, we introduced novel computations based on vector-norm analysis to prove that the required number of time-steps needed to reach a given accuracy $\epsilon > 0$ can be reduced by an exponential-with-n factor, compared to operator norm analysis. These results pave the way for the development of efficient quantum numerical schemes for other PDE problems such as the Vlasov equation or the Fokker-Planck equation, and for the development of novel numerical analysis for other types of quantum numerical schemes. In particular, the vector-norm scaling is expected to extend to non-periodic PDEs, although a different quantum numerical scheme would be required, since simple QFT-based diagonalization may no longer be applicable.

References

1. Kassal, I., et al.: Polynomial-time quantum algorithm for the simulation of chemical dynamics. Proc. Natl. Acad. Sci. **105**(48), 18681–18686 (2008)
2. Kivlichan, I.D., et al.: Bounding the costs of quantum simulation of many-body physics in real space. J. Phys. A: Math. Theor. **50**(30), 305301 (2017)
3. Babbush, R., et al.: Exponentially more precise quantum simulation of fermions in the configuration interaction representation. Quantum Sci. Technol. **3**(1), 015006 (2017)
4. Low, G.H., Chuang, I.L.: Optimal Hamiltonian simulation by quantum signal processing. Phys. Rev. Lett. **118**(1), 010501 (2017)
5. Low, G.H., Chuang, I.L.: Hamiltonian simulation by qubitization. Quantum **3**, 163 (2019)
6. Babbush, R., et al.: Quantum simulation of chemistry with sublinear scaling in basis size. NPJ Quantum Inf. **5**(1), 92 (2019)
7. Zhang, X.M., Li, T., Yuan, X.: Quantum state preparation with optimal circuit depth: implementations and applications. Phys. Rev. Lett. **129**(23), 230504 (2022)
8. Yuan, P., Zhang, S.: Optimal (controlled) quantum state preparation and improved unitary synthesis by quantum circuits with any number of ancillary qubits. Quantum **7**, 956 (2023)
9. Sun, X., et al.: Asymptotically optimal circuit depth for quantum state preparation and general unitary synthesis. IEEE Trans. Comput.-Aided Des. Integr. Circuits Syst. **42**(10), 3301–3314 (2023)
10. Zylberman, J., Debbasch, F.: Efficient quantum state preparation with walsh series. Phys. Rev. A **109**(4), 042401 (2024)
11. Zylberman, J., et al.: Efficient quantum circuits for non-unitary and unitary diagonal operators with space-time-accuracy trade-offs. In: ACM Transactions on Quantum Computing (2024)
12. Moosa, M., et al.: Linear-depth quantum circuits for loading fourier approximations of arbitrary functions. Quantum Sci. Technol. **9**(1), 015002 (2023)
13. Rosenkranz, M., et al.: Quantum state preparation for multivariate functions. Quantum **9**, 1703 (2025)
14. An, D., Fang, D., Lin, L.: Time-dependent unbounded Hamiltonian simulation with vector norm scaling. Quantum **5**, 459 (2021). ISSN: 2521-327X. http://dx.doi.org/10.22331/q-2021-05-26-459
15. Nielsen, M.A., Chuang, I.L.: Quantum Computation and Quantuminformation. Cambridge university press, Cambridge (2010)
16. Welch, J., et al.: Efficient quantum circuits for diagonal unitaries without ancillas. New J. Phys. **16**(3), 033040 (2014)
17. Brassard, G., et al.: Quantum amplitude amplification and estimation. arXiv preprint quant-ph/0005055 (2000)
18. Grinko, D., et al.: Iterative quantum amplitude estimation. NPJ Quantum Inf. **7**(1), 52 (2021)
19. Rall, P., Fuller, B.: Amplitude estimation from quantum signal processing. Quantum **7**, 937 (2023)
20. Giurgica-Tiron, T., et al.: Low depth algorithms for quantum amplitude estimation. Quantum **6**, 745 (2022)
21. Shang, Z.X., Zhao, Q.: Estimating quantum amplitudes can be exponentially improved. arXiv preprint arXiv:2408.13721 (2024)
22. Zylberman, J., et al.: Trotter-based quantum algorithm for solving transport equations with exponentially fewer time-steps. arXiv preprint arXiv:2508.15691 (2025)

Fully Quantum Algorithm for the 1-Dimensional Linear Lattice Boltzmann Method

Mohammed Bediche[1,2,3]($\boxtimes$) (iD), Matthijs van Waveren[1] (iD), Denis Ricot[2], and Pierre Sagaut[3] (iD)

[1] CS Research Lab, 3, Boulevard Thomas Gobert, 91120 Palaiseau, France
`mohammed.bediche@cs-soprasteria.com`,
`matthijs.vanwaveren@compilaflows.com`
[2] CS GROUP, 22, Avenue Galilée, 92350 Le Plessis-Robinson, France
`denis.ricot@cs-soprasteria.com`
[3] Aix Marseille Univ, Centrale Med, CNRS, M2P2 Laboratory, 13013 Marseille, France
`pierre.sagaut@univ-amu.fr`

Abstract. A fully quantum algorithm for solving the one-dimensional linear advection-diffusion equation using the Lattice Boltzmann method as a numerical procedure is presented in this work. We start by presenting a state of the art of the current usage of quantum algorithms for solving ordinary and partial differential equations. We then describe two algorithms for the one-dimensional Lattice Boltzmann method with two degrees of freedom. The first one is an existing hybrid quantum-classical algorithm with measurements at each time step, and the second one is our improved version, viz. a fully quantum algorithm where only one measurement is needed at the end of the algorithm. The fully quantum algorithm is first executed on a quantum simulator and then compared with a classical approach. Subsequently, the fully quantum algorithm is run on a quantum system with 133 qubits to investigate the effect of noise and the depth of the circuit on the output state. We find fluctuations in the final result due to the decoherence noise of the qubits.

Keywords: Quantum computing · Quantum algorithms · Fluid dynamics · Lattice Boltzmann Method · Linear partial differential equation

1 Introduction

Differential equations constitute the foundational framework for the mathematical modeling of physical systems. However, executing numerical simulations of nonlinear ordinary and partial differential equations (ODEs/PDEs) on classical computers remains a formidable challenge, primarily due to the extensive computational time and data processing resources required. In response to these limitations, quantum programming has emerged as a promising alternative, garnering increasing attention over recent decades. Certain computational problems that are intractable for classical machines can be addressed theoretically using quantum computers, for example, Shor's factoring algorithm, which underpins the concept of quantum supremacy. A key advantage of

F. Barbaresco and F. Gerin (Eds.): QUEST-IS 2025, CCIS 2744, pp. 264–273, 2026.
https://doi.org/10.1007/978-3-032-13855-2_24

quantum computing lies in its potential to exponentially expand the dimensionality of the state vector—hence the number of grid points in a simulation—as the number of qubits increases, thereby enabling the manipulation of high-dimensional state spaces. Quantum algorithms, owing to their inherently linear structure, are particularly well-suited for solving linear differential equations. Several such algorithms offer a speed-up over their classical counterparts [1, 2]. Nevertheless, progress in developing quantum algorithms for nonlinear differential equations has been comparatively limited, and these approaches have yet to outperform established classical techniques. Broadly speaking, the numerical solution of PDEs can be approached via two principal methods. The first involves discretizing the system using integration techniques such as finite difference or finite element methods, implemented with either a HHL-based quantum algorithm [1, 3] or a VQA algorithm [4]. The second strategy consists in transforming the problem into a Schrödinger-type equation, which can then be addressed using Hamiltonian simulation techniques [5].

In the field of fluid mechanics and the simulation of turbulent flows, the partial differential equations (PDEs) that are generally considered are the Navier–Stokes equations. For industrial flow configurations, simplified turbulence models must be employed within the system of equations, which leads to non-negligible discrepancies with respect to the actual physics. The potential capability of quantum simulation to address very large-scale models while maintaining the shortest possible computation times is of great interest for a wide range of physical domains, including electromagnetic computations, structural analysis, acoustics, materials, combustion, and many others. For fluid mechanics simulations, an alternative to solving the Navier–Stokes equations is the use of the Lattice Boltzmann Method (LBM) [6]. This numerical method enables the simulation of various convection–diffusion phenomena, both linear and nonlinear. Several academic and commercial software packages based on LBM are available, including ProLB, developed by CS GROUP within the framework of an industrial–academic partnership.

In this work, we present a quantum algorithm for the 1-dimensional linear advection-diffusion equation solved with the Lattice Boltzmann Method. We start by following the approach of L. Budinski [7], who wrote a hybrid quantum-classical algorithm that resembles the classical algorithm of the Lattice-Boltzmann Method. One of the drawbacks of this approach is the need of repeated measurements and reinitialization at each time step. This increases the complexity of the algorithm and hinders any potential for quantum advantage in the long run. We improve on the hybrid classical-quantum algorithm by proposing a fully quantum algorithm for the one-dimensional Lattice Boltzmann formulation D1Q2. We simulate the fully quantum algorithm on the Qiskit quantum simulator, and we execute it on the 133-qubit IBM quantum computer `ibm_torino`. The results from the computers have the same evolution of the flow as the classical results and the results from the quantum simulator. The results from the quantum computer however exhibit fluctuations due to the decoherence noise in the qubits.

2 Lattice Boltzmann Method

The advection-diffusion equation solved with the Lattice Boltzmann Method is the subject of this paper.

$$\frac{\partial \phi}{\partial t} + \frac{\partial (u_i \phi)}{\partial x_i} = \frac{\partial}{\partial x_i}\left(D \frac{\partial \phi}{\partial x_i}\right) \tag{1}$$

where ϕ is the depended variable (mass, momentum, energy, species, etc.), t is time, x_i is a Cartesian coordinate, u_i is the fluid velocity in the i-direction and D is the diffusion coefficient. The above equation is valid for general advection and diffusion phenomena, including both steady and unsteady situations.

In the Lattice Boltzmann Method, instead of solving the nonlinear equation, we solve an equivalent problem with a higher dimension. However, this formulation still contains a non-linear collision term that is not easily implemented on a quantum computer. Previous works have either neglected the non-linear part of this operator [7, 8] or proceeded by linearizing it using Carleman linearization [9]. The non-linear part is zero if the advection velocity is independent of the concentration and can be neglected if the mass transport is dominated by diffusion.

In the Lattice Boltzmann method, a particle density function $f(x, c, t)$ is used to describe the probability of finding one distribution of particles within the elemental cube $(x, x + dx)$ with a velocity comprised within the interval $(c, c + dc)$ at the time t. The discretized Boltzmann equation under the Bhatnagar-Gross-Krook (BGK) approximation in the phase space [10] is given by:

$$\frac{\partial f_\alpha(\vec{x}, t)}{\partial t} + c_{\alpha,i}\frac{\partial f_\alpha(\vec{x}, t)}{\partial x_i} = -\frac{f_\alpha(\vec{x}, t) - f^{eq}{}_\alpha(\vec{x}, t)}{\tau} \tag{2}$$

The index i refers to the discretization in space and the index α defines the discretization in the velocity components. These distribution functions are linked to the macroscopic variables by:

$$\rho(\vec{x}, t) = \sum_\alpha f_\alpha(\vec{x}, t); \; \rho\,\vec{u}(\vec{x}, t) = \sum_\alpha \vec{c_\alpha}.f_\alpha(\vec{x}, t) \tag{3}$$

The algorithm of the Lattice Boltzmann Method is a four-step process: i) initialization, ii) collision, iii) propagation and iv) macro-states calculation [11]. We develop these steps in the next section.

2.1 Fully Quantum Algorithm for the Linear D1Q2

We describe a fully quantum algorithm that uses the Lattice Boltzmann Method to solve the linear advection-diffusion equation in a one-dimensional space. This algorithm builds upon the work of Budinski [7] where he presented a hybrid classical-quantum algorithm for the linear D1Q2 model with a relaxation time equals to 1. In the D1Q2 model, the index α in (2) and (3) has values 1 and 2 with only two velocity components, $e_1 = -e_2 = \frac{\Delta x}{\Delta t}$. Additionally, we consider $\Delta x = \Delta t$. The Boltzmann equation becomes:

$$f_\alpha(x + e_\alpha \Delta t, t + \Delta t) = f^{eq}{}_\alpha \tag{4}$$

The linear equilibrium distribution functions are given by (5) with $\omega_\alpha = 0.5$:

$$f_\alpha^{eq}(\vec{x}, t) = \omega_\alpha \rho(\vec{x}, t)(1 + e_{\alpha,i} u_i) + o(u_i) = A_\alpha \rho(\vec{x}, t) \tag{5}$$

A_α are square diagonal matrices of size $N \times N$, with N the number of lattice points. We consider the lattice points $x_0, x_1, \ldots, x_{N-1}$ each describing a position of the distribution function on this lattice. To encode the model into our quantum algorithm, $log_2(N)$ qubits are needed with one additional qubit used to encode information about the two degrees of freedom. One ancilla qubit is added to aid calculations.

We encode the density at $t = 0$ in the normalized state vector of Eq. (6). We applied the reverse iterative procedure proposed by Shende et al. [15], which is implemented in the built-in function `initialize` in Qiskit.

$$|\psi_0\rangle = |0\rangle_a \sum_{i=0}^{2N-1} \frac{\rho_i}{\psi} |i\rangle \tag{6}$$

The following circuit in Fig. 1 shows the original hybrid quantum circuit published by Budinski [7]. It represents the 4 steps needed in the quantum Lattice Boltzmann algorithm. The dashed vertical lines separate the four sections of the algorithm. We first describe this circuit before describing our improvement to convert it to a fully quantum algorithm.

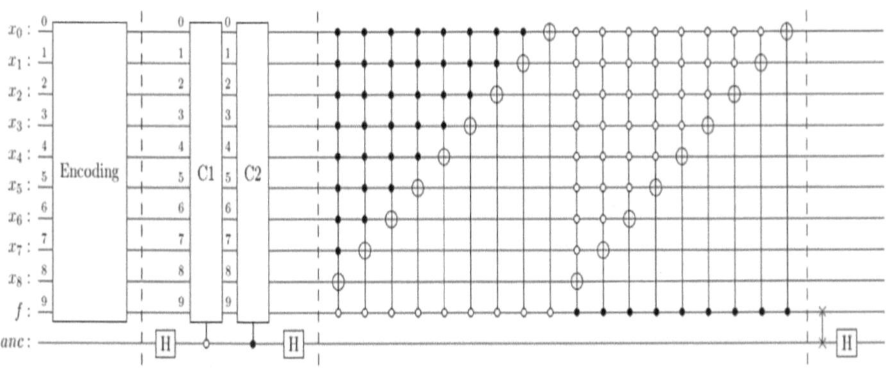

Fig. 1. The quantum circuit for one timestep. The leftmost section is the encoding step. The section with C1 and C2 is the collision step. The section to the right of it is the propagation step. The final rightmost section is the macro-states calculation step.

Collision. For the collision step, the Linear Combination of Unitaries approach [11] is used by applying the following set of operators:

$$|\psi_1\rangle = (H_a^+ \otimes I_q)(|0\rangle\langle 0|_a C_1 + |1\rangle\langle 1|_a C_2)(H_a \otimes I_q)|\psi_0\rangle \tag{7}$$

where C_1 and C_2 are unitary diagonal operators constructed from the matrices A_1 and A_2 and are described in [7]. The state vector post-collision is:

$$|\psi_1\rangle = \frac{1}{||\psi||}\left(|0\rangle_a \sum_{i=0}^{2N-1} f_{\alpha,i}|i\rangle + i|1\rangle_a \sqrt{I - A^2} \sum_{i=0}^{2N-1} \psi_{\alpha,i}|i\rangle\right) \tag{8}$$

Propagation. The propagation step is carried out using the quantum walk procedure, taking advantage of the fact that the quantum state is already in superposition. This is done using a series of multiple-controlled NOT, a C-NOT gate and a NOT gate. The difference between the two directions of propagation is the state of the control qubit. These operators can be seen in Fig. 1. The state vector $|\psi_2\rangle$ after propagation is:

$$|\psi_2\rangle = \hat{L}\hat{R}|\psi_1\rangle \tag{9}$$

Macro-states Calculations. To calculate the particle density and the velocity, We first followed Budinski's algorithm and applied a SWAP gate between the ancilla qubit and the N-1'th qubit followed by a Hadamard. The final state is:

$$|\psi_3\rangle = \frac{1}{\sqrt{(2)}\|\psi\|}\left(|0\rangle_a|0\rangle_{N-1}\sum_{i=0}^{N-1}(f_1+f_2)|i\rangle + |1\rangle_a|0\rangle_{N-1}\sum_{i=0}^{N-1}(f_1-f_2)|i\rangle + |0\rangle_a|1\rangle_{N-1}(C1-C2)\sum_{i=0}^{N-1}\psi_{\alpha,i}|i\rangle\right) \tag{10}$$

In the hybrid scheme, the resulting state vector is measured, and classical post-processing is performed to extract the density and the velocity. These are then input in the calculation of $|\psi_0\rangle$ of the next timestep. Each timestep requires thus executing a quantum circuit, measuring the resulting state vector, and performing classical post-processing to extract the density and the velocity. This procedure incurs significant overhead. For bigger systems where the Hilbert space has a large dimension, the algorithm behaves worse than a classical algorithm. A fully quantum algorithm is needed to exploit the full potential of quantum machines.

Fully Quantum Algorithm. To build a quantum algorithm that avoids repeated measurements, we start by the quantum state $|\psi_3\rangle$ calculated previously. The state vector shows that the information about f_1+f_2 and f_1-f_2 is already stored in the final state vector just before the measurements. We now develop the expression of the equilibrium distribution function and show that it contains these two quantities. It is easy to show that:

$$|\psi_0\rangle = |0\rangle_a\rho(\vec{x}, t+\Delta t) = f_1(\vec{x}, t+\Delta t) + f_2(\vec{x}, t+\Delta t);$$
$$\rho\,\vec{u}(\vec{x}, t+\Delta t) = f_1(\vec{x}, t+\Delta t) - f_2(\vec{x}, t+\Delta t) \tag{11}$$

From (5) and (11), we derive the following expressions:

$$f^{eq}_1(x, t) = \tfrac{1}{2}\big(\rho(\vec{x}, t) + \rho\,\vec{u}(\vec{x}, t)\big);$$
$$f^{eq}_2(x, t) = \tfrac{1}{2}\big(\rho(\vec{x}, t) - \rho\,\vec{u}(\vec{x}, t)\big) \tag{12}$$

This means that it is as if we had the result of the collision at the time step $t+1$ encoded in the state obtained after the calculations at the instant t. We just need to rearrange the states again to separate these two values. To do that, we modify the circuit of Fig. 1 by adding a SWAP gate between the sixth qubit in the q-register and the ancilla qubits and apply a Hadamard gate on qubit $N-1$. We obtain the state in Eq. (13). $|\lambda\rangle_q$ is an additional state that holds no useful information to us.

$$|\psi_3\rangle = \frac{1}{\|\psi\|}\left(|0\rangle_a|0\rangle_{N-1}\sum_{i=0}^{N-1}f^{eq}_{1,i}|i\rangle + |0\rangle_a|1\rangle_{N-1}\sum_{i=0}^{N-1}f^{eq}_{2,i}|i\rangle + |1\rangle_a|\lambda\rangle_q\right) \tag{13}$$

The state $|1\rangle_a|\lambda\rangle_q$ is for the other amplitudes obtained that has no exploitable information in this algorithm which we didn't write for simplicity. We see that we do not need to switch to a classical algorithm to prepare the system for future time steps, and that we can stay in the quantum space. Consequently, we can directly follow up by the propagation and the macro-states calculations, which drastically optimizes the time needed for the algorithm. The overall architecture of the algorithm is given in Fig. 2 where $|\psi_4\rangle$ is the state vector obtained after the quantum post-processing steps which allows us to bypass the intermediate measurement steps and classical processing in the hybrid algorithm to execute the next run.

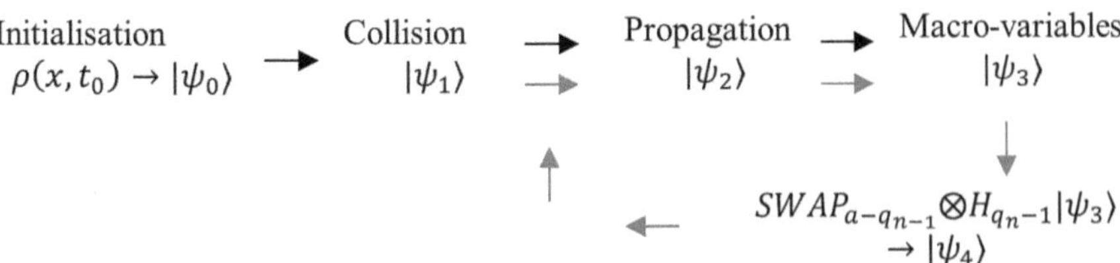

Fig. 2. The architecture of the quantum algorithm. The black arrows represent the first run of the algorithm and the grey ones the additional post treatment and the processes repeated in the m time loops afterwards for a total time of $t = m\Delta t$.

Our method eliminates the need for intermediate classical feedback. Instead, the required updates are achieved entirely within the quantum circuit by appending only two additional quantum gates at the end of each run. This results in a more efficient computation.

Complexity. To calculate the complexity of the quantum algorithm in Fig. 2, we first divide it into two parts: (a) a first forward part with the four major steps: encoding, collision, propagation and macro-states calculation and (b) a second loop part containing the quantum post-processing, the propagation and the calculation of the macroscopic variables. This second part is the one which is looped over once for each timestep.

In the first forward part, encoding the input variables into the state vector requires $2 \times 4^n - (2n+3) \times 2^n + 2n$ CNOT gates with n being the number of qubits according to Table 1 in [15]. For the collision step, we require $O(\log_2 m)$ ancilla qubits for m directions of propagation. The LCU scales as $O(1)$ in n, since the number of unitaries does not grow with the system size. The multi-controlled NOT gates used in the propagation step scale as $O(n^2)$ in the number of CNOT gates [16]. The macro-states calculations require 3 CNOT gates and one Hadamard gate so its complexity is negligible.

For the second loop part of our algorithm, we require $O(t(n)^2)$ CNOT gates with t being the number of timesteps. We see that the depth of the second part of the circuit in terms of CNOTs is linear in t and quadratic in n. Note that the number of qubits.

$n = \log_2 N + \log_2 m$, with N the number of grid points and m the number of directions of propagation.

3 Results

3.1 Execution on a Quantum Simulator

We ran the simulations using a centered Gaussian flow as input with periodic boundary conditions. The domain is discretized using 512 regular mesh points with 1000-time steps. The encoding of the initial values of the concentration was done using the instruction `initialize` available on Qiskit, which encodes an arbitrary quantum gate into the register of n qubits. The results of the fully quantum algorithm matched the ones obtained using the classical one, as shown in Fig. 3. Two time-instances, t = 200 and t = 600, were considered to show the evolution of the flow.

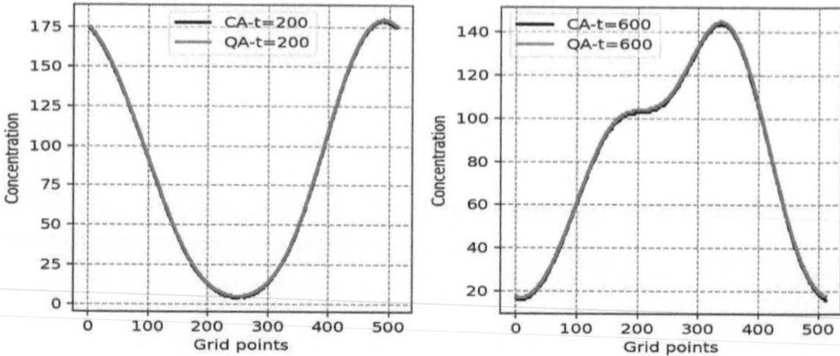

Fig. 3. Results of the fully quantum algorithm on a quantum simulator. CA corresponds to the Classical Algorithm, and QA to the Quantum Algorithm.

The original hybrid version of the algorithm was also compared with the classical one giving matching results. However, since the hybrid algorithm requires several measurement operations for each time step, it requires a longer time to execute.

3.2 Performance of the Fully Quantum Algorithm on a Quantum Computer

The fully quantum algorithm was then tested on the 133-qubit quantum machine `ibm_torino`. For every gate present in the quantum algorithm, a decomposition step is necessary to implement the gate on the quantum machine. For `ibm_torino`, there are 7 elementary gates which are the CZ, Id, RX, RZ, RZZ, SX and X. Every customized gate in the algorithm needs to be decomposed into these elementary gates before the algorithm is executed on the machine.

The quantum computer returns a histogram, which is the probability distribution obtained following the measurement operations at the end of the algorithm. Hence, we cannot directly access the quantum state. However, since the useful information of the quantum state only has a real part, it is possible to reconstruct the profile of the density as a function of the mesh points from the probability distribution. Due to the presence of noise on the quantum hardware, it was deemed necessary to reduce the size of the systems to 8 mesh points in order to decrease the depth of the algorithm. Multiple timesteps were also computed but as expected the output became noisier the longer the simulation.

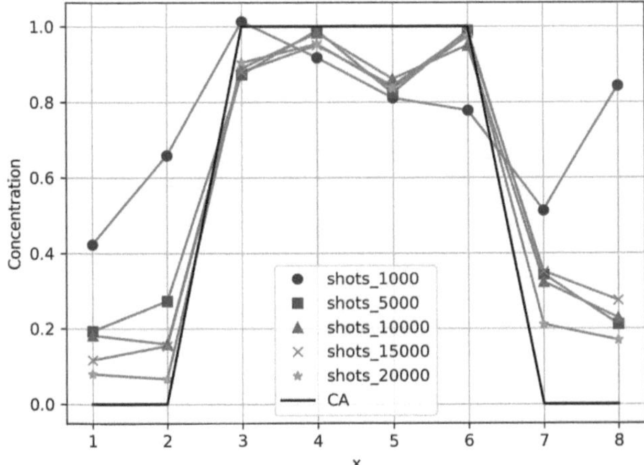

Fig. 4. Concentration results obtained from `ibm_torino`. The horizontal axis has the meshpoints.

Figure 4 represents the result obtained from `ibm_torino`, and it shows that we were successful in obtaining the same evolution of the flow as the results on the classical algorithm. We also note that increasing the number of shots from 1000 to 20000 led to better results. The values of the concentration from the run on the quantum machine fluctuate and differ from the values of the classical ones, but they have the same qualitative evolution of the flow. Our hypothesis is that this difference is due to the decoherence noise in the qubits, because larger systems and longer simulations lead to more noise. The noise in the transmon qubits arises mainly from the effects of dielectric loss and its manifestation through the interaction with two-level systems (TLS) [13]. To overcome the problem of noise in the qubits, quantum error mitigation and quantum error correction techniques are utilized. The basic idea of quantum error correction is the introduction of redundancy in the qubits, which means that one logical error-free qubit is represented by dozens of physical noisy qubits. Brayyi et al. [14] for example present an end-to-end quantum error correction protocol that implements fault-tolerant memory on the basis of a family of low-density parity-check codes. We expect that once the error mitigation and correction techniques have been implemented in production machines, we will be able to simulate larger systems.

4 Discussion

The goal of this work was to show that we can build a fully quantum algorithm for the one-dimensional linear advection-diffusion equation using the lattice Boltzmann method as a numerical procedure. The focus was more on avoiding measurements as much as possible and not as much on optimizing the steps and the gates in the algorithm. We were inspired by the hybrid quantum-classical algorithm of L. Budinski [7]. This hybrid quantum-classical algorithm is limited in its scope for quantum speedup due to the classical measurement bottleneck. We developed a fully quantum version of the algorithm by replacing the measurement step with a state vector rearrangement step. We tested this algorithm both on the Qiskit quantum simulator and on the 133-qubit IBM quantum computer `ibm_torino`. The results from the quantum computer have the same evolution of the flow as the results from the quantum simulator. The results from the quantum computer exhibit fluctuations due to the noise in the qubits. Our future work is to reduce these fluctuations by decreasing the depth of the circuits by optimizing the encoding, collision and propagation steps. We are also working on extending this algorithm to 2D systems and adding non-linearity to the model.

Acknowledgements. This study was partly funded by the French state through CIFRE.

Disclosure of Interests. The authors have no competing interests to declare that are relevant to the content of this article.

Appendix

The quantum algorithm was implemented using a Jupyter notebook. The version of Python that was used is 3.12.7. The API Qiskit version 1.2.4 was used with:

- Qiskit-aer: version 0.15.1
- Qiskit-ibm-runtime: version 0.30.0

The initial density is a centered Gaussian distribution, and the initial uniform velocity field is 0.2. The code is not open source, thus we cannot publish it on GitHub.

References

1. Montanaro, A., Pallister, S.: Quantum algorithms and the finite element method. Phys. Rev. A **93**(2), 032324 (2016). https://doi.org/10.1103/PhysRevA.93.032324
2. Childs, A.M., Liu, J.-L., Ostrander, A.: High-precision quantum algorithms for partial differential equations. Quantum **5**, 574 (2021). https://doi.org/10.22331/q-2021-11-10-574
3. Liu, J.-P., Kolden, H.Ø., Krovi, H.K., Loureiro N.F., Trivisa, K., Childs, A.M.: Efficient quantum algorithm for dissipative nonlinear differential equations. In: Proceedings of the National Academy of Sciences, vol. 118, no. 35, pp. 1–6 (2021). https://doi.org/10.1073/pnas.2026805118
4. Lubasch, M., Joo, J., Moinier, P., Kiffner, M., Jaksch, D.: Variational quantum algorithms for nonlinear problems. Phys. Rev. A. https://doi.org/10.1103/PhysRevA.101.010301
5. Jin, S., Liu, N., Yu, U.: Quantum simulation of partial differential equations via Schrödingerization. Phys. Rev. Lett. **133**, 230602 (2024). https://doi.org/10.1103/physrevlett.133.230602
6. Chen, S., Doolen, G.D.: Lattice Boltzmann Method for fluid flows, Ann. Rev. Fluid Mech. **30** (1998). https://doi.org/10.1146/annurev.fluid.30.1.329
7. Budinski, L.: Quantum algorithm for the advection–diffusion equation simulated with the Lattice Boltzmann method. Quan. Inf. Process. 20(2), 1–17 (2021). https://doi.org/10.1007/s11128-021-02996-3
8. Kocherla, S., et al.: Fully Quantum algorithm for Lattice Boltzmann methods with application to partial differential equations. AVS Quant. Sci. **6**(3), 033806 (2024). https://doi.org/10.1116/5.0217675

9. Itani W., Succi S.: Analysis of Carleman Linearization of Lattice Boltzmann. Fluids **7**(1), 24 (2022). https://doi.org/10.3390/fluids7010024

10. Bhatnagar, P.L., Gross, E.P., Krook, M.A.: Model for collision processes in Gases. I. small amplitude processes in charged and neutral one-component systems. Phys. Rev. **94**(3), 511–525 (1954). https://doi.org/10.1103/PhysRev.94.511

11. Succi, S.: Lattice Boltzmann Method for Fluid Dynamics and Beyond. Oxford University Press, Oxford (2001)

12. Low, G.H., Chuang L.L.: Hamiltonian Simulation by Qubitization. Quantum **3**, 163 (2019). https://doi.org/10.22331/q-2019-07-12-163

13. Murray, C.E.: Material matters in superconducting qubits. Mater. Sci. Eng.: R: Rep. **146**,100646 (2021). ISSN 0977–796X. https://doi.org/10.1016/j.mser.2021.100646

14. Brayyi, S., et al.: High-threshold and low-overhead fault-tolerant quantum memory. Nature **627**, 778–782 (2024). https://doi.org/10.1038/s41586-024-07107-7

15. Shende, V.V., Bullock, S.S., Markov, I.L.: Synthesis of quantum-logic circuits. IEEE Trans Comput Aided Design **25**(6), 1000–1010 (2006). https://doi.org/10.1109/TCAD.2005.855930

16. Barenco, A., et al.: Elementary gates for quantum computation. Phys. Rev. A **52**, 3457 (1995). https://doi.org/10.1103/PhysRevA.52.3457

Tucker Iterative Quantum State Preparation

Carsten Blank[(✉)] [iD] and Israel F. Araujo [iD]

Data Cybernetics ssc GmbH, 86899 Landsberg am Lech, Germany
`blank@data-cybernetics.com`
`https://data-cybernetics.com`

Abstract. Quantum state preparation is a fundamental component of quantum algorithms, particularly in quantum machine learning and data processing, where classical data must be encoded efficiently into quantum states. Existing amplitude encoding techniques often rely on recursive bipartitions or tensor decompositions, which either lead to deep circuits or lack practical guidance for circuit construction. In this work, we introduce Tucker Iterative Quantum State Preparation (Q-Tucker), a novel method that adaptively constructs shallow, deterministic quantum circuits by exploiting the global entanglement structure of target states. Building upon the Tucker decomposition, our method factors the target quantum state into a core tensor and mode-specific operators, enabling direct decompositions across multiple subsystems.

Keywords: Quantum state preparation · Entanglement · Circuit decomposition · Tensor decomposition · Tucker decomposition

1   Introduction

Quantum computing has emerged as a rapidly advancing field with significant recent progress in both theoretical and experimental domains. Current quantum hardware resides in the so-called Noisy Intermediate-Scale Quantum (NISQ) era, characterized by devices that operate with a limited number of qubits and are susceptible to noise and decoherence. These limitations pose substantial challenges in the engineering, control, and deployment of quantum systems. In particular, NISQ devices lack full fault tolerance and require sophisticated quantum error mitigation or correction strategies. Additional complications arise in the design and implementation of quantum circuits, especially concerning the reliable execution of quantum gates. A critical bottleneck in practical quantum computation is the task of encoding classical data into quantum states – a process referred to as *quantum state preparation*, or more generally, data encoding and loading.

Quantum state preparation is the process of encoding classical data into quantum states, which serve as inputs for quantum algorithms in applications such as quantum machine learning and data processing tasks [5,16,21]. Multiple

F. Barbaresco and F. Gerin (Eds.): QUEST-IS 2025, CCIS 2744, pp. 274–283, 2026.
https://doi.org/10.1007/978-3-032-13855-2_25

approaches have been developed to address the quantum state preparation problem. Among the most widely used are *Amplitude Encoding* [1–4], *Qubit Encoding* (also referred to as *Angle Encoding*) [9], and *Hamiltonian Encoding* [6,21].

In this work, we introduce a novel amplitude encoding method named *Tucker Iterative Quantum State Preparation* (Q-Tucker). This technique aims to provide a deterministic approach to state preparation by generating quantum circuits whose depth is adaptively determined by the entanglement structure of the target quantum state [1].

Entanglement has long been recognized as a fundamental resource in quantum computing, enabling capabilities that are unattainable in classical computation. Nevertheless, quantum features such as entanglement and superposition can, to a certain extent, be simulated on classical hardware. In practice, the preparation of an n-qubit quantum state – represented by a complex vector of $N = 2^n$ amplitudes – for execution on current NISQ-era quantum processing units (QPUs) often depends on classical pre-processing.

Delegating these preparatory computations to classical resources becomes increasingly challenging as the dimensionality of the quantum state grows, a scenario commonly encountered in quantum machine learning and quantum data processing tasks. A well-established strategy for mitigating this complexity involves decomposing the target state into approximately low-entangled subsystems. Even approximate decompositions can offer substantial advantages, such as reduced quantum circuit depth and, in many cases, improved fidelity when compared to exact state initialization. As a result, efficiently characterizing and exploiting the entanglement structure of quantum states is a critical objective in the design of scalable quantum algorithms and state preparation protocols.

1.1   Related Work

Several previous studies have explored low-rank quantum state preparation by leveraging hierarchical decompositions of multipartite quantum systems. Araujo et al. introduced an iterative scheme based on successive applications of the Schmidt decomposition to identify and disentangle bipartitions within a quantum system [1]. In their method, at each iteration, a suitable bipartition is selected, and the entanglement is reduced until no further disentanglement is possible, yielding a compact, low-rank representation of the original quantum state.

Regarding quantum state encoding, a variety of approaches have been proposed to map the N complex-valued amplitudes of an arbitrary quantum state onto a physical quantum circuit. One of the earliest deterministic techniques was presented by Grover et al., who proposed a method to construct quantum superpositions corresponding to efficiently integrable probability distributions [10]. While conceptually straightforward, this method requires $\mathcal{O}(N)$ quantum operations, rendering it impractical for large-scale quantum systems.

An alternative family of methods exploits tensor network structures, particularly the matrix product state (MPS) representation, to facilitate efficient state encoding. Schön et al. proposed a sequential scheme to construct MPS representations via one- and two-qubit quantum gates, enabling scalable circuit generation for a broad class of quantum states [20]. Owing to their strong approximation capabilities, MPS methods have become widely adopted in quantum many-body physics. Building upon this foundation, subsequent works have integrated parameterized tensor-network models to reduce quantum circuit depth and to support hybrid quantum-classical training schemes [19]. In some instances, iterative application of so-called *disentangler* operators to MPS representations has been proposed to further improve state compression and circuit efficiency.

A conceptually distinct approach based on the Tucker decomposition has been proposed by Protasov et al. [22]. In contrast to methods relying on the Schmidt decomposition or singular value decomposition (SVD), the Tucker decomposition generalizes these techniques by enabling one-versus-all decompositions across multiple modes. While the authors advocate for the use of the Tucker decomposition in quantum state preparation [22], their approach does not specify a selection criterion for determining the appropriate decomposition among the many non-unique possibilities. Finally, the work provides no concrete procedure for mapping the decomposed representation into a quantum circuit, leaving a critical gap between theoretical formulation and practical implementation.

1.2   Objectives and Organization

The primary objective of this work is to address the limitations of existing state preparation techniques by introducing a method based on tensor decomposition that is naturally suited for multipartite quantum systems. Unlike prior approaches that rely on recursive bipartitions – often resulting in complex hierarchical structures – our method circumvents these constraints by directly exploiting the global structure of the quantum state.

The remainder of the paper is organized as follows. Section 2 introduces the Tucker Iterative Process, detailing its decomposition strategy, fidelity control mechanisms, and the circuit synthesis procedure. Section 3 presents an analysis of the method's convergence. Section 4 reports the experimental results. Finally, Sect. 5 concludes the paper with a summary of the findings and future research directions.

2   Tucker Iterative Quantum State Preparation

This section introduces a hardware-efficient method for encoding classical data into a quantum device using an iterative tensor decomposition strategy. Using a multipartite-specific decomposition based on the Tucker tensor representation, the method addresses the limitations of previous methods [1] and eliminates the need for complex hierarchical bipartitions. The quantum circuit is constructed from the Tucker representation, with a performance metric guiding the iterative refinement process (see Fig. 1).

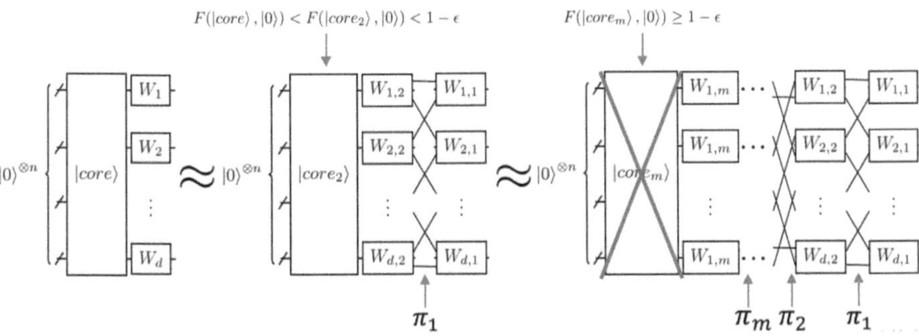

Fig. 1. Quantum circuit representation of the Tucker Iterative Process for circuit compression. The initial quantum state is decomposed into a core vector $|core\rangle$ and local hardware-efficient unitary (or isometry) factors W_i. At each iteration, the fidelity $F(|core_j\rangle, |0\rangle)$ is evaluated to determine whether the core can be approximated by the ground state. If the fidelity exceeds the threshold ($F \geq 1 - \epsilon$), the core is removed (indicated by the red cross). Between iterations, logical qubit swaps π_j are applied to optimize the placement of the factors. These swaps are guided by minimizing the bond dimension between qubit partition blocks – each corresponding to a factor W_i – which reflects the entanglement between subsystems.

Any vector $v \in \mathbb{C}^N$ can be represented as a m-tensor $T \in \mathbb{C}^{d_1} \otimes \ldots \otimes \mathbb{C}^{d_m}$, for an arbitrary $m > 1$, with the constraint $\prod_i d_i = N$. Given such a tensor T, a representation or decomposition can be determined. One such representation is the Tucker decomposition:

$$T = G \times_1 W^{(1)} \times_2 W^{(2)} \times_3 \ldots \times_m W^{(m)},$$

where $\times_i$ denotes the mode-i product, $G \in \mathbb{C}^{p_1} \otimes \ldots \otimes \mathbb{C}^{p_m}$ is the core tensor (with $p_i \leq d_i$), and the factor matrices $W^{(i)} \in \mathbb{C}^{d_i \times p_i}$. This decomposition is naturally suited for quantum computing since the tensor decomposition of T directly corresponds to a quantum state decomposition:

$$|\psi\rangle = (W^{(1)} \otimes \ldots \otimes W^{(m)})|G\rangle,$$

where we denote the operators $W^{(i)}$ using the same symbol as the factor matrices, even if their dimensions do not match ($p_i \neq d_i$), in which case columns are padded with basis vectors. The core tensor G is interpreted as a quantum state $|G\rangle$ via vectorization.

A key feature is that the factor matrices can be interpreted as unitary operators acting on separate subspaces (hence, the Tucker format is also known as the *subspace format*). From a quantum computing perspective, these unitary operations can be executed in parallel – if hardware permits – making the depth of any resulting quantum circuit determined by the longest individual factor circuit $W^{(i)}$. The geometric entanglement of the core tensor is never greater than that of the original state and can, in many cases, be significantly lower. This property enables a recursive or iterative approach that terminates. If recursion halts after

r steps, the resulting decomposition is:

$$|\psi\rangle = W_1 \ldots W_r |0\rangle.$$

Each iteration j of the process yields a factorization $W_j = W_{(j,1)} \otimes \ldots \otimes W_{(j,m)}$. If the resulting core tensor – or quantum state $|G_r\rangle$ – can be expressed as a product state (i.e., composed solely of operations that can be absorbed into the adjacent factors), or if the fidelity between the core state and the true ground state exceeds a predefined threshold, then the core state is adopted as the ground state. In this case, it requires no additional quantum operations. Each factor $W_{(j,i)}$ is implemented as a unitary or isometry operator $W^{(i)} \in \mathbb{C}^{d_i \times p_i}$. If a full unitary operator is not required ($p_i < d_i$), then isometries can be used – an important detail that enables circuit simplification and motivates the method described below.

It is well known that the tensorization of a vector $v \in \mathbb{C}$ into a tensor $T \in \mathbb{C}^{d_1} \otimes \ldots \otimes \mathbb{C}^{d_m}$ is not unique. Prior work has shown that different sequences of logical swap operations can lead to different entanglement behaviors [1]. This result extends to tensorization itself, resulting in a search space defined by the number of modes m, the dimension tuple $\mathbf{d} = (d_{i_1}, \ldots, d_{i_m})$, and a permutation $\pi \in \mathrm{Sym}([m])$ that encodes the swap pattern. Each *tensorization configuration* is thus characterized by a tuple $(m, \mathbf{d}, \pi)$. Since the optimal configuration $(m, \mathbf{d}, \pi)$ is not unique, it is possible to seek those that balance approximation quality with implementation complexity.

In general, there exists a configuration $(m, \mathbf{d}, \pi)$ such that the entanglement in the core tensor is smaller than in the original tensor T. This allows the method to be applied recursively until a product state is reached, which can be trivially initialized. Alternatively, methods such as that of Rudolph & Grover [10], or their concrete implementations [4,14], can be employed.

The search can be simplified by taking into account the coupling map limitations of the target device. On currently available quantum hardware, efficient two-qubit operations are typically restricted to neighboring qubits. Under this constraint, we can fix $m = n/2$, leading to $d_i = 4$. Since the dimension of these subsystems is fixed, the cost of each SVD is constant. This eliminates the scalability bottleneck often associated with tensor decompositions such as Tucker, where SVD costs scale cubically with the input dimension.

Even with such simplifications, the search process still depends on logical qubit swaps, which are costly. Identifying the partition with the minimal bond dimension – that is, the one producing the core vector with the least entanglement – requires a combinatorial search. The complexity of this search is determined by the size of the partition blocks, $n_i = \log_2 d_i$, and grows according to the binomial coefficient $\binom{n}{n_i}$. In the worst-case scenario, where all blocks have equal size, the complexity becomes $\mathcal{O}(2^n)$. For large systems, the computational overhead of evaluating the bond dimension at each step becomes prohibitive, making entanglement quantification infeasible.

To address this, heuristic strategies and physically motivated constraints are essential to guide the search process. These include limiting the partition search space to locality-preserving cuts, prioritizing low-overlap subsystems, or leveraging prior knowledge about the structure of the target state (e.g., symmetries or sparsity). Such strategies drastically reduce the search space and enable approximate yet effective decomposition, balancing computational tractability with circuit efficiency.

3 Convergence Guarantee of Q-Tucker

Here we formalize the correctness of the iterative procedure described in Sect. 2. Throughout, an *oracle* provides a partition $\mathcal{P} = \{B_1, \ldots, B_m\}$ (a qubit set per block) that maximizes "inner" entanglement within blocks (equivalently, minimizes inter-block entanglement or bond dimension). Given this $\mathcal{P}$, we run a Tucker step and then *fix the gauge* ("monotone gauge") by choosing the first column of each block operator to maximize the block-product overlap with the current state. Tucker/HOSVD and the non-uniqueness of factors up to rotations inside singular subspaces are classical facts [7,12].

Notation and Per-step Objective. Let $|G_{j-1}\rangle$ be the core state before iteration j and $|0\rangle = \bigotimes_i |0\rangle_{B_i}$ the block-wise ground state. For a partition $\mathcal{P}$, define the multipartite product-overlap

$$\alpha_{\mathcal{P}}(|\phi\rangle) \ := \ \max_{\|u_i\|=1} \big|\langle u_1 \otimes \cdots \otimes u_m \mid \phi\rangle\big|,$$

which is called the *entanglement eigenvalue*, and corresponds to the closest separable state [24]. The *monotone gauge* sets the first column of each block unitary so that $W^{(i)}|0\rangle_{B_i} = u_i^\star$ where $(u_i^\star)$ attains the maximum above.

Lemma 1 (monotone-gauge identity). *For the (oracle-provided) partition $\mathcal{P}_j$ used at iteration j,*

$$\langle 0 \mid G_j\rangle \ = \ \alpha_{\mathcal{P}_j}(|G_{j-1}\rangle), \qquad F_j \ = \ \alpha_{\mathcal{P}_j}(|G_{j-1}\rangle)^2.$$

Proof. By construction, $(\otimes_i W_j^{(i)\dagger})|G_{j-1}\rangle = |G_j\rangle$ and $W_j^{(i)}|0\rangle_{B_i} = u_i^\star$; hence $\langle 0 \mid G_j\rangle = \langle \otimes_i u_i^\star \mid G_{j-1}\rangle$, which is the maximizing value.

Corollary 1 (per-step monotonicity for any partition). *For any partition $\mathcal{P}$,*

$$\alpha_{\mathcal{P}}(|G_{j-1}\rangle) \ \geq \ \big|\langle 0 \mid G_{j-1}\rangle\big|,$$

since $|0\rangle$ is a feasible block-product vector. Therefore, with the monotone gauge on $\mathcal{P}_j$,

$$F_j \ = \ \alpha_{\mathcal{P}_j}(|G_{j-1}\rangle)^2 \ \geq \ \big|\langle 0 \mid G_{j-1}\rangle\big|^2 \ = \ F_{j-1}.$$

This mirrors the monotonicity of ALS/HOOI-style block updates for Tucker models [8, 12, 23].

Role of the Oracle. The oracle's partition, which maximizes inner entanglement, is used to *accelerate* progress: it typically enlarges the cut-based ceiling for block-product overlaps (see below). However, monotonicity $F_j \geq F_{j-1}$ holds *for any partition* when the monotone gauge is applied.

Cut Ceiling and Strict Improvement. For a partition $\mathcal{P}$, let $\beta(\mathcal{P}) := \min_C \lambda_{\max}(C)$ be the minimum, over block cuts C, of the largest Schmidt coefficient of $|G_{j-1}\rangle$ across C. For any bipartition, the largest Schmidt coefficient equals the maximal overlap with a product state across that cut [15, 17]; therefore any block-product is bounded by

$$F_j = \alpha_{\mathcal{P}_j}(|G_{j-1}\rangle)^2 \leq \beta(\mathcal{P}_j).$$

Partitions maximizing inner entanglement tend to increase $\beta(\mathcal{P}_j)$, raising a tight upper bound on F_j. A *strict* increase $F_j > F_{j-1}$ is guaranteed whenever the "worst" cut for $\mathcal{P}_j$ is unique and its top Schmidt vectors factor across blocks, in which case $\alpha_{\mathcal{P}_j}^2 = \beta(\mathcal{P}_j)$ and the larger ceiling is attained.

Stalls and Block-size Expansion. Fix a maximum block size k (so each block unitary acts on $\leq k$ qubits). If $F_j = F_{j-1}$ persists, we say the process *stalls at level k*. This occurs because, with fixed k, each step (via the monotone gauge) already attains the *best possible* block-product overlap within the current model class, i.e., $F_j = \alpha_{\mathcal{P}_j}(|G_{j-1}\rangle)^2$. When the state's overlap is capped by the *cut ceiling* for the chosen partition ($\alpha_{\mathcal{P}_j} = \beta(\mathcal{P}_j)$), or when all admissible partitions yield the same ceiling (or the worst-cut top Schmidt vectors fail to factor across blocks), no strict improvement is possible. Increasing k strictly enlarges the feasible set and (weakly) raises the ceilings β, so the limiting fidelity cannot decrease; allowing k to reach n (a single block) makes $F = 1$ attainable.

Theorem 1 (global convergence, stall-and-grow). *At each iteration j, (i) receive only the partition $\mathcal{P}_j$ from the oracle; (ii) run a Tucker step on $\mathcal{P}_j$ and apply the monotone gauge to form $|G_j\rangle$; (iii) if stalled at block size k, increase k. Then (F_j) is non-decreasing and converges for fixed k. Moreover, under stall-and-grow with k allowed to reach n (one block), the process attains $F = 1$ (exact preparation) in finitely many expansions.*

Proof. Monotonicity and boundedness ($0 \leq F_j \leq 1$) yield convergence for fixed k. Enlarging k cannot reduce the achievable limit; with $k = n$ any pure state can be mapped to $|0\rangle$ by a single block unitary [15], whence $F = 1$.

Summary. Q-Tucker's per-iteration fidelity *never decreases* when the monotone gauge is enforced after the oracle's partition. Oracle partitions that maximize inner entanglement typically improve the *rate* of progress by enlarging the cut ceiling. If the process stalls at a given block size, raising the block size restores progress; allowing it to reach the full register guarantees exact convergence.

4   Numerical Results

To evaluate the utility of the method, we apply it to the MNIST 784 dataset [13], compressing the circuit with a maximum factor size of two. Each digit is embedded into 10 qubits, and each iteration introduces five two-qubit general unitaries. These decompose into a circuit of approximate depth 13-14 with respect to the gate set $\{R_x, R_y, R_z, \mathrm{CX}\}$. Consequently, the circuit depth scales linearly with the number of iterations. For example, after six iterations the depth is approximately 78, while after 260 iterations it reaches about 3500, without applying any circuit optimization.

For the digit "zero" (index 0 of [13]), a clear fidelity-depth trade-off is observed: increasing the number of iterations reduces the fidelity loss, but simultaneously increases the circuit depth. This relationship is illustrated in Fig. 2. The standard Qiskit [18] initializer, which employs the isometries method [11], is used as a baseline. Any circuit depth exceeding this baseline is marked as *No Use*, as it provides no practical advantage. Approximations below this threshold – even those associated with relatively high fidelity loss – may still offer sufficient quality for certain applications.

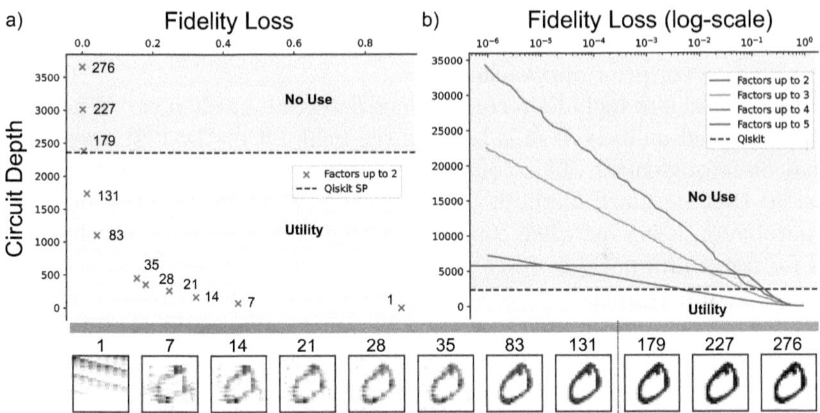

Fig. 2. This study examines the relationship between iteration count, circuit depth, and fidelity loss in the Q-Tucker method. As illustrated by the scatter plot, the method achieves rapid convergence – requiring fewer iterations – when an acceptable fidelity loss is targeted, highlighting its potential utility.

To examine the effect of the factor size, values of 2, 3, 4, and 5 were compared on the same digit (cf. Fig. 2). Since the synthesis of general three-, four-, and five-qubit unitaries is expensive, the method does not scale efficiently in this regime, highlighting the need for improved multi-qubit unitary synthesis techniques. Nevertheless, larger factor sizes exhibit a clear advantage in terms of convergence speed (not shown in the figure). For the digit "zero", convergence below a precision of 10^{-6} is achieved after 553 iterations with factor size

2, 163 with factor size 3, 51 with factor size 4, and only 8 with factor size 5. These results indicate that advances in unitary synthesis could yield substantial improvements for the iterative Tucker decomposition.

5   Comments and Conclusion

Empirical tests on real-world data show that even when Q-Tucker fails to fully converge – an infrequent scenario – it still produces high-quality approximations, often matching the best achievable results within a given classical computational budget. In our implementation, the number of iterations is capped at the square of the number of qubits. This parameter, which directly influences circuit depth, can be tuned to match the constraints of the target device and application. Within this limit, the method consistently delivers efficient decompositions compatible with available classical resources.

A key strength of the approach lies in its reliance on SVDs over two-qubit subsystems. Since these subsystems have fixed dimensionality, each SVD incurs a constant cost. This avoids the scalability bottleneck typical of tensor decompositions like Tucker, where SVD costs scale cubically with input size. Additionally, the heuristic used to search for a good configuration tuple has linear complexity with respect to the number of state amplitudes. As a result, the overall asymptotic cost of the method is linear in the number of amplitudes – a significant improvement over prior approaches [1].

The method also includes a convergence detection mechanism. When convergence is deemed unlikely, it falls back on the Schmidt method [1], incorporating partial disentanglement. This concept of partial disentanglement enables compressions that standard methods often overlook. However, the fallback remains comparatively slow, even when combined with greedy heuristics, highlighting the need for faster alternatives.

In summary, Q-Tucker achieves the best possible approximation allowed by the classical resources at hand, offering two key guarantees: (1) The approximation improves proportionally with increased classical resources. (2) The algorithm always returns a result – even for large, complex states where only a coarse approximation may be feasible.

Acknowledgments. We acknowledge funding of the Federal Ministry of Research, Technology and Space Project Number 13N17157 (Q-ROM).

Disclosure of Interests. Carsten Blank is the co-founder and CEO of data cybernetics. Furthermore, a patent application covering the methods described herein has been filed and is pending; details are under confidentiality restrictions.

References

1. Araujo, I.F., Blank, C., Araújo, I.C.S., da Silva, A.J.: Low-rank quantum state preparation. IEEE Trans. Comput. Aided Des. Integr. Circuits Syst. **43**(1), 161–170 (2024)

2. Araujo, I.F., Park, D.K., Ludermir, T.B., Oliveira, W.R., Petruccione, F., Da Silva, A.J.: Configurable sublinear circuits for quantum state preparation. Quantum Inf. Process. **22**(2), 123 (2023)
3. Araujo, I.F., Park, D.K., Petruccione, F., da Silva, A.J.: A divide-and-conquer algorithm for quantum state preparation. Sci. Rep. **11**(1), 6329 (2021)
4. Bergholm, V., Vartiainen, J.J., Möttönen, M., Salomaa, M.M.: Quantum circuits with uniformly controlled one-qubit gates. Phys. Rev. A **71**, 052330 (2005)
5. Blank, C., Park, D.K., Rhee, J.K.K., Petruccione, F.: Quantum classifier with tailored quantum kernel. NPJ Quantum Inf. **6**(1), 1–7 (2020)
6. Cade, C., Mineh, L., Montanaro, A., Stanisic, S.: Strategies for solving the fermi-hubbard model on near-term quantum computers. Phys. Rev. B **102**, 235122 (2020)
7. De Lathauwer, L., De Moor, B., Vandewalle, J.: A multilinear singular value decomposition. SIAM J. Matrix Anal. Appl. **21**(4), 1253–1278 (2000)
8. De Lathauwer, L., De Moor, B., Vandewalle, J.: On the best rank-1 and rank-(r1 ,r2 ,. . .,rn) approximation of higher-order tensors. SIAM J. Matrix Anal. Appl. **21**(4), 1324–1342 (2000)
9. Grant, E., et al.: Hierarchical quantum classifiers. NPJ Quantum Inf. **4**(1), 65 (2018)
10. Grover, L., Rudolph, T.: Creating superpositions that correspond to efficiently integrable probability distributions. arXiv (2002)
11. Iten, R., Colbeck, R., Kukuljan, I., Home, J., Christandl, M.: Quantum circuits for isometries. Phys. Rev. A **93**(3), 032318 (2016)
12. Kolda, T.G., Bader, B.W.: Tensor decompositions and applications. SIAM Rev. **51**(3), 455–500 (2009)
13. LeCun, Y., Cortes, C., Burges, C.: Mnist handwritten digit database. ATT Labs (2010). http://yann.lecun.com/exdb/mnist **2**
14. Möttönen, M., Vartiainen, J.J., Bergholm, V., Salomaa, M.M.: Quantum circuits for general multiqubit gates. Phys. Rev. Lett. **93**(13), 130502 (2004)
15. Nielsen, M.A., Chuang, I.L.: Quantum Computation and Quantum Information: 10th Anniversary Edition. Cambridge University Press (2010)
16. Park, D.K., Petruccione, F., Rhee, J.K.K.: Circuit-based quantum random access memory for classical data. Sci. Rep. **9**(1), 3949 (2019)
17. Plbnio, M.B., Virmani, S.: An introduction to entanglement measures. Quantum Inf. Comput. **7**(1), 1–51 (2007)
18. Qiskit contributors: Qiskit: an open-source framework for quantum computing (2023). https://doi.org/10.5281/zenodo.2573505
19. Ran, S.J.: Encoding of matrix product states into quantum circuits of one- and two-qubit gates. Phys. Rev. A **101**, 032310 (2020)
20. Schön, C., Hammerer, K., Wolf, M.M., Cirac, J.I., Solano, E.: Sequential generation of matrix-product states in cavity qed. Phys. Rev. A **75**, 032311 (2007)
21. Schuld, M., Petruccione, F.: Supervised Learning with Quantum Computers. Quantum Science and Technology, Springer, Cham (2018). https://doi.org/10.1007/978-3-319-96424-9
22. Stanislav, P., Marina, L.: Faster quantum state decomposition with Tucker tensor approximation. Quantum Mach. Intell. **5**(2), 25 (2023)
23. Tseng, P.: Convergence of a block coordinate descent method for nondifferentiable minimization. J. Optim. Theory Appl. **109**(3), 475–494 (2001)
24. Wei, T.C., Goldbart, P.M.: Geometric measure of entanglement and applications to bipartite and multipartite quantum states. Phys. Rev. A **68**(4) (2003)

Quantum Approaches to the Minimum Edge Multiway Cut Problem

Ali Abbassi[1,2]($\boxtimes$), Yann Dujardin[1], Eric Gourdin[1], Philippe Lacomme[3], and Caroline Prodhon[2]

[1] Orange Research, Chatillon, France
{yann.dujardin,eric.gourdin}@orange.com
[2] LIST3N, University of Technology of Troyes, Troyes, France
ali.abbassi@utt.frorange.com, caroline.prodhon@utt.fr
[3] LIMOS - UMR CNRS 6158, ISIMA - Université Clermont Auvergne,
Clermont-Ferrand, France
philippe.lacomme@isima.fr

Abstract. We investigate the minimum edge multiway cut problem, a fundamental task in evaluating the resilience of telecommunication networks. This study benchmarks the problem across three quantum computing paradigms: quantum annealing on a D-Wave quantum processing unit, photonic variational quantum circuits simulated on Quandela's Perceval platform, and IBM's gate-based Quantum Approximate Optimization Algorithm (QAOA). We assess the comparative feasibility of these approaches for early-stage quantum optimization, highlighting trade-offs in circuit constraints, encoding overhead, and scalability. Our findings suggest that quantum annealing currently offers the most scalable performance for this class of problems, while photonic and gate-based approaches remain limited by hardware and simulation depth. These results provide actionable insights for designing quantum workflows targeting combinatorial optimization in telecom security and resilience analysis.

Keywords: Quantum annealing · Multiway cut · Variational quantum circuits · QAOA · Photonic computing

1 Introduction

The *Minimum Edge Multiway Cut* (MEMC) problem arises in tasks where specific terminal nodes must be separated to satisfy specific resilience criteria for a network or unauthorized communication [7]. Given an undirected graph and a set of k terminals, the objective is to identify a minimum-weight set of edges whose removal ensures that no two terminals lie in the same connected component. For $k = 2$, the problem reduces to a classical min-cut, solvable in polynomial time via max-flow algorithms [14]. For $k \geq 3$, the problem is NP-hard [9]. We evaluate the MEMC problem using three quantum computing paradigms: (i)

F. Barbaresco and F. Gerin (Eds.): QUEST-IS 2025, CCIS 2744, pp. 284–293, 2026.
https://doi.org/10.1007/978-3-032-13855-2_26

quantum annealing on D-Wave hardware, (ii) the Quantum Approximate Optimization Algorithm (QAOA) implemented with IBM Qiskit, and (iii) variational photonic quantum circuits simulated using Quandela's Perceval platform. Each paradigm operates on a shared QUBO formulation of MEMC, enabling consistent problem modeling across solvers [8].

The QUBO model is mapped to an Ising formulation for execution on the D-Wave platform and is employed as the cost Hamiltonian for QAOA and photonic variational circuits. D-Wave serves as a natural baseline due to its native support for QUBO-based quantum annealing and available hybrid solvers [10]. We report numerical results on small graph instances, including annealing behavior, QAOA convergence trends, and bitstring sampling accuracy in photonic circuits.

This study aims to characterize the behavior of different quantum optimization approaches on a unified problem encoding, highlighting trade-offs in scalability, model expressiveness, and hardware compatibility [2]. The paper is structured as follows: Sect. 2 defines the MEMC problem and recalls key complexity results. Section 3 presents the QUBO formulation used across all quantum paradigms. Section 4 introduces the three quantum optimization approaches: quantum annealing, QAOA, and photonic variational circuits. Section 5 reports numerical results obtained on representative instances. Section 6 concludes with a discussion of current limitations and directions for future work.

2 Problem Statement

We consider the *Minimum Edge Multiway Cut* (MEMC) problem, a classical NP-hard graph partitioning problem with applications in network interdiction, secure communications, and containment strategies [20].

Input. An undirected connected graph $G = (V, E)$, a non-negative cost function $C : E \to \mathbb{R}_{\geq 0}$, and a set of $k \geq 2$ terminals $T = \{t_1, \ldots, t_k\} \subseteq V$.

Goal. Find a minimum-cost set of edges $F \subseteq E$ such that in the graph $G' = (V, E \setminus F)$, no two terminals lie in the same connected component.

The problem generalizes the classic *minimum s–t cut* (when $k = 2$), solvable in polynomial time via max-flow algorithms [14]. For $k \geq 3$, the problem becomes NP-hard [9], and remains hard even under restrictions such as bounded degree or treewidth [6,22]. While fixed-parameter tractable algorithms exist for certain parameters [4,21], the decision variant–testing whether a cut of cost at most K exists–is NP-complete in general. Greedy heuristics offer a $2 - \frac{2}{k}$ approximation ratio [9], but exact solutions for larger instances often require exponential-time methods or relaxation techniques. We consider the optimization variant and reformulate it as a Quadratic Unconstrained Binary Optimization (QUBO), which serves as a common ground for deploying quantum solvers.

3 QUBO Formulation

We adopt the QUBO formulation for the MEMC problem from [18], designed for efficient embedding on quantum hardware. Given a graph $G = (V, E)$, each vertex $u \in V$ is assigned to one of the terminals $t \in T$ via binary variables $x_{u,t} \in \{0,1\}$, where $x_{u,t} = 1$ if node u is assigned to terminal t. This defines a partition of the graph vertices. We hence consider the following Hamiltonian:

$$H(x) = \alpha \left(\sum_{u \in V} \left(1 - \sum_{t \in T} x_{u,t} \right)^2 + \sum_{\substack{t,t' \in T \\ t \neq t'}} x_{t,t'} \right)$$
$$+ \sum_{\{u,v\} \in E} \sum_{t \in T} \sum_{\substack{t' \in T \\ t \neq t'}} C(\{u,v\}) x_{u,t} x_{v,t'} \tag{1}$$

The first term ensures that each node is assigned to exactly one terminal, and the second term prevents conflicting terminal overlaps (no terminal is assigned to another terminal).

The final term accumulates the cost of cutting edges that connect vertices assigned to different terminals. The scalar α regulates the strength of the constraint terms relative to the objective.

4 Quantum Optimization

We describe the different quantum optimization workflows considered in this study and briefly recall their theoretical soundness and encoding strategies [1,2].

4.1 Quantum Annealing

Quantum annealing (QA) is a continuous-time optimization method based on the quantum adiabatic theorem [11]. The system evolves under a time-dependent Hamiltonian

$$H(t) = A(t)H_0 + B(t)H_p, \quad t \in [0, T], \tag{2}$$

where H_0 is a transverse-field driver and H_p encodes the cost function. Provided $A(t), B(t)$ vary slowly enough and the spectral gap remains non-negligible, the system ends near the ground state of H_p.

Our QUBO formulation is mapped to an Ising energy function via $x_i = (1 - z_i)/2$ with $z_i \in \{-1, 1\}$ [15]:

$$E(z) = \sum_{i<j} J_{ij} z_i z_j + \sum_i h_i z_i, \tag{3}$$

where $J_{ij} = \frac{1}{4} Q_{ij}$ and $h_i = -\frac{1}{4} \sum_j (Q_{ij} + Q_{ji})$. This defines an Ising spin glass whose ground state solves the original problem.

The Ising cost function is promoted to a diagonal quantum Hamiltonian:

$$H_p = \sum_{i<j} J_{ij} Z_i Z_j + \sum_i h_i Z_i, \tag{4}$$

where Z_i is the Pauli-Z operator on qubit i. On D-Wave systems, this Hamiltonian is implemented using programmable couplings and local fields over a sparse hardware graph [10]. Minor embedding is used to satisfy hardware connectivity constraints by linking physical qubits into chains, enabling annealing toward a minimum-energy configuration that corresponds to low cost solution.

4.2 Quantum Approximate Optimization Algorithm

The Quantum Approximate Optimization Algorithm (QAOA) [12] is a variational quantum algorithm for solving discrete optimization problems using alternating Hamiltonian dynamics on shallow circuits. It operates on a cost Hamiltonian H_C, derived from the QUBO model, and a mixer Hamiltonian H_M. The QAOA_p ansatz prepares the state

$$|\psi(\boldsymbol{\gamma}, \boldsymbol{\beta})\rangle = \prod_{\ell=1}^{p} e^{-i\beta_\ell H_M} e^{-i\gamma_\ell H_C} |\psi_0\rangle,$$

where $\boldsymbol{\gamma}, \boldsymbol{\beta} \in \mathbb{R}^p$ are tunable parameters and $|\psi_0\rangle = |+\rangle^{\otimes n}$ is the initial uniform superposition. In practice, H_C is implemented using single-qubit R_Z gates; the standard mixer $H_M = \sum_i X_i$ corresponds to global R_X rotations.

The parameters are optimized by minimizing the energy expectation value $\langle \psi(\boldsymbol{\gamma}, \boldsymbol{\beta}) | H_C | \psi(\boldsymbol{\gamma}, \boldsymbol{\beta}) \rangle$ using classical optimizers such as COBYLA. The QAOA structure can be interpreted as a Trotterized approximation of evolution under $H = H_C + H_M$, with a "bang-bang" control profile [5].

While standard QAOA explores the full Hilbert space, constrained extensions like the Quantum Alternating Operator Ansatz (QAOA-A) [17] use custom mixers to enforce problem constraints. These were not used in our experiments; instead, constraints are embedded directly in the QUBO energy landscape.

4.3 Variational Circuits on Photonic Quantum Computers

Photonic variational quantum circuits offer an alternative paradigm for solving combinatorial optimization problems using linear optical components [3, 19]. These circuits operate in the discrete-variable regime, where Fock states–photon number states–form the computational basis. Unitary transformations are implemented via passive interferometers built from beam splitters and phase shifters. Classical optimization is used to adjust circuit parameters to minimize a QUBO-derived cost function.

Each Fock state $|\boldsymbol{n}\rangle = |n_0, n_1, \ldots, n_{M-1}\rangle$ encodes the number of photons in M modes, with $\sum_j n_j = P$ photons in total. For binary optimization, encodings such as dual-rail and parity are used. In dual-rail encoding, a logical qubit spans two modes:

$$|0\rangle = |1\rangle_0 \otimes |0\rangle_1, \quad |1\rangle = |0\rangle_0 \otimes |1\rangle_1.$$

We use a parity-based mapping [24], where each bit is computed as $x_i = n_i \bmod 2$, enabling direct evaluation of QUBO objectives: $H(x) = x^\top Q x$.

The circuit applies a unitary transformation $U(\boldsymbol{\theta}, \boldsymbol{\phi})$ to the input Fock state $|\boldsymbol{n}_{\text{in}}\rangle$. This unitary is decomposed into parameterized beam splitters and phase shifters:

$$\text{BS}(\theta, \phi) = \begin{pmatrix} \cos(\theta) & -e^{i\phi}\sin(\theta) \\ e^{-i\phi}\sin(\theta) & \cos(\theta) \end{pmatrix}, \quad \text{PS}(\varphi) = \begin{pmatrix} 1 & 0 \\ 0 & e^{i\varphi} \end{pmatrix}.$$

These gates evolve the quantum state to:

$$|\psi_{\text{out}}\rangle = U(\boldsymbol{\theta}, \boldsymbol{\phi})|\boldsymbol{n}_{\text{in}}\rangle.$$

Measurements in the Fock basis yield output configurations $\boldsymbol{n}_{\text{out}}$, which are mapped to bitstrings $x \in \{0, 1\}^N$ via parity. The QUBO energy is estimated as:

$$\mathbb{E}_{x \sim p(\boldsymbol{\theta}, \boldsymbol{\phi})}[H(x)] = \sum_x p(x) \cdot x^\top Q x.$$

Minimizing this expected value over $(\boldsymbol{\theta}, \boldsymbol{\phi})$ forms the variational optimization loop. We simulate small MEMC instances using `Perceval` [19] under idealized conditions and evaluate the circuit's ability to sample near-optimal bitstrings.

5   Numerical Analysis

Preliminary results confirm that *small-scale instances* of the MEMC problem can be solved to optimality across all three quantum workflows–quantum annealing (QA), gate-based QAOA, and photonic variational circuits (VQC). In cases with $k = 2$ and $|V| \leq 10$, all platforms return the optimal solution, demonstrating that constrained combinatorial problems are tractable in low-dimensional quantum settings (Table 1).

Figure 1 compares cut costs obtained by simulated annealing, quantum annealing, and D-Wave's hybrid solver across increasing graph sizes. For small graphs, all solvers yield the same optimal values. However, quantum annealing performance deteriorates sharply starting from the $(10, 70)$ instance, where QA incurs significantly higher cut costs. This degradation is likely due to *embedding overheads, chain break errors, or insufficient annealing schedule expressivity* in larger QUBO encodings. Figure 2 shows runtime scaling on D-Wave hardware. QA remains efficient (under $1\,\text{s}$) for small graphs but exhibits a two-order-of-magnitude spike on the $(10, 70)$ instance, likely reflecting increased overhead for embedding and retries. In contrast, the hybrid solver maintains stable runtime (10–$20\,\text{s}$) across all sizes, while consistently achieving better solution quality. These observations highlight a key trade-off between computational time and output accuracy, with hybrid workflows offering robustness as instance size increases.

Figure 3 shows the QAOA circuit structure for a depth-$p = 1$ instance on a small MEMC graph. The ansatz alternates between entangling ZZ-cost layers and parametrized single-qubit mixers. While the structure is amenable to gate-based simulation, circuit width and entanglement depth increase quadratically with instance size, limiting simulation feasibility to $n \lesssim 14$ qubits.

Table 1. Numerical summary

Backend	Small instances ($k = 2$, $|V| \leq 10$)	Larger instances ($k = 2$, $10 < |V| < 70$)	Convergence/notes
D-Wave QA	Optimal (100%)	Gap $\approx$ 20–40% (topology-dependent)	Chain breaks observed, avg. chain length ~ 8; hybrid solver recommended for stability.
D-Wave Hybrid	Optimal (100%)	Near-optimal; typical gap $\lesssim$ 10–15%	Longer runtime; classical post-processing mitigates embedding effects.
QAOA (sim.)	Optimal bitstring sampled with $\sim 40\%$ probability at $p = 1$; no improvement for larger iterations.	Gap $\sim$15–30% at fixed p; no systematic improvement for larger p (barren plateau)	Converges in $\sim$10–20 iterations; decreasing probability of sampling best bitstring at $p = 1$, effect of BP for medium to large instances.
Photonic (sim.)	Optimal sampled with >90% probability.	Gap $\sim$20–40%; infeasible to simulate larger systems with SLOS backend	$\sim$300–1500 parameter updates; Nelder–Mead more stable than COBYLA.

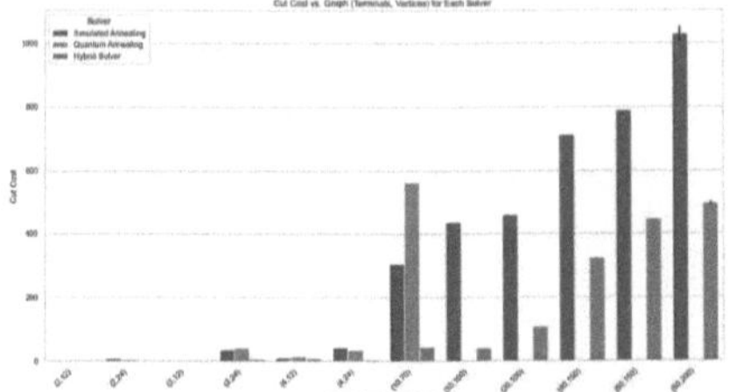

Fig. 1. Hybrid vs Quantum Annealing Cut Cost

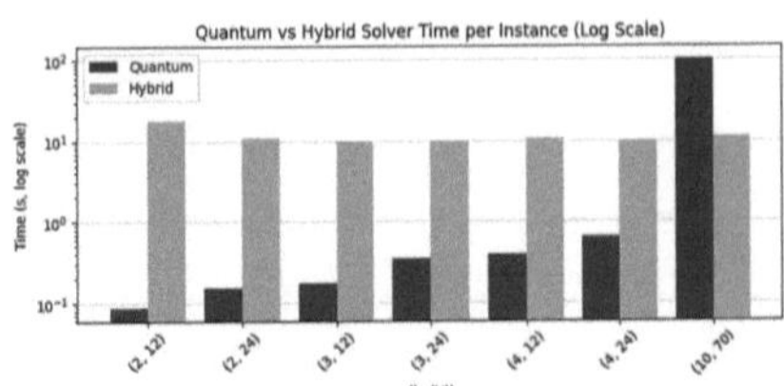

Fig. 2. D-Wave Runtime vs Instance Size

Figure 4 reports the optimization trajectory of the expected cost across 25 classical steps Despite operating in a noise-free environment, the energy landscape remains rugged, leading to unstable convergence and local minima. This reflects known challenges such as barren plateaus in variational algorithms. While convergence is eventually reached, QAOA failed to scale to larger MEMC instances ($k \geq 3$ and $|V| \geq 15$) in our setting. Nevertheless, it remains more expressive than photonic workflows in terms of circuit design and model capacity.

Photonic simulations were conducted using Perceval's `GenericInterferometer` and the SLOS backend, combined with classical optimizers. For each parameter configuration, $N = 10000$ measurement shots estimated the QUBO cost. In Fig. 6, a toy instance ($|V| = 4$, $k = 2$) demonstrates strong sampling concentration on the optimal bitstring. Figure 5 shows convergence of QUBO energy using multiple optimizers, with Nelder–Mead yielding greater stability.

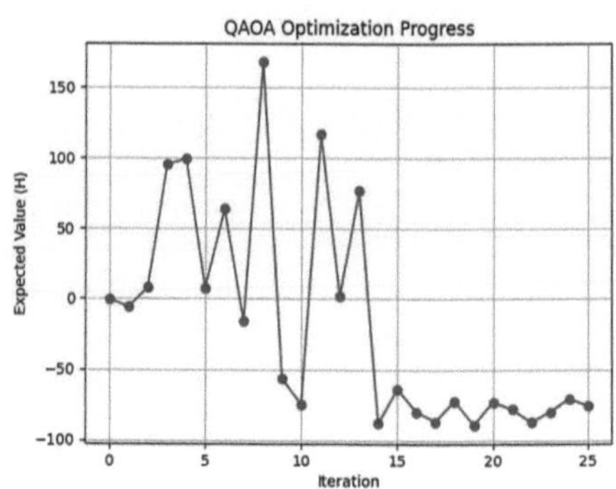

Fig. 3. QAOA Circuit Structure ($p = 1$)   **Fig. 4.** QAOA Training Convergence

Nonetheless, scalability of the photonic workflow remains limited: performance degrades significantly for problems with more than 10 variables or $k \geq 3$. Embedding effects, photon loss, and decoherence are not modeled here. Future directions include testing with qudit encodings or hybrid analog encodings to address mode limitations.

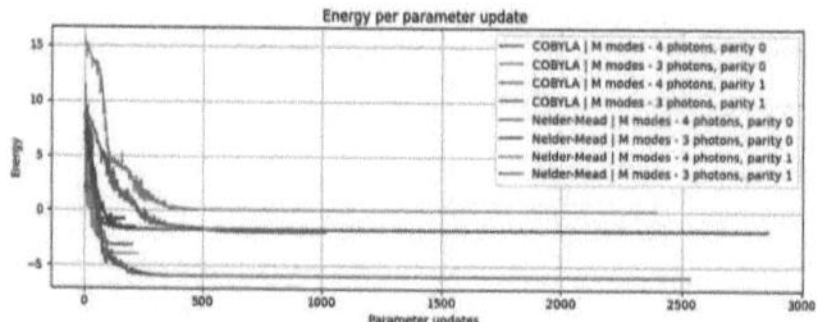

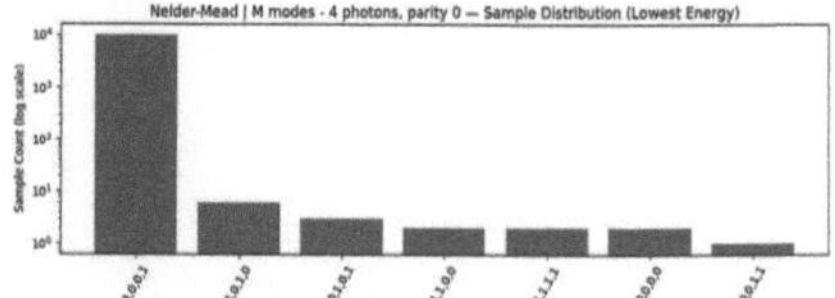

Fig. 5. Photonic Energy Convergence   **Fig. 6.** Photonic Bitstring Frequencies

6 Conclusion

We benchmarked the MEMC problem across three quantum paradigms under the same QUBO encoding. For small instances ($|V| \leq 10$, $k \leq 2$), all methods–quantum annealing, QAOA, and photonic variational circuits–recovered optimal solutions, confirming baseline compatibility. Among these, D-Wave showed the highest scalability, solving larger graphs with reasonable solution quality despite embedding overhead. The QUBO formulation, designed with annealing in mind, favored sparse, pairwise interactions, reinforcing D-Wave's suitability. QAOA was limited by circuit depth and barren plateaus. Despite convergence in simulation, performance did not extend to medium-scale problems. More advanced designs (e.g., constraint-preserving mixers [17], parameter concentration [5]) remain to be tested. Photonic simulations reliably captured optimal bitstrings on toy instances but degrade rapidly beyond $\sim$10 variables. Current parity encodings and the exponential state space are major constraints. Hardware-oriented enhancements (e.g., qudit encodings) are promising.

Overall, Quantum annealing remains the most mature option at scale. Future work will target more expressive encodings, noise-resilient QAOA circuits, and physical deployment of photonic models.

7 Supplementary Material

Table 2. System properties and algorithmic parameters across the three backends.

D-Wave Advantage (eu-central-1)	
QPU	Advantage_system5.4, Pegasus topology (5614 working qubits)
Temperature	16.4 mK
Supported Problems	QUBO, Ising
Annealing Time	[0.5, 2000] μs (default 20 μs)
Embedding	Automatic minor-embedding, avg. chain length
Coupling Ranges	$J \in [-2, 1]$, $h \in [-18, 15]$
Reads / Postproc	Multiple reads, majority-vote unembedding
Quandela Perceval (Photonic VQC, simulated)	
Backend	SLOS shot-based simulator
Ansatz	Generic interferometer (beam splitters + phase shifters)
Encoding	Parity-mapped Fock states ($j = 0, 1$), full/reduced occupancy
Parameters	Phases $\{\theta_i, \phi_{tr,i}, \phi_i\}$ uniform in $[0, 2\pi]$
Optimizers	Powell, COBYLA, Nelder–Mead (max. 1000 it.)
Loss Function	Variational energy $\langle \psi(\boldsymbol{\theta})|H|\psi(\boldsymbol{\theta})\rangle$
Samples	10 000 per evaluation
Postproc	Parity conversion, histogramming of distributions

(continued)

Table 2. (*continued*)

IBM Qiskit QAOA (gate-based, simulated)	
Backend	Qiskit AerSimulator (statevector + 4000 shots)
Ansatz	QAOAAnsatz, depth $p = 1$ (U_C, U_M)
Initialization	$\beta_0 = \pi/2$, $\gamma_0 = \pi/4$
Optimizer	COBYLA, bounds $[0, \pi]$
Objective	$\langle \psi(\beta, \gamma)|H|\psi(\beta, \gamma)\rangle$
Compilation	Transpilation, preset pass manager (opt. level 0)
Sampling	All qubits measured, normalized distributions

Reproducibility and Scope. All implementations use public SDKs (Ocean, Qiskit, Perceval), with parameter ranges listed in Table 2. Classical baselines are not included here; they remain essential for competitiveness analysis but outside the scope of this feasibility study.

Contributions and Scope. This work provides one of the first cross-paradigm studies of the *Minimum Edge Multiway Cut* (MEMC) using a single QUBO formulation across three modalities: D-Wave annealing, gate-based QAOA, and photonic variational circuits. The main contributions are: (i) a unified encoding enabling consistent benchmarking of embedding overheads and workflow constraints; (ii) a qualitative analysis of scalability bottlenecks

(iii) an application-driven link between MEMC and telecom-network resilience;

Unlike prior work focused on native device problems (e.g., Max-Cut on hardware graphs), we benchmark a combinatorial problem with direct industrial motivation. The target is feasibility and methodology, not quantum advantage.

Telecom Resilience. Finding minimum sets of "critical" edges (such that a simultaneous breakdown of these edges would imply a shutdown of a service) is essential for designing "resilient" telecom networks with low probability of failure. This is a hard task since there is a trivial reduction from the NP-hard problem "min edge multiway cut" to this problem.

References

1. Abbas, A., et al.: Challenges and opportunities in quantum optimization. Nat. Rev. Phys. (2024)
2. Bochkarev, A., Heese, R., Jäger, S., Schiewe, P., & Schöbel, A.: Quantum computing for discrete optimization: a highlight of three technologies (2024). arXiv:2409.01373
3. Bombardelli, M., Fleury, G., Lacomme, P., Vulpescu, B.: Foundations of photonic quantum computation (2025). arXiv preprint arXiv:2509.04266

4. Bousquet, N., Daligault, J., Thomassé, S.: Multicut is FPT. In: Proceedings of the 43rd ACM Symposium on Theory of Computing (STOC), pp. 459–468 (2011)

5. Brady, L.T., Van Heck, B., Keating, C.T.: Bang-bang control and quantum approximate optimization. Quantum **5**, 500 (2021)

6. Calinescu, G., Karloff, H., Rabani, Y.: Multicuts in unweighted graphs and digraphs with bounded degree and bounded tree-width. J. Algorithms **48**(2), 333–359 (2003)

7. Costa, M.-C., Coudert, D., Jaumard, B.: Minimal multicut and maximal integer multiflow: a survey. Eur. J. Oper. Res. **162**(1), 55–69 (2005)

8. Cruz-Santos, W., Salas, C., López-Sánchez, J.C.: A QUBO formulation of minimum multicut problem instances in trees for D-Wave quantum annealers. Sci. Rep. **9**, 1721 (2019)

9. Dahlhaus, E., Johnson, D.S., Papadimitriou, C.H., Seymour, P.D., Yannakakis, M.: The complexity of multiterminal cuts. SIAM J. Comput. **23**(4), 864–894 (1994)

10. D-Wave Systems Inc. (2024). Hybrid workflow solvers. Technical documentation

11. Farhi, E., Goldstone, J., Gutmann, S., Sipser, M.: Quantum computation by adiabatic evolution (2000). arXiv:quant-ph/0001106

12. Farhi, E., Goldstone, J., Gutmann, S.: A quantum approximate optimization algorithm (2014). arXiv:1411.4028

13. Flamini, F., Spagnolo, N., Sciarrino, F.: Photonic quantum information processing: a review. Rep. Prog. Phys. **82**(1), 016001 (2018)

14. Garg, N., Vazirani, V.V., Yannakakis, M.: Approximate max-flow min-(multi)cut theorems and their applications. SIAM J. Comput. **25**(2), 235–251 (1996)

15. Glover, F., Kochenberger, G., Du, Y.: A tutorial on formulating and using QUBO models (2019). arXiv:1811.11538

16. Guo, J., Niedermeier, R.: Complexity and exact algorithms for vertex multicut in interval and bounded treewidth graphs. Eur. J. Oper. Res. **186**(2), 542–553 (2008)

17. Hadfield, S., Wang, Z., O'Gorman, B., Rieffel, E., Venturelli, D., Biswas, R.: From the quantum approximate optimization algorithm to a quantum alternating operator ansatz. Algorithms **12**(2), 34 (2019)

18. Heidari, S., Dinneen, M.J., Delmas, P.: An equivalent QUBO model to the minimum multi-way cut problem. CDMTCS Research Reports, Technical report CDMTCS-565 (2022)

19. Heurtel, C., Arrazola, J.M., Delgado, A., Thiel, V., Wright, L.G., Orieux, A.: Perceval: a software platform for discrete-variable photonic quantum computing. Quantum Sci. Technol. **8**, 014006 (2023)

20. Magnouche, Y., Mahjoub, A.R., Martin, S.: The multi-terminal vertex separator problem: branch-and-cut-and-price. Discret. Appl. Math. **286**, 168–189 (2020)

21. Marx, D., Razgon, I.: Fixed-parameter tractability of multicut parameterized by the size of the cutset. In: Proceedings of the 43rd ACM Symposium on Theory of Computing (STOC), pp. 469–478 (2011)

22. Papadopoulos, C.: Restricted vertex multicut on permutation graphs. Discret. Appl. Math. **160**(12), 1791–1797 (2012)

23. Venegas-Andraca, S.E., Barrera, J.F., Boyer, M.L., Carrasco, A.E.: A cross-disciplinary introduction to quantum annealing-based algorithms. Contemp. Phys. **59**(2), 174–196 (2018)

24. Brádler, K., Wallner, H.: Certain properties and applications of shallow bosonic circuits. arXiv:2112.09766 [quant-ph] (2021). https://doi.org/10.48550/arXiv.2112.09766

Systematic Improvement of the quantum approximate optimisation algorithm for Combinatorial Optimisation Using Quantum Subspace Expansion

Yann Beaujeault-Taudière$^{(\boxtimes)}$

Independent researcher, Orsay, France
`yann.beaujeault-t@protonmail.com`

Abstract. The quantum approximate optimisation algorithm (QAOA) is one of the flagship algorithms used to tackle combinatorial optimisation on graphs problems using a quantum computer, and is considered a strong candidate for early fault-tolerant advantage. In this work, I study the enhancement of the QAOA with a generator coordinates method (GCM), and achieve systematic performances improvements in the approximation ratio and fidelity for the Maximum Independent Set on Erdös-Rényi graphs. The cost-to-solution of the present method and the QAOA are compared by analysing the number of logical CNOT and T gates required for either algorithm. Extrapolating on the numerical results obtained, it is estimated that for this specific problem and setup, the approach surpasses QAOA for graphs of size greater than 75 using as little as eight trial states.

Keywords: quantum approximate optimisation algorithm · quantum subspace expansion · combinatorial optimisation

1 Introduction

Among the myriad fundamental and applied use cases that quantum computing is expected to disrupt, combinatorial optimisation on graphs stands a peculiar role, finding applications across several fields. The possibility to achieve quantum advantage for solving NP-hard instances has been the focus of extensive experimental and theoretical research in the recent years, within the scope of both analogue and digital approaches. The quantum approximate optimisation algorithm (QAOA) [1,2] has emerged as one of the most promising variational algorithms to tackle graph problems. Several variations have been put forth, aiming for instance at reducing the circuit depth by using more sophisticated mixers, or the number of evaluations of the cost function during the optimisation (see [2] for a review). The existing flavours of QAOA all add refinements of varying complexity to the original ansatz. In this work, an alternative and

F. Barbaresco and F. Gerin (Eds.): QUEST-IS 2025, CCIS 2744, pp. 294–302, 2026.
https://doi.org/10.1007/978-3-032-13855-2_27

complementary route is followed, combining the QAOA ansatz with a generator coordinate method (GCM), which is a standard tool for approximating the solution to eigenvalues problems on classical processors.

This paper is organised as follows. Sections 2, 3, 4 respectively present the general aspects of the MIS problem, of the QAOA, and of the GCM. Section 5 defines a simplified metric that can be used to determine whether applying the QSE protocol on top of a circuit optimised by QAOA leads to an improvement in the time-to-solution. The evaluation of the so-called Hamiltonian kernels, which represent the most challenging quantity to evaluate on a quantum processing unit (QPU), is discussed in Sect. 6 ; three different procedures are briefly analysed. Section 7 presents numerical results.

2   Maximum Independent Set

The maximum independent set is a classic problem of graph theory [3]. Given a graph $G = (V, E)$, where V is the set of vertices and E is the set of edges, the objective is to find the largest set of independent vertices, that is, the largest set where no two vertices are linked by an edge. The MIS problem is generally NP-hard, which means that there is no known classical algorithm that can find the optimal solution in polynomial time. There exist however efficient heuristic algorithms that can get close to the optimal solution, provided the MIS in question belongs to the appropriate complexity class [4]. The MIS problem can be formalised as finding the minimum of the cost Hamiltonian

$$\hat{H}_C = -\sum_{i \in V} \hat{n}_i + \sum_{(i,j) \in E} \hat{n}_i \hat{n}_j, \tag{1}$$

where the number operator $\hat{n}_i = (1 - Z_i)/2$ corresponds to rejecting ($\langle \hat{n}_i \rangle = 0$) or including ($\langle \hat{n}_i \rangle = 1$) the node i in the solution. Expanding (1) in terms of the Pauli matrices Z_i puts the Hamiltonian in a form suited for quantum simulation.

3   quantum approximate optimisation algorithm

The quantum approximate optimisation algorithm (QAOA) [1,2] is one of the leading algorithms for solving combinatorial optimisation problems [2]. In its original version, QAOA makes use of two non-commuting operators $\hat{H}_C$ and $\hat{H}_M$, called the cost and mixer Hamiltonians, respectively. Together, they define the alternating parametric unitary of depth L

$$U(\boldsymbol{\gamma}, \boldsymbol{\beta}) = \prod_{\ell=1}^{L} e^{-i\beta_\ell \hat{H}_M} e^{-i\gamma_\ell \hat{H}_C}, \tag{2}$$

and one seeks to minimise the following expectation value:

$$\begin{aligned} C(\boldsymbol{\gamma}, \boldsymbol{\beta}) &= \langle \phi | U^\dagger(\boldsymbol{\gamma}, \boldsymbol{\beta}) \hat{H}_C U(\boldsymbol{\gamma}, \boldsymbol{\beta}) | \phi \rangle \\ &\equiv \langle \Phi_0 | \hat{H}_C | \Phi_0 \rangle, \end{aligned} \tag{3}$$

In order to lower the energy, the mixer Hamiltonian should not commute with $\hat{H}_C$. It should furthermore allow reaching the ground state of the cost Hamiltonian, which can be formalised as the requirement that the nested commutators of the cost and mixer produce an algebra whose exponential map contains at least one path transforming $|\phi\rangle$ into the proper ground state. These two conditions can be fulfilled regardless of the initial state by the simple yet efficient choice of a coherent rotation mixer:

$$\hat{H}_M = -\sum_{i=0}^{N-1} X_i, \tag{4}$$

where X_i is the Pauli X operator acting on qubit i.

4   Generator Coordinate Method

The generator coordinate method (GCM) [5,6] is a general method to obtain ground and excited states properties of a Hamiltonian. The GCM is based upon expanding the eigenstates of the cost Hamiltonian on a set of (usually non-orthogonal) generating states $\{|\chi_k(\boldsymbol{q}_k)\rangle\}_{k=0,\dots,K-1}$ according to

$$|\Psi_j\rangle = \sum_{k=0}^{K-1} f_{k,j}\,|\chi_k(\boldsymbol{q}_k)\rangle, \tag{5}$$

the parameters $\boldsymbol{q}_k$ being referred to as the generator coordinates. One has full freedom in choosing the generating states; for instance, this formulation encompasses real-time quantum subspace expansion (QSE) [11] as the special case $\boldsymbol{q}_k = t_k$ and $|\chi_k(t_k)\rangle = e^{-i\hat{H}_C t_k}\,|\Phi_0\rangle$. The GCM approach has been recently applied on quantum computers in order to access properties of strongly coupled fermionic systems, such as the one-body Green's functions [15,16], ground and excited states building on a unitary coupled clusters ansatz [17,18] or coherent one-body rotations in the case of Hamiltonians possessing permutation invariance [19].

Given the expansion (5), approximate eigenstates of $\hat{H}_C$ can be obtained by applying the Rayleigh-Ritz variational principle to the cost function

$$C_{\mathrm{GCM}}^{(j)}(\{f_{k,0}\}) = \sum_{kk'} f_{k,j}^* f_{k',j} \langle \chi_k|\hat{H}_C|\chi_{k'}\rangle, \tag{6}$$

with the normalisation

$$\sum_{kk'} f_{k,j}^* f_{k',j} \langle \chi_k|\chi_{k'}\rangle = 1, \tag{7}$$

leading to a generalised eigenvalue equation

$$\mathcal{H}\,|\Psi\rangle = E\mathcal{S}\,|\Psi\rangle, \tag{8}$$

where $\mathcal{H}$ and $\mathcal{S}$ are the so-called Hamiltonian and overlap kernels, whose elements read

$$\mathcal{H}_{kk'} = \langle \chi_k | \hat{H}_C | \chi_{k'} \rangle, \tag{9}$$

$$\mathcal{S}_{kk'} = \langle \chi_k | \chi_{k'} \rangle. \tag{10}$$

In practice, the kernels (9)–(10) estimated on the QPU, and (8) is solved classically.

Due to the non-orthogonality of the generating states, the overlap matrix may contain small eigenvalues which spoil the numerical resolution of Eq. (8), even when the matrix elements are computed exactly [12,13]. This instability is exacerbated in the presence of noise. A simple prescription is to project out the states that are quasi-degenerate with the rest of the basis, by removing the eigenvalues that are smaller than a given threshold $\varepsilon_{\mathrm{cut}}$. This has been used successfully in the nuclear physics context since over three decades [12,13], delivering robust and reliable results for virtually any reasonable choice of the threshold. The convergence and stability properties of this approach were studied in detail in [11].

5   When Is GCM a Good FTQC Algorithm?

Whether an initial state algorithm should be employed rather than another is measured by whether it reduces the overall computational resources, that is, runtime, number of gates, and/or number of logical qubits required. Among these metrics, the number of gates plays the most critical role, as it essentially dictates the balance between space and time tradeoffs during the execution on the QPU. Focusing on this quantity alone is thus a good first first-order estimate of an algorithm's performances, and in the following discussion, I use the term "cost" to refer to the number of gates required to run the algorithms.

The fidelity between the ground state(s) and the state prepared by a given algorithm "X" is written $F^{(X)}$. The cost-to-fidelity ratio of the GCM is favourable when the fractional increase in the fidelity outweighs the fractional increase in computational cost. This is achieved if the following condition is met:

$$\frac{F^{(\mathrm{GCM})}}{F^{(\mathrm{QAOA})}} \gtrsim 4(K-1)\left(1 + \frac{1}{L'}\right) \times f, \tag{11}$$

where f is a scaling factor depending on which of the subcircuits introduced in Sect. 6 is used to extract the matrix elements of the kernels (see [20]), and $L' \leq L$ is the number of layers retained after the optimisation of the QAOA angles.

From (11), it can readily be argued that under some assumptions, there will always exist some critical graph size beyond which the GCM is always favourable over the plain QAOA, in the case of a fixed depth. Indeed, while there are no rigorous results on the practical scaling of the QAOA fidelity with the number of nodes, a simple empirical fit of the form $F^{(X)}(N) = \beta_X(1 + e^{N\alpha_X})^{-1}$ captures the most salient expected features of the fidelity, namely its exponential decay from one to zero as N grows, and its plateau effect for small N. As the left-hand side of (11) grows exponentially, while the right-hand side scales polynomiallly, the GCM will deliver a better cost-to-fidelity ratio than QAOA for larger graphs. In Sect. 7, this simple metric is used to extrapolate the QAOA and GCM results in the special case of Erdös-Rényi graphs with fixed average density.

6   Computation of the Kernels

The matrix elements $\mathcal{A}_{kk'} \equiv \langle \Psi_k | \hat{A} | \Psi_{k'} \rangle$ can be evaluated using two Hadamard tests, as shown in Fig. 1.

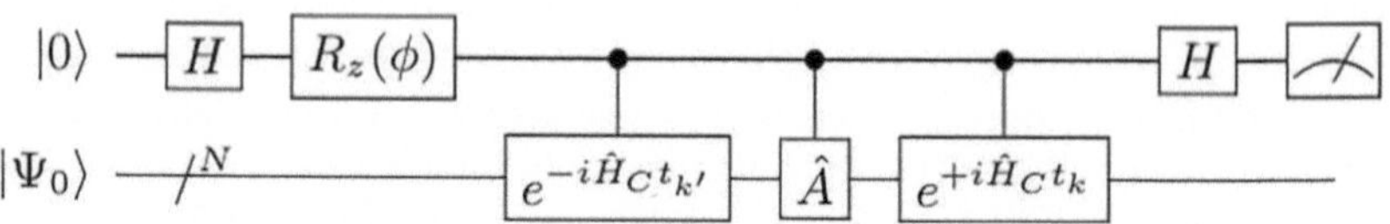

Fig. 1. Circuit for the evaluation of $\mathcal{A}_{kk'}$ with Hadamard tests. The real part of $\mathcal{A}_{kk'}$ is obtained by setting ϕ to zero. The imaginary part is obtained using $\phi = -\pi/2$, which corresponds to the modified Hadamard test. Both parts are recovered as the difference between the probabilities of measuring the ancillary qubit in the state zero and one.

While evaluating $\mathcal{S}_{kk'}$ poses little challenge, estimating the kernels $\mathcal{H}_{kk'}$ requires more care, as the cost Hamiltonian is not a unitary operator. This can be addressed in several ways (see [20] for a detailed discussion and derivation):

- By estimating the expectation values of the Pauli strings $\{Z_n, Z_n Z_{n'}\}$ directly. With this approach, $N/2$ two-qubits expectation values can be estimated simultaneously by grouping the operators into commuting sets. This approach furthermore avoids the use of extra CNOT and T gates beyond the ones required to encode the states. Because $\mathcal{O}(\sqrt{\rho}N)$ rounds of measurements are performed, each with approximately $\sqrt{\rho}N$ terms evaluated, the total number of shots scales as $(\sqrt{\rho}N)^3$.
- By using real time evolution and forming linear combinations. In that case, one estimates $\langle \Psi_k | e^{-i\hat{H}_C t_j} | \Psi_{k'} \rangle = \mathcal{S}_{kk'} - it_j \mathcal{H}_{kk'} + \mathcal{O}(t_j^2)$ for several propagation times $\{t_j\}_{j=1,\cdots,p}$. The kernels $\mathcal{S}_{kk'}, \mathcal{H}_{kk'}$ can then be recovered to precision $\mathcal{O}(t^p)$ by forming linear combinations of the p expectation values. The principal disadvantage of this approach is that it requires large values of p in order to estimate matrix elements to high precision, which in turns increases the final variance.
- By encoding $\hat{H}_C$ into a larger space using the linear combination of unitaries (LCU) approach [14]. This approach involves $\mathcal{O}(\log_2(N))$ ancillary qubits. The LCU is a probabilistic algorithm, and close to the optimal solution, $p_{\text{success}} \sim (\log_2(N)/N)^4$.

Which approach is optimal depends on several parameters, including the availability of ancillary qubits, the required precision to which the unitaries must be realised, which algorithm is used to decompose them on the elementary gate set, and the scaling of the off-diagonal moments $\langle \Psi_k | \hat{H}_C^p | \Psi_{k'} \rangle$. The three approaches are analysed and compared in more depth in [20].

7 Numerical Results

7.1 Parameter Setting

For all the numerical applications shown below, the number of QAOA layers is set to $L = 20$, and only one layer is optimised at a time in order to keep the runtime reasonable. That is, once the ℓth layer's parameters $(\gamma_\ell, \beta_\ell)$ have been optimised, they are fixed before moving on to the optimisation of the $(\ell + 1)^{\text{th}}$ layer. The angles are optimised using Scipy's implementation of the limited-memory Broyden-Fletcher-Goldfarb-Shanno (L-BFGS-B) gradient descent algorithm [7,8]. Finally, after the L layers, the optimal depth $L' \leq L$ is chosen simply by selecting the depth that leads to the lowest cost function. All simulations are carried out using PennyLane library [21], using a noiseless statevector simulator.

For the QSE stage, the evolution times are taken equally spaced in the $[-\pi(1 - 1/K); \pi(1 - 1/K)]$ interval. This gives both $\mathcal{H}$ and $\mathcal{S}$ a Toeplitz structure, such that the number of matrix element to estimate is linear in K. When diagonalising the overlap kernel, states with eigenvalue below $\varepsilon_{\text{cut}} = 10^{-3}$ are discarded. These parameters are chosen somewhat arbitrarily, and several problem-specific improvements are possible. As the QSE is guaranteed to deliver improved solution with respect to the QAOA, the simple choice for the values of the generator coordinates nonetheless leads to substantially better solutions.

7.2 Erdös-Rényi Graphs

To showcase and benchmark the capabilities of the QSE more systematically, I apply the machinery on Erdös-Rényi (ER) graphs of average density $\rho = 0.5$. These graphs are NP-hard in general [4]. In particular, they cannot be addressed using analogue quantum processors when the number of nodes becomes large, as the probability that a random ER graph with fixed non-zero density is a unit-ball graph approaches zero in the large N limit [9]. Although for small graph size, the probability of generating a disconnected graph is non-negligible, this probability decreases exponentially fast as N grows and ρ remains fixed and strictly nonzero. For ER graphs, there is therefore no practical need to identify whether a graph is connected or not [9, 10].

The QAOA-plus-QSE workflow is applied to graphs of size $N = 2$ to 10. In addition to the approximation ratio and the fidelity, the error in the particle number (i.e. in the Hamming weight) and parity are also shown on Fig. 2.

The QAOA achieves rather poor approximation ratios, reaching 0.5 for $N = 10$. Naturally, the fidelity degrades to even lower values, namely 0.15 at $N = 10$. Furthermore, the error in both the Hamming weight and the parity of the QAOA

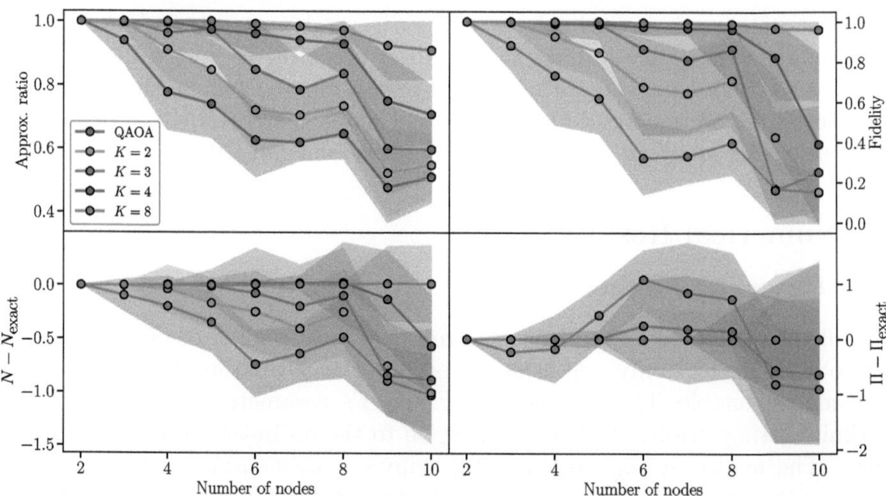

Fig. 2. Approximation ratio (top left), fidelity (top right), error in the Hamming weight/particle number (bottom left), and error in the parity (bottom right). On all subplots, the blue curve corresponds to the QAOA results, while the other curves represent the results obtained when adding a QSE on top, with $K = 2, 3, 4, 8$ trial states. For each number of nodes, 14 random Erdös-Rényi graphs are generated. The shaded areas correspond to one standard deviation. (Color figure online)

solutions is of order one, indicating that QAOA fails to evolve the initial state towards the correct symmetry subspace.

Adding the QSE step resolves all the issues of the QAOA optimisation, and increasing the number of trial states systematically improves all the relevant performance metrics. Notably, for $N = 10$, the approximation ratio is almost double. The performance of the QSE is even more striking for the fidelity, which remains significantly higher that the QAOA solutions, with a value of 0.96 for $N = 10$ nodes. As a consequence of the high fidelity, the Hamming weight (and therefore the parity) of the QSE solution quickly converge to the exact values.

Fitting the fidelities of the QAOA and the QSE to Fermi-Dirac distributions, one can extract the critical graph size N_* beyond which the QSE presents a favourable time-to-solution as the smallest N for which

$$\frac{\beta_{\mathrm{QAOA}}}{1 + e^{N\alpha_{\mathrm{QAOA}}}} \frac{1 + e^{N\alpha_{\mathrm{QSE}}(K)}}{\beta_{\mathrm{QSE}}(K)} > 2K(\sqrt{\rho}N)^3, \tag{12}$$

assuming the Hadamard plus Pauli method is used. In that specific setting and for $K = 8$, the QAOA-plus-QSE surpasses QAOA for $N > 75$, and increasing K should further reduce N_*. It should be stressed that this number implicitly depends on the choice of the QAOA initial state and mixer, on the number of layers, on the optimisation algorithm and strategy, on the type of graphs, and on the optimisation problem itself. As such, the precise value obtained here bears little meaning, and the only merit of such extrapolation is to illustrate the advantages of the QSE over QAOA for graphs with size of practical interest.

Acknowledgments. I thank Denis Lacroix and Jing Zhang for useful discussions. I also wish to warmly thank E. Beaujeault-Taudière for enthusiastic comments at all stages of this work.

Disclosure of Interests. I declare no competing interests.

References

1. Farhi, E., Goldstone, J., Gutmann, S.: A quantum approximate optimization algorithm (2014). arXiv:1411.4028 [quant-ph]
2. Blekos, K., Brand, D., Ceschini, A., Chou, C., Li, R., Pandya, K., Summer, A.: A review on quantum approximate optimization algorithm and its variants. Phys. Rep. **1068**, 1–66 (2024). https://doi.org/10.1016/j.physrep.2024.03.002
3. West, D.: Introduction to Graph Theory, 2nd Edition. Prentice Hall (2000)
4. Dalyac, C., Henry, L., Kim, M., Ahn, J., Henriet, L.: Exploring the impact of graph locality for the resolution of MIS with neutral atom devices (2023). arXiv: arXiv:2306.13373 [quant-ph]
5. Hill, D., Wheeler, J.: Nuclear constitution and the interpretation of fission phenomena. Phys. Rev. **89**, 1102–1145 (1953). https://doi.org/10.1103/PhysRev.89.1102
6. Ring, P., Schuck, P.: The Nuclear Many-Body Problem. Springer, Berlin, Heidelberg (1980)
7. Byrd, R., Lu, P., Nocedal, J., Zhu, C.: A limited memory algorithm for bound constrained optimization. SIAM J. Sci. Comput. **16**, 1190–1208 (1995). https://doi.org/10.1137/0916069
8. Zhu, C., Byrd, R., Lu, P., Nocedal, J.: Algorithm 778: L-BFGS-B: Fortran subroutines for large-scale bound-constrained optimization. ACM Trans. Math. Softw. **23**, 550–560 (1997). https://doi.org/10.1145/279232.279236
9. Cazals, P., et al.: Identifying hard native instances for the maximum independent set problem on neutral atoms quantum processors (2025). arXiv:2502.04291 [quant-ph]
10. Kim, K., Kim, M., Park, J., Byun, A., Ahn, J.: Quantum computing dataset of maximum independent set problem on king's lattice of over hundred Rydberg atoms. Sci. Data. **11**, 111 (2024). arXiv:2311.13803 [quant-ph]
11. Epperly, E., Lin, L., Nakatsukasa, Y.: A theory of quantum subspace diagonalization (2023). arXiv:2110.07492
12. Bonche, P., Dobaczewski, J., Flocard, H., Heenen, P., Meyer, J.: Analysis of the generator coordinate method in a study of shape isomerism in 194Hg. Nuclear Phys. A **510**, 466-502 (1990). https://www.sciencedirect.com/science/article/pii/037594749090062Q
13. Marević, P.: Towards a unified description of quantum liquid and cluster states in atomic nuclei within the relativistic energy density functional framework. (Université Paris-Saclay,2018). http://www.theses.fr/2018SACLS358/document, Thèse de doctorat dirigée par Khan, Elias Structure et réactions nucléaires Université Paris-Saclay (ComUE) (2018)

14. Childs, A., Wiebe, N.: Hamiltonian simulation using linear combinations of unitary operations. Quantum Inf. Comput. **12** (2012). arXiv:1202.5822 [quant-ph]
15. Jamet, F., Agarwal, A., Rungger, I.: Quantum subspace expansion algorithm for Green's functions (2022). arXiv:2205.00094
16. Umeano, C., Jamet, F., Lindoy, L., Rungger, I., Kyriienko, O.: Quantum subspace expansion approach for simulating dynamical response functions of Kitaev spin liquids (2024). arxiv:2407.04205
17. Zheng, M., Peng, B., Wiebe, N., Li, A., Yang, X., Kowalski, K.: Quantum algorithms for generator coordinate methods (2022). arXiv:2212.09205
18. Zheng, M., Peng, B., Li, A., Yang, X., Kowalski, K.: Unleashed from constrained optimization: quantum computing for quantum chemistry employing generator coordinate method (2024). arXiv:2312.07691
19. Beaujeault-Taudiere, Y., Lacroix, D.: Solving the Lipkin model using quantum computers with two qubits only with a hybrid quantum-classical technique based on the generator coordinate method (2023). arXiv:2312.04703
20. Beaujeault-Taudière, Y.: Systematic improvement of the quantum approximate optimisation algorithm for combinatorial optimisation using quantum subspace expansion (2025). arxiv:2506.18594
21. Bergholm, V., et al.: PennyLane: automatic differentiation of hybrid quantum-classical computations (2022). arxiv:1811.04968

Session: 11 Quantum Algorithms, Computing; Simulation – Hardware and Architectures

The Current Landscape of Quantum Hardware Development - An Overview

Siddharth R. Chander[(✉)][iD]

California Institute of Technology, Pasadena, CA 91125, USA
schander@caltech.edu

Abstract. Quantum computing has developed since the 1980s, with significant progress in its theoretical and practical applications. A critical aspect of this field is quantum hardware development, which supports research and real-world applications. One notable example of quantum computing's potential is cryptography, where the RSA protocol has been employed to secure browsers and other internet applications. The private key of the RSA protocol is based on two prime numbers that are so large that even supercomputers cannot factor them to their prime factors in a reasonable amount of time. In 1994, Caltech alumnus Peter Shor proposed Shor's Algorithm, which exploits the unique properties of quantum computers to factorize large numbers quickly and efficiently [1, pp. 5–8]. Implementing this algorithm on quantum hardware would compromise the security of the RSA protocol. Quantum computing has been touted as revolutionary, but understanding the progress of different quantum hardware types is vital. This paper aims to analyze the types of quantum hardware and applications they are best suited for, presenting a comprehensive look into most quantum hardware in development. By understanding the current state of quantum hardware, we can gain valuable insights into the potential applications of quantum computing.

1 Introduction

Quantum computing represents a paradigm shift from classical computation, leveraging quantum mechanical phenomena like superposition and entanglement to solve problems intractable for classical systems. Since Peter Shor's 1994 algorithm demonstrated quantum computing's potential to break widely-used encryption, the field has accelerated dramatically.

This paper examines the current state of quantum hardware development, categorizing platforms by commercial maturity and research focus. We analyze superconducting quantum processors (currently leading commercial deployment), trapped ion systems (notable for high fidelity), photonic approaches (offering room-temperature operation), and neutral atom arrays (showing recent rapid advancement). We then explore emerging platforms including semiconductor spin qubits, topological systems, NMR quantum computing, and diamond-based processors. For each approach, we discuss operating principles, current performance metrics, scalability challenges, and potential applications.

F. Barbaresco and F. Gerin (Eds.): QUEST-IS 2025, CCIS 2744, pp. 305–317, 2026.
https://doi.org/10.1007/978-3-032-13855-2_28

2 Superconducting Quantum Processors

Superconducting quantum processors utilize Josephson junctions to create artificial atoms serving as qubits, operating at cryogenic temperatures (10–15 mK) where superconductivity eliminates electrical resistance. Qubit states correspond to different energy levels manipulated via microwave pulses, leveraging the unique property of superconductors where resistivity drops to near zero at optimal temperatures [3]. The system is quantized from classical to quantum mechanics using amplitude complexity, modifying wave phases through superconductors. Processor initialization involves fine-tuning electrical elements by classically adjusting capacitance and inductance, then modifying the total energy following quantum mechanical laws including Kirchoff's circuit laws and quantum Hamiltonian properties [4].

Fabrication employs lithography to etch metals and pattern materials for precise junctions and components. These processors uniquely use Josephson Junctions where two thin superconductive wires separated by a few-atom-thick insulator create superconducting currents that activate qubits [5]. Operational control involves sending specific-frequency microwave pulses to each qubit for designated times to apply unitary operators. Current systems face significant challenges including decoherence from environmental interactions requiring cooling to near absolute zero, crosstalk between qubits and control lines introducing noise, fabrication difficulties in creating high-quality low-loss circuits, control complexity in signal generation and error correction, and interfacing issues with classical systems requiring efficient signal conversion between microwave, optical, and electrical domains [7–9] (Fig. 1).

Note: Schematic of an IBM Superconducting Quantum Processor. Note that the processor is only the bottommost component of this figure, with the rest being cooling and interactive equipment [6].

Despite these challenges, superconducting processors represent the most mature quantum technology with commercial systems available through cloud platforms. Notable implementations include Google's Sycamore demonstrating quantum advantage, IBM's Q System One, Rigetti's Aspen, and Intel's Horse Ridge. These systems show significant potential in cryptography, optimization, machine learning, and physics applications. Various superconducting qubit types exist including fixed-frequency and tunable qubits, with the latter allowing individual qubit frequency adjustment [10].

3 Trapped Ion Quantum Processors

Trapped ion quantum processors confine atomic ions (typically Yb+ or Ca+) using electromagnetic fields, utilizing their electronic states as qubits. These systems currently lead in gate fidelities and coherence times, making them strong contenders for fault-tolerant quantum computation. The process begins with trapping and cooling ions using electromagnetic fields to create confining potential wells, typically arranged linearly. Laser cooling techniques reduce kinetic

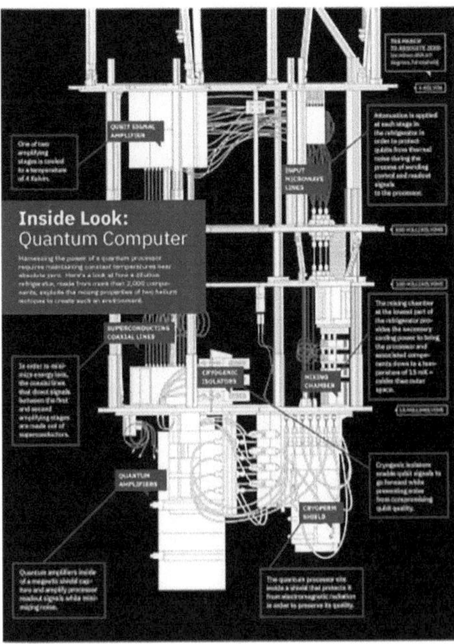

Fig. 1. Inside Look of Superconducting Quantum Processor.

energy, bringing ions to their ground state for enhanced qubit stability against external noise [11,12].

State initialization prepares qubits in well-defined states, typically the ground state, accomplished through laser pulses or microwave radiation that excite or de-excite ions to specific energy levels. Quantum operations employ carefully engineered laser beams coupling different internal energy levels, with precise control achieved by manipulating laser intensity, frequency, and duration. Entanglement generation leverages strong inter-ion interactions through designed laser pulses that create correlated systems where individual ion states cannot be described independently, enabling complex quantum computations [12] (Fig. 2).

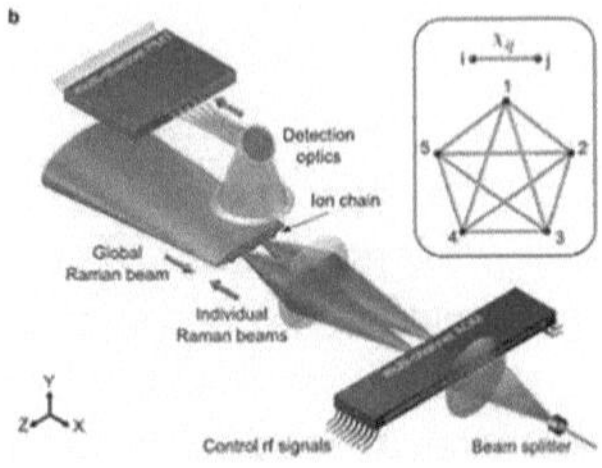

Fig. 2. Diagram of how a generic trapped ion quantum processor works, not considering how one interacts and implements programs on the processor [13].

Measurement extracts information by mapping quantum states to classical states through separate laser beams that ionize atoms and produce detectable electrical currents. Quantum error correction combats decoherence through techniques like encoding multiple qubits into single logical qubits and implementing error-detecting codes, enabling fault-tolerant computing. Recent progress includes IonQ's 2021 commercial trapped ion computer availability, newer systems employing surface code error protection, and Tsinghua University's 2023 programmable quantum phononic processor combining trapped ion and photonic approaches that offers improved scalability for complex problems [14].

Trapped ion processors leverage unique properties including well-defined energy levels and strong interactions to achieve quantum advantage. The technical processes involving trapping, cooling, state initialization, entanglement, and error correction all play crucial roles in system functionality. Ongoing development continues to produce more powerful systems with enhanced capabilities for practical quantum computing applications.

4 Photonic Quantum Processors

Photonic quantum processors utilize photons as quantum information carriers, encoding qubits in optical properties such as polarization, path encoding, or time bins. These systems leverage quantum interference and measurement-induced nonlinearities to perform quantum computations based on quantum entanglement principles where linked particles share correlated states regardless of separation distance, allowing noise-resistant information transmission [15].

Photonic processors work by initializing photons in superposition states then controlling them using optical elements that flip states or create entanglement between photons. Calculations employ quantum algorithms operating on photons, with results read out through optical measurement elements. Since photons cannot interact in vacuum, additional media facilitate photon interactions within the processor architecture. Key advantages include light-speed operation enabling faster calculations than classical computers, inherent scalability requiring less effort for multi-qubit setups compared to other platforms, room-temperature operation eliminating cryogenic requirements, natural networking capability through fiber optic compatibility, and low decoherence from weak environmental interactions [16] (Fig. 3).

Note: Simplified diagram of how a photonic quantum processor works. Note the photon emitters, circuits, and the detectors [15].

Significant limitations include susceptibility to noise causing quantum property loss and calculation errors, extreme manufacturing complexity preventing mass production, and high fabrication costs. Despite these challenges, photonic processors have advanced beyond pure research to commercial applications. Recent progress includes Xanadu's Borealis processor demonstrating quantum advantage using 216 squeezed-light qubits and Gaussian boson sampling, development of new optical elements enabling more complex calculations, improved

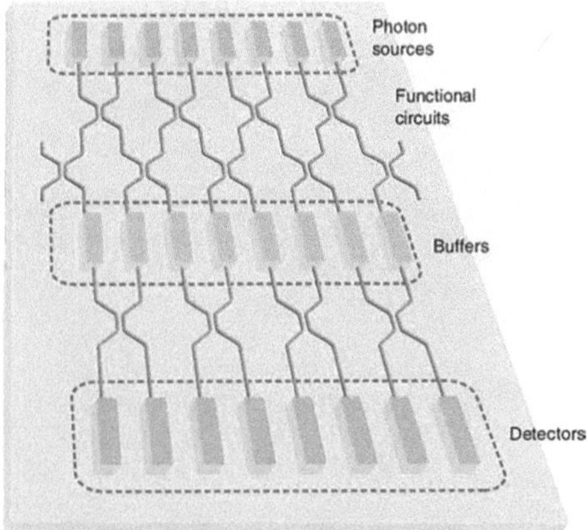

Fig. 3. Overview of a Photonic Quantum Processor.

manufacturing methods creating more reliable and affordable systems, and Chinese research teams demonstrating photonic systems solving specialized problems 10^{14} times faster than classical supercomputers [16].

Primary applications include quantum simulation of molecular vibrations and bosonic systems, Gaussian boson sampling for graph problems and machine learning, quantum networking and quantum key distribution, and specialized optimization problems. The technology continues developing toward greater power and affordability, increasing accessibility for researchers and businesses while accelerating new quantum application development.

5   Neutral Atom Quantum Processors

Neutral atom quantum processors utilize arrays of individual atoms (typically rubidium or cesium) trapped in optical tweezers and cooled to microkelvin temperatures. These systems encode qubits in atomic energy levels and leverage Rydberg interactions to create entanglement between atoms, offering unique advantages in scalability and connectivity [17]. The first step involves laser cooling atoms to microkelvin temperatures using repeated photon absorption and emission to reduce kinetic energy. Once cooled, atoms are trapped using optical tweezers or magnetic fields that create confining potential wells at standing wave nodes where potential energy is minimized [17].

Qubits are typically encoded in the spin states of atoms, where up and down states represent classical 0 and 1 values, with lasers manipulating these states through precise rotations. Multiple atoms can be arranged into reconfigurable arrays to represent multiple qubits. Quantum operations include rotations for state manipulation, entanglement through Rydberg blockade effects that create correlated quantum states, and measurement operations that collapse quantum states to classical values [17] (Fig. 4).

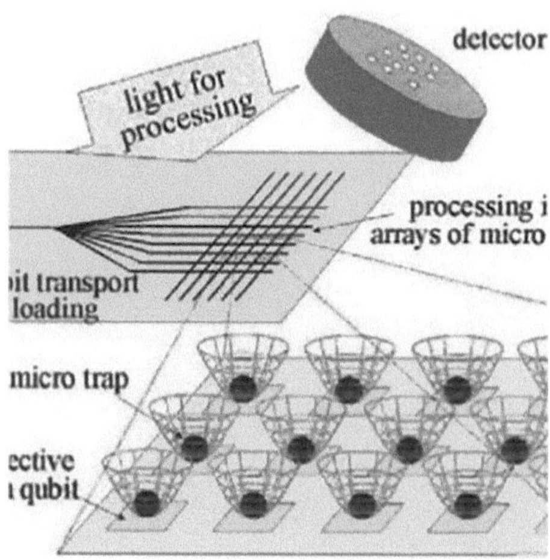

Fig. 4. Neutral Atom Processor Diagram.

Note: Schematic of how a generic neutral atom quantum processor works [18].

Recent advances include QuEra's 256-qubit Aquila processor (2022) and 280-qubit system (2023), PASQAL's 200-qubit quantum processing unit, and Atom Computing's 100-qubit system with nuclear spin qubits. These systems have demonstrated quantum error correction, high-fidelity two-qubit gates (99.5%) using Rydberg interactions, and dynamic qubit rearrangement during computation [19]. Neutral atom processors excel in quantum simulation of many-body systems and optimization problems, with commercial systems available through cloud services.

Despite progress, significant challenges remain. Decoherence presents particular difficulties as neutral atoms are highly sensitive to environmental interactions, making it challenging to maintain coherent states for extended periods [19]. Precise control of individual atoms remains difficult, and scaling to larger qubit arrays is constrained by decreasing coupling strength with distance. Additionally, these systems require specialized equipment and facilities, making them expensive to build and operate [20].

The future outlook for neutral atom quantum processors remains promising due to their exceptional scalability, long coherence times, and flexible qubit arrangements. With continued research addressing control precision, error reduction, and system integration, neutral atom systems are progressing toward 1000+ qubit demonstrations within 2–3 years and may achieve quantum advantage for specific optimization problems before universal quantum computers [19, 20].

6   Semiconductor Spin Qubits

Semiconductor spin qubits leverage the quantum properties of electron or hole spins confined in quantum dots within semiconductor structures, primarily sili-

con. These qubits benefit from compatibility with existing semiconductor manufacturing infrastructure, offering a promising path toward scalable quantum
computing [21]. Significant progress has been made with researchers achieving
99.9% single-qubit gate fidelities in isotopically enriched silicon-28, 99.8% two-
qubit gate fidelities demonstrating entanglement, coherence times exceeding 10 s
through dynamical decoupling techniques, and operation temperatures up to
1.5 K, reducing cooling requirements. Intel's 2023 demonstration of a 12-qubit
array with all-to-all connectivity represents the largest silicon spin qubit system
to date, with their Horse Ridge cryogenic controller enabling scalable control of
qubit arrays [21].

Spin qubits are implemented using electron spins in quantum dots defined by
electrostatic gates, hole spins in germanium-silicon heterostructures, or donor
atoms (phosphorus) implanted in silicon substrates. Qubit control is achieved
through ESR techniques for microwave manipulation of electron spins, electric
dipole spin resonance for electric field control of hole spins, and exchange coupling for mediating two-qubit operations through controlled interactions [21].
The primary advantage lies in CMOS fabrication compatibility, enabling potential integration with classical control electronics. Recent developments include
3D integration of qubits and control circuitry, cryogenic CMOS controllers operating at 4 K, and quantum dot arrays with pitch sizes compatible with advanced
lithography [22] (Fig. 5).

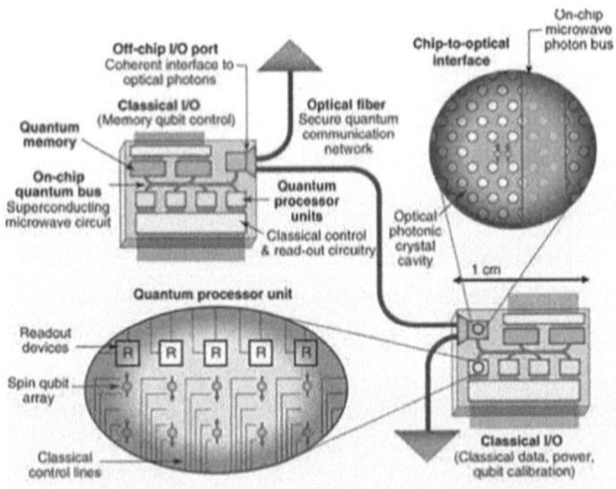

Fig. 5. User Interface with Silicon Spin Quantum Processor.

Note: How one can work and use a silicon spin quantum processor [22].

While spin qubits show exceptional coherence properties and fabrication compatibility, challenges remain in maintaining uniformity across large qubit arrays,
scaling control electronics for thousands of qubits, improving readout fidelity
and speed, and implementing error correction in dense arrays. The field is progressing rapidly toward 50–100 qubit demonstrations within the next 2–3 years,
with several companies and research institutions pursuing different architectural
approaches. The compatibility with existing semiconductor infrastructure makes

spin qubits particularly attractive for eventual mass production and integration with classical computing systems [23].

7   Topological Quantum Processors

Topological quantum computing represents a fundamentally different approach to quantum information processing, encoding information in non-local quantum states that are intrinsically protected from local noise and decoherence [24]. These systems utilize exotic quasiparticles—particularly Majorana zero modes—that emerge in topological materials where the bulk acts as an insulator while the surface conducts electricity [25]. Significant experimental advances include Microsoft's Quantum Lab demonstrating preliminary evidence of Majorana zero modes in semiconductor-superconductor nanowires (2023) and QuTech researchers demonstrating braiding operations with topological defects in quantum circuits. The most promising development comes from Microsoft and Quantinuum's 2023 collaboration, demonstrating active protection of logical qubits using topological codes that reduced error rates by 800x compared to physical qubits [26].

Topological qubits are implemented through Majorana-based approaches utilizing semiconductor-superconductor heterostructures to host Majorana zero modes, surface code implementations using anyonic braiding operations in two-dimensional systems, and photonic topological systems implementing protected states in photonic waveguides and resonators [27]. Key advantages include intrinsic error protection, potential for higher-temperature operation, and reduced error correction overhead, as the topological protection mechanism means local perturbations cannot destroy quantum information. Significant challenges remain in reliable creation and detection of Majorana particles, developing fabrication techniques for complex heterostructures, implementing practical braiding operations, and ensuring material purity and interface quality for semiconductor-superconductor systems [28] (Fig. 6).

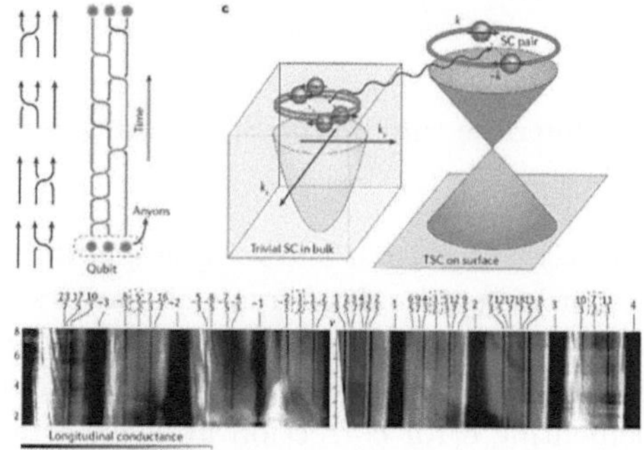

Fig. 6. Topological Qubit Representation.

Note: Representation of topological qubits from the inside and on the surface [28].

While still in the research phase, topological quantum computing has progressed from theoretical concept to experimental demonstration. Most experts estimate practical topological quantum processors remain 5–10 years from realization, with Microsoft's Azure Quantum ecosystem continuing to invest heavily in topological approaches. The field is rapidly evolving with new material systems including ternary compounds and van der Waals heterostructures showing promise for hosting robust topological states. Recent theoretical work also suggests hybrid approaches combining topological protection with conventional qubit architectures may provide near-term benefits [28].

8 Nuclear Magnetic Resonance Quantum Processors

Nuclear magnetic resonance (NMR) quantum computing utilizes the spin states of atomic nuclei within molecules as qubits, operating through precise radiofrequency pulses in strong magnetic fields [29]. Unlike other quantum computing approaches, NMR employs ensemble systems with billions of identical molecules rather than isolated qubits, providing inherent redundancy. NMR quantum computation leverages well-established magnetic resonance techniques where nuclear spins (typically ^{1}H or ^{13}C) encode quantum information, with quantum gates implemented through carefully crafted RF pulse sequences that manipulate spin states, while readout occurs through induction detection of the bulk magnetization [30].

While NMR systems cannot scale to large qubit counts due to signal strength limitations, they remain invaluable research tools. Recent work has focused on quantum control techniques, quantum simulation, and benchmarking quantum algorithms, with researchers demonstrating advanced quantum error correction protocols and quantum process tomography with up to 12 qubits in custom-designed molecules [31]. NMR quantum computing excels in quantum simulation of molecular dynamics and complex quantum systems, with recent applications including protein folding simulations, quantum thermodynamics experiments, and studies of quantum chaos. NMR techniques have also contributed significantly to developing quantum control methods now used across quantum computing platforms [32] (Fig. 7).

Note: Visual of how states are created and affected in a NMR system [32].

Although not pursuing quantum advantage, NMR quantum computing provided the first experimental demonstrations of quantum algorithms including Shor's algorithm in 2001 and continues to serve as a testbed for quantum information processing concepts. Its techniques for precise quantum control have influenced the development of all subsequent quantum computing architectures, maintaining relevance through ongoing contributions to quantum simulation and control methodology development [33].

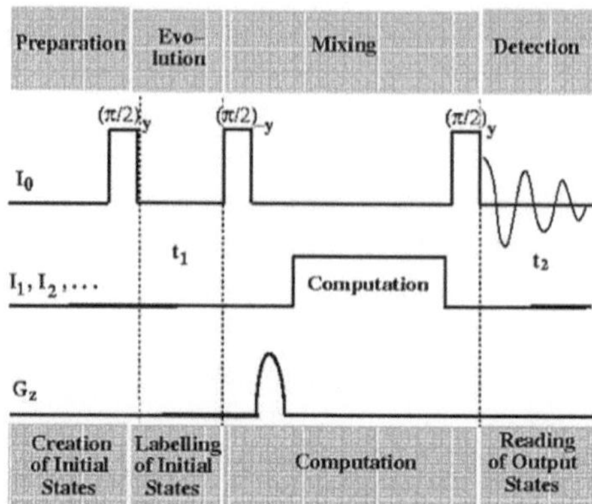

Fig. 7. NMR Computation Processes Diagram.

9 Diamond Quantum Processors

Diamond quantum processors utilize nitrogen-vacancy (NV) centers—lattice defects where nitrogen atoms replace carbon atoms—as qubits [34]. These systems uniquely operate at room temperature while maintaining coherence times exceeding milliseconds, making them particularly valuable for quantum sensing and networking applications. Recent advances include demonstrations of multi-qubit entanglement and quantum registers with 10+ qubits using nearby carbon-13 nuclear spins, with researchers achieving quantum control fidelities exceeding 99% for single-qubit operations and developing integrated photonic interfaces for scalable quantum networks [35].

Key advantages include room-temperature operation, long coherence times, and optical addressability, with NV centers serving dual purposes for quantum computation and ultra-sensitive sensing of magnetic, electric, and temperature fields. Challenges include precise NV center placement difficulties, with current fabrication techniques producing non-uniform qubit distributions, slower quantum gate speeds compared to superconducting systems, and significant materials engineering challenges for scaling beyond tens of qubits [35] (Fig. 8).

Note: A diagram of where the NV centers and synthetic diamonds are in a Diamond quantum processor [35].

Applications focus on quantum sensing including nanoscale MRI and magnetometry, quantum networks as quantum memory nodes, and tests of fundamental quantum phenomena. Commercial quantum sensors based on NV centers are already available for materials characterization and biomedical applications. Research focuses on improving NV center formation precision, enhancing spin-photon interfaces, and developing hybrid systems combining diamond qubits with other quantum platforms. Advances in diamond growth and nanofabrication may enable larger-scale quantum processors within the next decade [35].

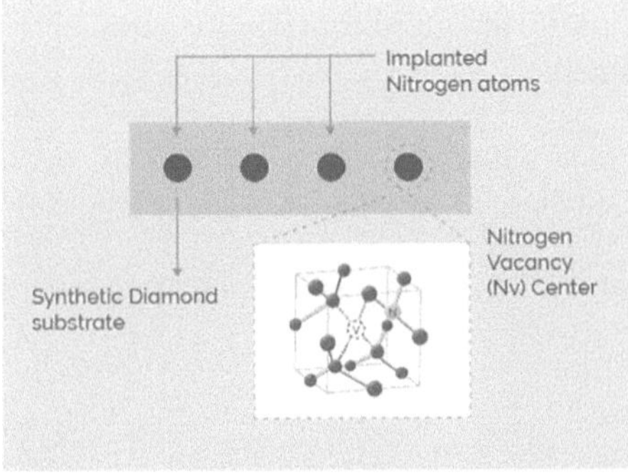

Fig. 8. Diamond Quantum Processor Schematic - Simplified.

10   Conclusion

Quantum computing represents a rapidly developing field with transformative potential across numerous industries, with quantum hardware development serving as the critical foundation for realizing this potential. This review has comprehensively examined the current landscape of quantum processors, including superconducting, trapped ion, photonic, neutral atom, semiconductor spin, topological, nuclear magnetic resonance, and diamond-based systems. Each platform demonstrates unique advantages and limitations: superconducting processors represent the most mature technology with demonstrated quantum advantage; trapped ion systems offer exceptional coherence times and high fidelities; photonic approaches provide rapid operation and scalability but face noise susceptibility; neutral atom arrays show promising scalability and long coherence times; semiconductor spin qubits benefit from classical electronics compatibility and mass-production potential; topological systems promise inherent noise resistance though remain in early development; NMR processors offer operational simplicity despite scalability constraints; and diamond-based systems provide remarkable stability with room-temperature operation despite fabrication challenges.

The future of quantum computing appears increasingly promising as technology continues advancing toward more powerful and efficient quantum processors capable of solving problems intractable for classical computers. Several critical challenges must be addressed to achieve full potential: decoherence remains a fundamental obstacle as qubits maintain fragility to environmental interactions across all platforms; error correction presents complex implementation hurdles despite being essential for scalable quantum computation; scalability demands substantial advancement to achieve practical qubit counts for meaningful applications; and cost reduction is necessary to overcome current economic barriers to widespread adoption. Despite these challenges, considerable optimism persists within the research community, with continued development expected to yield solutions in the coming years. Quantum computing's potential to revolutionize numerous fields makes this an exceptionally promising and exciting area of technological development.

Acknowledgments. This paper was done individually by the author, Siddharth Chander, without any guidance or association with anyone else.

Disclosure of Interests. The author has no competing interesting.

References

1. Shor, P.W.: Algorithms for quantum computation: discrete logarithms and factoring. In: Proceedings 35th Annual Symposium on Foundations of Computer Science, pp. 124–134. IEEE (1994)
2. Arute, F., et al.: Quantum supremacy using a programmable superconducting processor. Nature **574**(7779), 505–510 (2019)
3. Krantz, P., et al.: A quantum engineer's guide to superconducting qubits. Appl. Phys. Rev. **6**(2), 021318 (2019)
4. Oliver, W.D., Welander, P.B.: Materials in superconducting quantum bits. MRS Bull. **38**(10), 816–825 (2013). https://doi.org/10.1557/mrs.2013.229
5. Clarke, J., Wilhelm, F.K.: Superconducting quantum bits. Nature **453**(7198), 1031–1042 (2008)
6. IBM Quantum. IBM Quantum System One (2023). https://www.ibm.com/quantum/system-one
7. Devoret, M.H., Schoelkopf, R.J.: Superconducting circuits for quantum information: an outlook. Science **339**(6124), 1169–1174 (2013)
8. Gambetta, J.M., et al.: Building logical qubits in a superconducting quantum computing system. NPJ Quantum Inf. **3**(1), 1–7 (2017)
9. Wendin, G.: Quantum information processing with superconducting circuits: a review. Rep. Prog. Phys. **80**(10), 106001 (2017)
10. Koch, J., et al.: Charge-insensitive qubit design derived from the Cooper pair box. Phys. Rev. A **76**(4), 042319 (2007)
11. Cirac, J.I., Zoller, P.: Quantum computations with cold trapped ions. Phys. Rev. Lett. **74**(20), 4091 (1995)
12. Monroe, C., Kim, J.: Scaling the ion trap quantum processor. Science **339**(6124), 1164–1169 (2013)
13. IonQ. How IonQ Trapped Ion Quantum Computers Work (2023). https://ionq.com/technology
14. Zhang, J., et al.: A programmable quantum phononic processor based on trapped ions. Nat. Phys. **19**(3), 345–350 (2023)
15. Xanadu. Photonic Quantum Computing (2023). https://www.xanadu.ai/technology
16. Madsen, L.S., et al.: Quantum computational advantage with a programmable photonic processor. Nature **606**(7912), 75–81 (2022)
17. Saffman, M., et al.: Quantum information with Rydberg atoms. Rev. Mod. Phys. **82**(3), 2313 (2010)
18. QuEra Computing. Neutral Atom Quantum Computers (2023). https://www.quera.com/technology
19. Browaeys, A., Lahaye, T.: Many-body physics with individually controlled Rydberg atoms. Nat. Phys. **16**(2), 132–142 (2020)

20. Bernien, H., et al.: Probing many-body dynamics on a 51-atom quantum simulator. Nature **551**(7682), 579–584 (2017)
21. Veldhorst, M., et al.: An addressable quantum dot qubit with fault-tolerant control fidelity. Nat. Nanotechnol. **10**(7), 589–592 (2015)
22. Silicon Quantum Computing. Our Technology (2023). https://www.siliconquantum.com.au/our-technology
23. Zwanenburg, F.A., et al.: Silicon quantum electronics. Rev. Mod. Phys. **85**(3), 961 (2013)
24. Nayak, C., et al.: Non-Abelian Anyons and topological quantum computation. Rev. Mod. Phys. **80**(3), 1083 (2008)
25. Kitaev, A.Y.: Fault-tolerant quantum computation by Anyons. Ann. Phys. **303**(1), 2–30 (2003)
26. Google AI Quantum Team: Observation of Majorana fermions in a superconducting InAs nanowire. Nature **556**(7700), 74–79 (2018)
27. Satzinger, K.J., et al.: Realizing topologically protected qubits in a superconducting circuit. Nature **595**(7868), 508–513 (2021)
28. Microsoft Quantum. Topological Qubits (2023). https://www.microsoft.com/en-us/quantum/topological-qubits
29. Vandersypen, L.M., Chuang, I.L.: NMR techniques for quantum control and computation. Rev. Mod. Phys. **76**(4), 1037 (2005)
30. Jones, J.A.: Quantum computing with NMR. Prog. Nucl. Magn. Reson. Spectrosc. **59**(2), 91–120 (2011)
31. Vandersypen, L.M., et al.: Experimental realization of Shor's quantum factoring algorithm using nuclear magnetic resonance. Nature **414**(6866), 883–887 (2001)
32. Laflamme, R., et al.: NMR quantum information processing. Quantum Inf. Comput. **2**(3), 166–176 (2002)
33. Biamonte, J., et al.: Quantum machine learning. Nature **549**(7671), 195–202 (2018)
34. Doherty, M.W., et al.: The nitrogen-vacancy colour centre in diamond. Phys. Rep. **528**(1), 1–45 (2013)
35. Quantum Diamond Technologies. NV Center Diamond Quantum Sensors (2023). https://www.quantumdiamondtech.com/technology

Coherence Development at IQM Quantum Computers

Matthew Steggles[(✉)], Arthur Rebello, Ivan Takmakov, Andrew Guthrie,
Lihuang Zhu, Azad Karis, Alejandro Gómez-Frieiro, Tuomas Mylläri,
Eelis Takala, Leonid Abdurakhimov, and Juha Hassel

IQM Quantum Computers, Keilaranta 19, 02150 Espoo, Finland
`matthew.steggles@meetiqm.com`

Abstract. We showcase recent developments in improving coherence
times of superconducting qubits at IQM Quantum Computers. We
demonstrate how optimized fabrication and design result in increases in
coherence times, enabling enhanced performance when integrated into
quantum processing units (QPUs). We also report current efforts to mitigate the detrimental effects on coherence of remaining two-level-system
(TLS) defects. Our results showcase state-of-the-art fabrication capabilities at IQM Quantum Computers, exhibiting T_1 and T_2 echo times up to
the millisecond level in test devices. These improvements, when applied
to QPU-compliant designs and combined with IQM's quantum computer
control software stack, yield consistently improved performance, enabling
increased gate fidelities and paving the way to building fault-tolerant
quantum computers.

Keywords: Superconducting Quantum Computing · Fabrication ·
Design · Coherence · TLS defects

1 Introduction

Coherence times are one of the most foundational performance indicators of
quantum processing units (QPUs). Long energy relaxation and coherence times
(hereafter 'coherences') enable high fidelities of single-qubit gates, two-qubit
gates, and readout [1]. Nevertheless, there are numerous decay channels that
substantially limit the relaxation times that must be dealt with to increase QPU
performance. The coherences of superconducting qubit platforms are frequently
limited by dielectric losses via so-called two-level-system (TLS) defects, a broad
class of microscopic defect states that couple to the qubit electric field via their
dipole moment [2–4].

While defect states are always present in any nanofabrication process, there
are several strategies which can be employed to decrease their effect. By
improving fabrication, design, and parking strategies we present here significant improvements in the lifetimes of transmon qubits, showing the benefits of
an integrated approach to coherence development.

F. Barbaresco and F. Gerin (Eds.): QUEST-IS 2025, CCIS 2744, pp. 318–325, 2026.
https://doi.org/10.1007/978-3-032-13855-2_29

2   Improvements from Fabrication and Design

Here we discuss how qubit coherence can be improved by decreasing the impact of the TLS bath. In the Sect. 2.1 we demonstrate high coherence by improving fabrication quality, which we attribute to decreasing the density of TLSs. Then, in the Sect. 2.2 we optimize the qubit design to be less sensitive to TLSs, which we attribute to shifting the distribution of coupling strengths towards weaker qubit-TLS coupling.

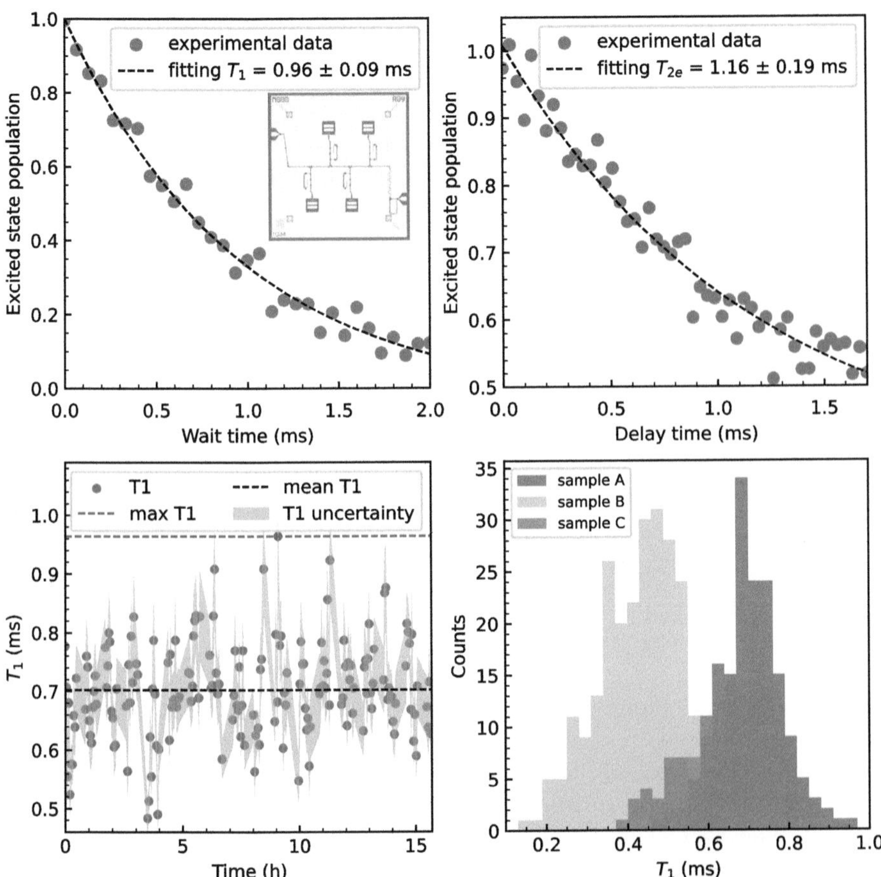

Fig. 1. Performance of isolated double-pad test qubits. (a) Best qubit T_1 trace. (b) Best qubit T_2 echo trace. (c) T_1 temporal fluctuations. (d) Histograms of T_1 distributions for best qubits from three different samples.

2.1   Fabrication Improvements

The most straightforward way to improve the relaxation times of a qubit is to improve the quality of the fabrication. We optimized multiple steps in the fabrication process for niobium-on-silicon isolated double-pad qubits, including silicon substrate cleaning, base-layer Nb film deposition, and Al/AlOx/Al Josephson junction double-angle evaporation. Fabrication was performed in IQM's internal clean room facilities.

The double-pad qubit is used as a reference design in order to compare more directly with the literature, as there is a significant wealth of papers showing high coherence with double-pad designs [7–10]. We note that millisecond coherence times have been obtained with different materials, such as tantalum [10], but keep the materials used constant to more effectively benchmark the fabrication process.

In Fig. 1 we show results from 3 test devices with the highest maximum T_1, highlighting the maximum T_1 time of 0.96 ± 0.09 ms and maximum T_2 echo time of 1.16 ± 0.19 ms. The fluctuation data presented also show mean performance $T_{1,\,\mathrm{mean}} > 0.7$ ms. Our maximum T_1 compares favourably with the literature: Table 1 compares this work to various high-coherence transmon qubits, all fabricated with the same materials.

Table 1. Comparison of highest niobium-on silicon transmon T_1 data published in academic literature. Maximum and median T_1 values are included where available.

	Max T_1	Median T_1	Qubit Frequency	Max Q-factor
This work, IQM [6]	0.96 ms	0.70 ms	3.6 GHz	22 M
R.T. Gordon et al. [7]	0.6 ms	0.40 ms	3.7 GHz	14 M
S. Kono et al. [8]	0.45 ms	N/A	4.8 GHz	14 M
M. Tuokkola et al. [9]	0.67 ms	0.50 ms	2.9 GHz	12 M

We close this section by noting that our maximum Q-factor of 22M is higher than for any niobium-on-silicon double-pad transmon published in the academic literature. We acknowledge that some transmons made of other materials have demonstrated Q-factors up to 25M [10] with tantalum-on-silicon. In that work, they conclude that they are limited by substrate losses. The conclusion that our maximum Q-factor is also near the substrate limit is supported by energy participation ratio simulations.

We are also aware of reports of higher T_1 [11] but due to the materials platforms and fabrication details not being disclosed in those sources we cannot make direct comparison.

2.2   Design Improvements

Another avenue for improving coherence is through optimizing the distribution of the qubit electric field through design. The qubit and its immediate environment consist of several components made of materials with differing loss factors [5]. By optimizing the energy participation ratio (EPR) of the qubit electric field in each of these regions we reduce the coupling of the qubit to the TLSs at lossy interfaces. The size of the interaction between a TLS and the qubit scales like $\boldsymbol{p}_{\mathrm{TLS}} \cdot \boldsymbol{E}_{\mathrm{qb}}$ [3,4], where $\boldsymbol{p}_{\mathrm{TLS}}$ is the TLS dipole moment and $\boldsymbol{E}_{\mathrm{qb}}$ the qubit electric field. Hence, reducing the field strength in lossy regions where the TLS density is high reduces the total coupling to the TLS bath. We have already seen this at work in the double-pad qubits; there the electric field is distributed largely away from the lossy interfaces, explaining the high coherence. However, there are challenges associated with scaling qubit counts with double pad qubits due to their large size, so instead we optimize the design of an IQM QPU-compliant qubit.

In Fig. 2 we show data comparing the T_1 performances of isolated QPU-compliant qubits before and after EPR optimization. The data indicate an improvement in mean T_1 of 19% following the geometry changes.

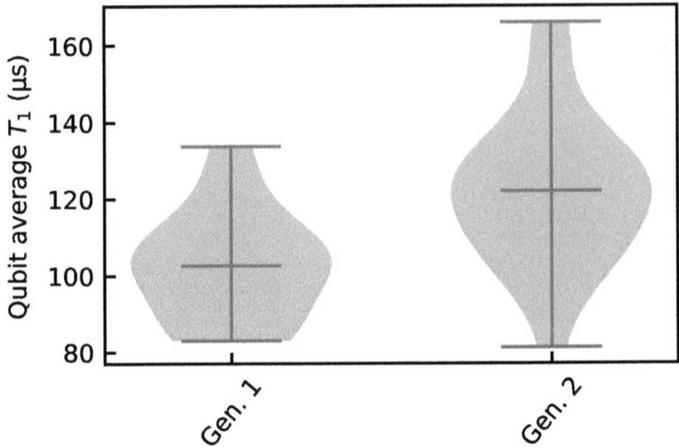

Fig. 2. QPU-compliant test device performances, demonstrating improved average performance. (a) A violin plot of mean T_1 values across 13 first generation qubits. (b) A violin plot of mean T_1 values across 22 s generation qubits.

The main benefits of this enhanced coherence are seen readily once this qubit design is incorporated into a QPU-compliant test device. When combined with advanced gate calibration techniques this qubit design has allowed for two qubit gate fidelities of over 99.9% fidelity [6,15].

3 Mitigating the Effects of TLS Defects

Even with improvements to fabrication quality and design, one cannot completely remove the effects of TLSs. With this in mind, we have studied different strategies to dynamically reduce the impact of near-resonant TLSs on the performance of the qubits.

3.1   Electric Field Tuning of TLS Defects

A promising approach to mitigating the impact of TLSs on superconducting qubits involves directly manipulating the defect frequencies using an external electric field. This methodology has been explored in the literature [11,13,14]. However, this technique requires careful consideration of the chip architecture to ensure compatibility as several implementations require dedicated control lines to manipulate the TLS frequency.

We demonstrate the feasibility of tuning TLS defects by introducing a bias tee following reference [13] and applying DC voltage to the qubit drive line. By performing AC-Stark TLS spectroscopy [12], we observed a TLS defect frequency shifting under applied voltage, illustrated in Fig. 3.

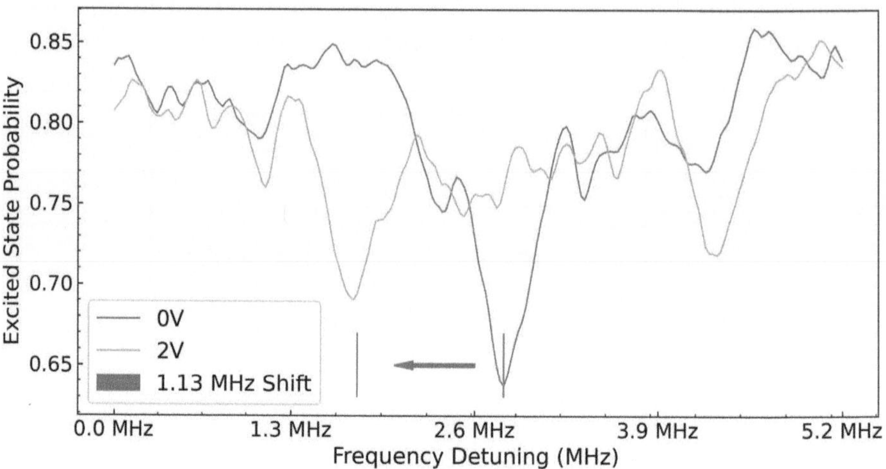

Fig. 3. Results of AC-Stark TLS spectroscopy at two different biasing voltages, showing the frequency shift of a TLS defect by over 1 MHz at 2 V.

With a biasing voltage of 2V, the TLS is seen to have been tuned by 1.13 MHz. However, the majority of TLSs are not tuned. An electromagnetics simulation demonstrates that the qubit geometry shields the surroundings, so only a small minority of TLSs are exposed to the biasing field. This highlights that trade-offs must be made; optimizing the design for resilience to TLS and for high coupling means TLS tuning is not viable at present in our devices, although other groups with different designs have employed it to good effect [13].

3.2   TLS Defect Avoidance

Alternatively, a different approach that does not require hardware modifications is based on qubit frequency parking optimization. Possible parking strategies based on particular types of optimizers were reported in [16,17]. We are investigating other optimization frameworks, and, in Fig. 4, we share results of optimizing coherence times with such an approach, improving T_1 by 14% on a 45 qubit square topology QPU. By utilizing advanced pulse shaping methodologies [18] and crosstalk avoidance [19], we can achieve fast high-fidelity gates on QPUs which are mostly coherence-limited, so optimizing coherence within this framework can directly improve QPU performance.

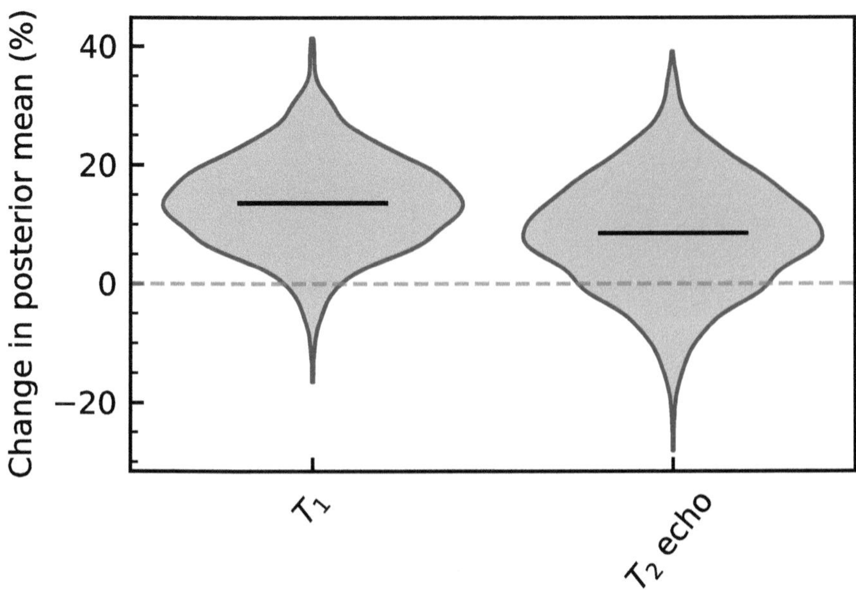

Fig. 4. Posterior distributions for the change in mean T_1 and T_2 echo on a 45 qubit QPU. The maximum *a posteriori* (MAP) estimate for the increase in coherence is 14%.

4 Outlook

In this paper we have summarized several strategies for improving coherence times at IQM Quantum Computers, demonstrating the role state-of-the-art fabrication and integrated approaches to design and software play in improving QPU performance. We are currently implementing these changes in larger scale QPUs to reap similar benefits and adapting what we have learned from TLS avoidance into our calibration methodologies.

Acknowledgments. We would like to acknowledge Marco Marin Suarez, Sampo Saarinen, Snigdha Kumar for measurement support; Lan-Husan (Lance) Lee, Seung-Goo Kim for sample fabrication; Tianyi Li, Liuqi Yu, Jeroen Verjauw, Vladimir

Milchakov, Jean-Luc Orgiazzi for technical discussions. This research has been partially supported by the Horizon Europe program through project OpenSuperQ+ of the EU Flagship on Quantum Technologies.

Disclosure of Interests. The authors are employed by IQM Quantum Computers.

References

1. Abad, T., Fernández-Pendás, J., Frisk Kockum, A., Johansson, G.: Universal fidelity reduction of quantum operations from weak dissipation. Phys. Rev. Lett. **129**, 150504 (2022)
2. Simmonds, R.W., Lang, K.M., Hite, D.A., Nam, S., Pappas, D.P., Martinis, J.M.: Decoherence in Josephson phase qubits from junction resonators. Phys. Rev. Lett. **93**, 077003 (2004)
3. Martinis, J.M., et al.: Decoherence in Josephson qubits from dielectric loss. Phys. Rev. Lett. **95**, 210503 (2005)
4. Müller, C., Cole, J.H., Lisenfeld, J.: Towards understanding two-level-systems in amorphous solids: insights from quantum circuits. Rep. Prog. Phys. **82**, 124501 (2019)
5. Wang, C., et al.: Surface participation and dielectric loss in superconducting qubits. Appl. Phys. Lett. **107**, 162601 (2015)
6. IQM Quantum Computers Press Release (2024). https://meetiqm.com/press-releases/iqm-quantum-computers-achieves-new-technology-milestones-with-99-9-2-qubit-gate-fidelity-and-1-millisecond-coherence-time/. Accessed 25 Aug 2025
7. Gordon, R.T., et al.: Environmental radiation impact on lifetimes and quasiparticle tunneling rates of fixed-frequency transmon qubits. Appl. Phys. Lett. **120**, 074002 (2022)
8. Kono, S., et al.: Mechanically induced correlated errors on superconducting qubits with relaxation times exceeding 0.4 ms. Nat. Commun. **15**, 3950 (2024)
9. Tuokkola, M., et al.: Methods to achieve near-millisecond energy relaxation and dephasing times for a superconducting transmon qubit. Nat. Commun. **16**, 5421 (2025)
10. Bland, M.P., et al.: 2D transmons with lifetimes and coherence times exceeding 1 millisecond, arXiv:2503.14798 (2025)
11. Dane, A., et al.: Performance Stabilization of High-Coherence Superconducting Qubits, arXiv:2503.12514 (2025)
12. Carroll, M., Rosenblatt, S., Jurcevic, P., Lauer, I., Kandala, A.: Dynamics of superconducting qubit relaxation times. NPJ Quantum Inf. **8**, 132 (2022)
13. Chen, L., et al.: Scalable and Site-Specific Frequency Tuning of Two-Level System Defects in Superconducting Qubit Arrays, arXiv:2503.04702 (2025)
14. Lisenfeld, J., et al.: Electric field spectroscopy of material defects in transmon qubits. NPJ Quantum Inf. **5**, 105 (2019)
15. Marxer, F., et al.: Above 99.9% Fidelity Single-Qubit Gates, Two-Qubit Gates, and Readout in a Single Superconducting Quantum Device, arXiv:2508.16437 (2025)
16. Google Quantum AI and Collaborators; Quantum error correction below the surface code threshold. Nature **638**, 920 (2025)
17. Jiang, T., et al.: Generation of 95-qubit genuine entanglement and verification of symmetry-protected topological phases, arXiv:2505.01978 (2025)

18. Hyyppä, E., et al.: Reducing leakage of single-qubit gates for superconducting quantum processors using analytical control pulse envelopes. PRX Quantum **5**, 030353 (2024)
19. Wesdorp, J., et al.: In preparation (2025)

ST Microelectronics/Quobly Path for Spin Based Large Scale Quantum Core

Tristan Meunier[1]([✉]), Nicolas Daval[1], Maud Vinet[1], Franck Arnaud[2], and Jean-Charles Barbé[1,3]

[1] Quobly, Grenoble, France
`tristan.meunier@quobly.io`
[2] STMicroelectronics, Crolles, France
[3] CEA-Leti, Grenoble, France

Abstract. Recent advancements in qubit manipulation in quantum dot arrays, as well as in the classical/quantum co-integration of FD-SOI spin-based quantum circuits shed light on an industrialization path for large-scale quantum computing. We present this path to designing and engineering good qubits using technology that is as close as possible to the most advanced industrial FD-SOI nodes. We then investigate qubit design accounting for the constraints arising from the established industrial fabrication process. More precisely, we repurpose the W vias and, in a single contact patterning step, define the gates that enable to define the electro-chemical potential of quantum dots (QDs), as well as the W vias that control the coupling barriers between adjacent QDs. We present simulation-based and experimental results on the individual coupling control of QDs in arrays fabricated on the industrial 28 nm FD-SOI technology. We present detailed wafer-level transfer characteristics of each barrier implemented on a 1x3 linear array at room temperature and at 2K. These results demonstrate that the vias behave like MOSFET gates, providing effective electrostatic control over the silicon channel. This validates the compatibility of the 28 nm FD-SOI industrial route with essential requirements for demonstrating a two-qubit gate.

Keywords: Spin qubit · semiconductor · Industrialization

1 Introduction

Silicon-based qubits are considered the most promising experimental system for scaling quantum computing. Due to the intrinsic compatibility with semiconductor industry, engineering quantum processors at large scale would benefit from the well-established (mature) semiconductor manufacturing. In particular, the ability to co-integrate qubits and transistors would be an asset for controlling large scale quantum circuits and keep their size comparable to classical processors. Among the various Si-based quantum technology platforms, FinFeET- and FD-SOI-based quantum technologies are the closest to advanced commercial CMOS technologies [1].

Demonstrating both hole and electron spin qubits [2], charge array control with tunnel coupling compatible with spin manipulation [3], and cryoelectronic circuits for

F. Barbaresco and F. Gerin (Eds.): QUEST-IS 2025, CCIS 2744, pp. 326–332, 2026.
https://doi.org/10.1007/978-3-032-13855-2_30

qubit control and read-out [4] has been key to validating the potential of the FD-SOI spin qubit technology. In particular, the last demonstration showed that the level of control in quantum dot arrays was sufficient to reach the last remaining milestone of the technology: the two-qubit gate in the form pulsed induced coherent exchange oscillations [5].

Precise control of the electrical parameters of a quantum dot array is essential for well controlled manipulation of qubits and, consequently, for reliable quantum computation [2]. Controlled exchange interactions can be achieved by independently tuning the tunnel coupling between adjacent QDs, which is a key requirement for implementing two qubit gates in spin-based quantum processors. Previous approaches in commercial FD-SOI linear arrays have relied on local back biasing [6] (Fig. 1 – top) or an extra front-gate [7] (Fig. 1 – Middle) to tune QD tunnel coupling. While these approaches are functional, they have the respective inconveniences of not being scalable, even in linear arrays, and of sacrificing one front-gate from two, which leads to a larger effective pitch for neighboring 2-qubit cells to have full control.

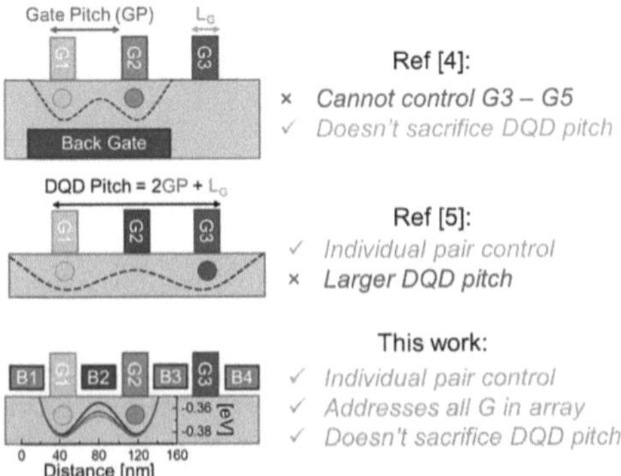

Fig. 1. Cross-section schematics of our design (bottom), depicting a 1x3 front gate array (G1 – G3), leveraging W vias as barrier gates, B1 – B4, as compared to previous work on foundry fabricated devices using the 22FDX® platform. In each case, the gate used for coupling control is shown in dark blue. Simulations of our design show the barrier height control by the bias applied on B2, between G1 and G2.

Here, we review our effort to design QD arrays that can be manufactured using known industrial fabrication methods, i.e. design rules compatible. The coupling control between adjacent QDs fabricated on 28 nm FD-SOI [8] by using vias to tune the barrier height is discussed. In comparison to previous designs, this integration approach allows in-situ tunnel coupling control over individual pairs of QDs in linear arrays while respecting a tighter pitch and reserving the FD-SOI back gate for tuning other important aspects of the device performance, such as mobility [9] and charge noise [10].

2 Device Geometry and Characterization

The devices measured were fabricated on the industrial 28 nm FD-SOI technology platform. Nanowires of various widths were defined by STI etching on an SOI wafer and were covered by a series of high-κ metal gates using standard gate-first processes

328   T. Meunier et al.

with a 90 nm pitch. These front gates (G1 – G3) are used to trap charges (electrons or holes) in the Si nanowire to form QDs (Fig. 1 - Bottom). Both the nanowire and the front gates were contacted by standard W vias, but additional vias were engineered to serve as barrier gates (B1 – B4) to tune the tunnel coupling between adjacent dots, inspired by our previous work [9] based on non-commercial FD-SOI technology.

Figure 2a shows a 2D map of $I_D(V_{barrier}, V_{GT})$ for a typical 1x3 qubit array (Fig. 1 – Bottom) at room temperature. The four barrier gates (vias) are swept together ($V_{barrier}$) while the front gates (G1-G3) are biased at different V_G to achieve the same V_{GT} ($V_{GT} = V_G - V_{TH,G}$). Figures 2b and 2c show how the I-V curves change with the voltages of the barrier and front gates, while Fig. 2d shows how the barrier's threshold voltage depends on the back gate voltage. Figure 2e-h show the impact of capacitive coupling of the barrier gate with the front and back gates at room temperature and at 2K.

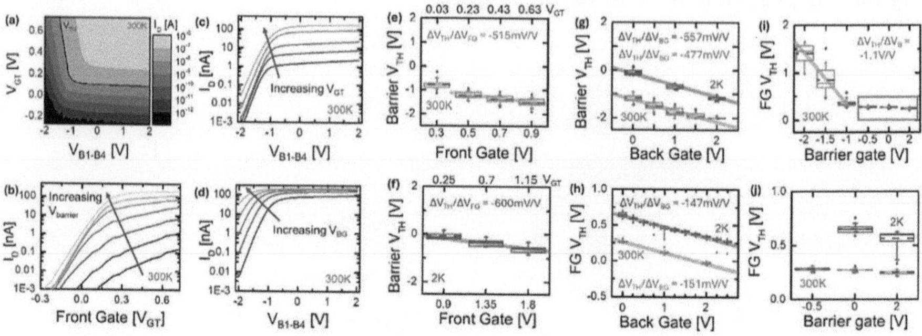

Fig. 2. (**a**) Typical $I_D(V_{Br}, V_{GT})$ sweeping all 4 barrier vias (B1–B4) in a 1x3 array with different front gate (G1–G3) V_{GT} values. (**b-c**) Sample I_D-V_{gate} with back gate = 0 V for (b) different $V_{barrier}$ while sweeping V_{GT} and (c) different V_{GT} while sweeping $V_{barrier}$. (**d**) $I_D(V_{barrier})$ for different back gate voltages (V_{GT} = 230 mV). (**e-h**) Full wafer statistics showing the impact of capacitive coupling between gates: (e-f) barrier's threshold voltage variation with V_{GT} (V_{BG} = 0 V) at (e) 300K and (f) 2K; Barrier's (g, V_{GT} = 0V) and Front gate (h, $V_{barrier}$ = 0V) threshold voltage variation with V_{BG} at 300K (red) and 2K (blue). (**i-j**) front gate threshold voltage variation with $V_{barrier}$ (V_{BG} = 0 V) at 300K (red) and 2K (blue). V_{DS} = 50 mV for all panels. V_{TH} calculated at I_D = 1nA. Lines overlayed to guide the eye for linear extrapolation. (Color figure online)

As barrier gate's threshold voltage is strongly affected by the operating V_{GT} (Fig. 2 e-f), to enable comparison of the individual barriers, the front gates were biased at near-threshold V_{GT} while the remaining barrier were biased at 0V. A MOS-like I_D-V_{Bi} of each barrier gate B_i was measured at room temperature on a typical die (see Fig. 3a). This was also collected at wafer level and at 300K [2K] displaying a mean threshold voltage over the 4 barrier gate $V_{TH,Bi}$ = −1.2V [−0.54 V] (Fig. 3b), mean I_{ON} = 92.85 nA, and a mean subthreshold swing (SS) of 233 mV/dec (Fig. 4d), with good matching between barrier gates. At 2K, the SS was visibly smaller but, due to current oscillations induced by the formation of sub-bands at cryogenic temperatures [11], an accurate SS extraction was not possible. We observe a ~ 250 mV interquartile range for $V_{TH,Bi}$ and 1V min-max range at both temperatures. While large, the EOT of the barrier gate is ~ 10–15 nm, leading to an electric field strength that is ~ 3–12x as weak as a typical high-κ front gate (EOT ~ 1.25–3.5 nm); the range we observe is therefore comparable to the ~ 300–400 mV range typically observed for a standard commercial front gate in a qubit arrays [8, 12].

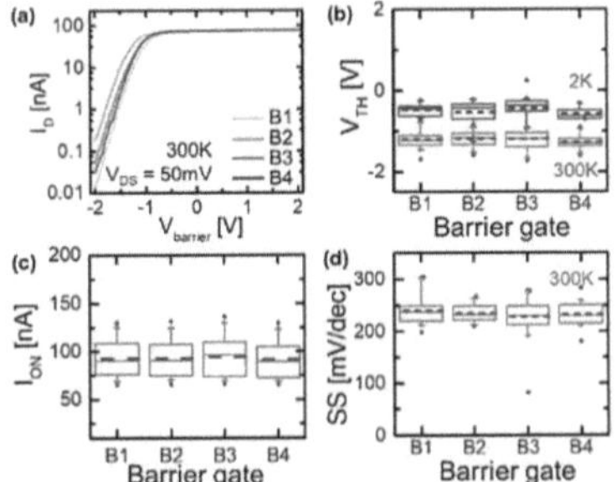

Fig. 3. (a) Typical $I_D V_B$ for each of the 4 barrier gate at 300K and $V_{GT} = 127mV$ showing MOSFET-like control over the Si channel **(b-d)** Boxplots of (b) $V_{TH,Bi}$ (c) I_{ON} and (d) SS_{Bi} for each of the 4 barrier vias, at $V_{GT} = 127mV$ [40mV] at 300K [2K] showing individual barrier gate matching at 300K (26 dies) and 2K (10 dies). V_{TH} calculated using linear extrapolation due to low voltages.

3   Quantum Dot Coupling Control

To test the effect of the barrier on the tunnel coupling control, we first performed wafer-level screening of various array designs (Design A-C). To do so, we collect $I_D V_G$ for each front gate at low $V_{DS} = 5$ mV and target regular patterns of Coulomb peaks (SET behavior) which have been predicted to be evenly spaced by 10–20 mV by simulation. As shown in Fig. 4, we observe an improvement in SET behavior by modifying the layout and characteristic size of the qubit array. In Design A we observe a purely MOSFET-like behavior, which improves slightly but shows poor Coulomb blockade in Design B and finally, good electrostatic definition of the QDs in Design C. The observation of some Coulomb peaks for Design B shows an improvement in quantum confinement when compared to Design A due to a 60% reduction of the front gate size. The undesired current in Design B, (MOSFET-like superposed to the peaks,) indicates that the structure is not operating properly in the Coulomb blockade regime and that uncontrolled carriers are flowing through the device. This behavior is improved for Design C, where the Coulomb peak heights are more even and the mean peak spacing is 15 mV which is in good agreement with simulations.

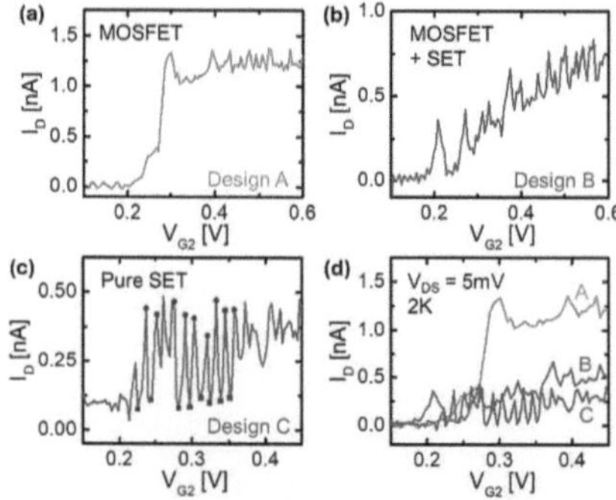

Fig. 4. Typical $I_D V_G$ at $V_{DS} = 5$ mV for three different qubit array designs measured during our screening protocol on the same die displaying (**a**) pure MOSFET (**b**) mixed MOSFET + SET and (**c**) pure SET characteristics. (**d**) all three $I_D V_G$ overlaid to show comparison. In design C, Coulomb peak heights and valleys are highlighted to calculate average peak distance.

For Design C, we show 2D maps of $I_D(V_{G2}, V_{G3})$ in Fig. 5. The existence of different sets of regular Coulomb peaks controlled by each of the two gates confirm that electrostatic QDs are defined by G2 and G3. The dots are found to be coupled with each other since honeycomb patterns can be distinguished [14]. For this design, we see over an 80% yield of the dies tested (20/24) spread across the center, middle and edge radii of the wafer. The yield indicates that a double QD system was identified through qualitative analysis of the 2D maps (*i.e.* visual inspection of patterns).

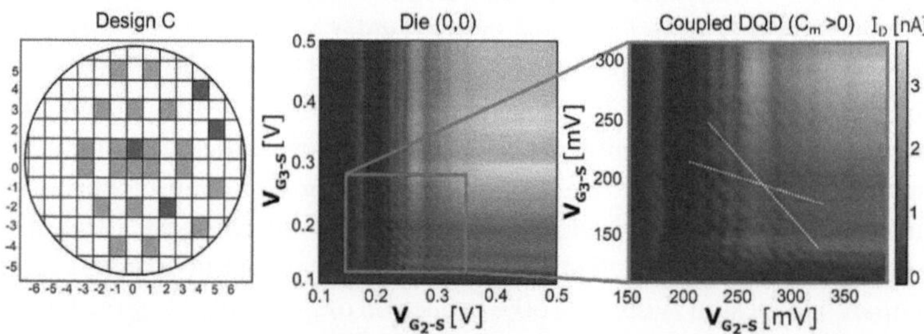

Fig. 5. Wafer-level 2D stability diagrams are collected at 2K for all neighboring front gate combinations to confirm SET behavior, collect information on their positions as well as verify the existence of coupling without the application of a barrier via bias. 80% of Design C devices displayed a DQD. Example of a coupled DQD shown for Die (0,0) defined by front gates G2 and G3.

In Fig. 6 we show simulations and experimental results for different regimes of coupling between two QDs, formed underneath front gates G1 and G2, for the same wafer, die and device as shown in Fig. 5. Simulations suggest, and experiment confirms tunnel coupling control between two neighboring QDs in our linear array design. The stability diagrams show $I_D(V_{G1}, V_{G2})$ collected for $V_{B2} = -0.5$ V, $+0.75$ V and $+1.5$ V, where the tunnel coupling varies from weak (observation of bias triangles) to strong. In the moderate coupling regime, honeycomb patterns are clearly observed, indicative of capacitive coupling.

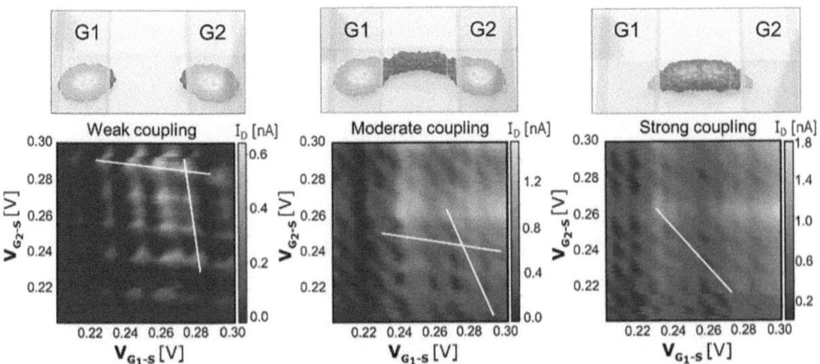

Fig. 6. (top) Simulation of envelope probability density, showing the effect of applied bias on the barrier gate via, B2. **(bottom)** Stability diagrams of G1 and G2 from Design C, with coupling controlled by the barrier gate B2, collected at 2K. White solid lines are added to guide the eyes.

4 Conclusions

We summarize our effort to design and operate quantum circuits using industrial 28 nm FD-SOI technology platform. We proposed the use of W vias as barrier gates in qubit devices and demonstrate the ability to control the tunnel coupling between two adjacent QDs operating in the many-electron regime. The vias act as proper electrostatic gates and control both conduction within the Si channel as well as allow for coupling control in the Coulomb blockade regime. Our design fulfills the requirement of individual coupling control needed for implementing quantum gates in spin-based quantum processors while maintaining an aggressive pitch. Our work demonstrates the first step towards an industrialization of spin qubit processors based on the existing semiconductor ecosystem.

Acknowledgments. We thank the Quobly characterization and engineering team for providing the data.

References

1. Meunier, T., Daval, N., Perruchot, F., Vinet, M.: Silicon spin qubits: a viable path towards industrial manufacturing of large-scale quantum processors. Eur. Phys. J. A **61**(3), 1–11 (2025)
2. Paz, B.C., et al.: FDSOI platform for quantum computing. In: 2024 IEEE International Electron Devices Meeting (IEDM), pp 1–4 (2024). https://doi.org/10.1109/IEDM50854.2024.10873521
3. Hamonic, P., et al.: Combining multiplexed gate-based readout and isolated CMOS quantum dot arrays. Nature Commun. **16**(1), 6323 (2025)

4. Schmidt, Q., et al.: Compact frequency multiplexed readout of silicon quantum dots in monolithic FDSOI 28nm technology. In: 2024 IEEE European Solid-State Electronics Research Conference (ESSERC), pp. 157–160 (2024). https://doi.org/10.1109/ESSERC62670.2024.10719580

5. Hamonic, P., et al.: Coherent exchange oscillation manuscript (submitted) (2022)

6. Bonen, S., Tripathi, S.P., McIntosh, J., Jager, T., Voinigescu, S.P.: Investigation of p- and n-type quantum dot arrays manufactured in 22-nm FDSOI CMOS at 2–4 K and 300 K. IEEE Electron Device Lett. **45**(10), 2025–2028 (2024). https://doi.org/10.1109/LED.2024.3435380

7. Amitonov, S.V., et al.: Commercial CMOS process for quantum computing: quantum dots and charge sensing in a 22 nm fully depleted silicon-on-insulator process (2024). arXiv: arXiv:2412.08422. https://doi.org/10.48550/arXiv.2412.08422

8. Planes, N., et al.: 28 nm FDSOI technology platform for high-speed low-voltage digital applications. In: 2012 Symposium on VLSI Technology (VLSIT), pp. 133–134 (2012). https://doi.org/10.1109/VLSIT.2012.6242497

9. Elbaz, G.E., et al.: Transport characterization and quantum dot coupling in commercial 22FDX (2025). arXiv: arXiv:2501.10146. 10.48550/arXiv:2501.10146

10. Spence, C., et al.: Probing low-frequency charge noise in few-electron CMOS quantum dots. Phys. Rev. Appl. **19**, 044010 (2023). https://doi.org/10.1103/PhysRevApplied.19.044010

11. Bédécarrats, T., et al.: A new FDSOI spin qubit platform with 40 nm effective control pitch. In: 2021 IEEE International Electron Devices Meeting (IEDM), pp. 1–4 (2021). https://doi.org/10.1109/IEDM19574.2021.9720497

12. Colinge, J.-P.: Quantum-wire effects in trigate SOI MOSFETs. Solid-State Electron. **51**(9), 1153–1160 (2007). https://doi.org/10.1016/j.sse.2007.07.019

13. Neyens, S., et al.: Probing single electrons across 300-mm spin qubit wafers. Nature **629**(8010), 80–85 (2024). https://doi.org/10.1038/s41586-024-07275-6

14. van der Wiel, W.G., et al.: Electron transport through double quantum dots. Rev. Mod. Phys. **75**(1) (2002). https://doi.org/10.1103/RevModPhys.75.1

Session: 12 Quantum Algorithms, Computing; Simulation – Specialized Applications and Security

A Hybrid Quantum-Classical Approach for Urban Drone Path Planning Insights from Quantum Computer Implementation

Hian Lee Kwa[(✉)], Bing Hong Teh, Teck Yoong Chai, and Yung Sze Gan

CortAIx Singapore, Thales Solutions Asia, Singapore, Singapore
{hianlee.kwa,binghong.teh,teckyoong.chai}@thalesgroup.com,
yungsze.gan@asia.thalesgroup.com

Abstract. As drone traffic and air taxi services expand globally, regulators face increasing pressure to manage complex airspace safely, while operators aim to optimise energy-efficient routes. Traditional computing may not scale to meet future path planning demands, making quantum computing a promising alternative. However, the variety of qubit modalities across quantum processing units (QPUs) raises questions about which platforms are most suitable for such applications. This paper presents a hybrid quantum-classical drone path planning approach that accounts for urban obstacles and its implementation on two QPUs: D-Wave's quantum annealer and Pasqal's neutral atom processor. We compare solution quality, time-to-solution, and hardware-specific trade-offs against a classical solver, providing insights into the scalability, embedding challenges, and the practical feasibility of different quantum computing platforms for urban air mobility.

Keywords: Drone Path Planning · Hybrid Quantum-Classical Algorithm · Neutral Atom Quantum Computers · Quantum Annealing · Quantum Computing · Urban Air Mobility

1 Introduction

The use of unmanned aerial vehicles (UAVs) in urban environments has recently started to grow rapidly, with applications ranging from building inspection and surveillance to parcel delivery. As the uptake of urban air mobility increases, regulators face mounting challenges in ensuring the safe and efficient management of airspace. At the same time, drone operators seek to minimise energy usage by optimising flight paths. A centralised planner that accounts for drone requirements and environmental constraints could address both needs. However, traditional computing approaches may struggle with scalability, a challenge that quantum computing could overcome.

Quantum computing in such applications can be achieved through hybrid quantum-classical frameworks that combine classical techniques with quantum algorithms. A common method involves converting real-world problems into

F. Barbaresco and F. Gerin (Eds.): QUEST-IS 2025, CCIS 2744, pp. 335–345, 2026.
https://doi.org/10.1007/978-3-032-13855-2_31

Quadratic Unconstrained Binary Optimisation (QUBO) form [3], which can be solved using quantum annealers, such as those provided by D-Wave, or gate-based processors using variational quantum algorithms (VQAs) [9].

In this paper, the contributions are as follows: (1) extending the method proposed by James and Raheb [4] to one that has been adapted to account for environmental obstacles, (2) implementing this approach using a D-Wave quantum annealer and a simulated neutral atom QPU from Pasqal, as well as benchmarking the resulting solutions for a single drone case against those produced by a classical algorithm, and (3) evaluating the two QPUs in terms of solution quality and generation time, and discuss their implementation, scalability, and the challenges involved in real-world deployment.

2   Background and Motivation

In classical computing, path planning for autonomous vehicles typically relies on graph search algorithms, which aim to identify a sequence of edges connecting a start node to a destination [7]. In practice, classical routing problems often map nodes to points on a fine grid, allowing near-continuous movement across a space. These approaches can be adapted to quantum computing since graph search problems are readily reformulated as Quadratic Unconstrained Binary Optimisation (QUBO) problems and can subsequently be solved via Quantum Annealing or other Variational Quantum Algorithms (VQAs) [1].

However when formulating the problem with quantum computing, nodes are typically mapped to key landmarks, such as drop-off locations or intersections, due to the limited qubit count and connectivity in current quantum processing units (QPUs) [6,14]. As a result, drones are often constrained to predefined routes and are unable to take advantage of variations in their flight capabilities, such as the speed of fixed-wing drones or the agility of multi-rotor drones.

To better exploit UAV manoeuvrability, an iterative approach that does not rely on fixed graphs is required. This enables drones to carry out control decisions anytime during the flight instead of at pre-determined waypoints. Similar quantum-based methods have been explored in contexts like robot swarms [2] and airport landing path generation [4]. However, these do not fully account for obstacles such as buildings and no-fly zones. In what follows, we formulate a QUBO problem that allows the optimiser to route drones from their starting points to their destinations while avoiding collisions with both environmental obstacles and other drones. This is done without mapping waypoints to graph nodes, making it applicable to any area with known obstacles, and easily scalable to accommodate an arbitrary number of drones.

3   QUBO Problem Formulation

To guide drones toward their destinations while avoiding obstacles, we use a combination of attractive and repulsive potential fields. The quantum optimiser attempts to minimise the total potential field experienced by each UAV at each time-step through choosing a speed and turning rate instruction the UAV is to carry out. This potential value is related to the distance of a drone to its destination and to other nearby drones. For the original formulation, the reader is directed to [4]. In what follows, we describe an extension of their QUBO formulation that accounts for the presence of buildings and other obstacles, which is done via classical computing methods.

The overall potential field guiding the drone is given by the equation:

$$P_i(x_i) = P_{\text{att},i}(x_i) + P_{\text{rep},i}(x_i). \tag{1}$$

where, $P_{\text{att},i}$ and $P_{\text{rep},i}$ is the attractive field and repulsive field experienced by a Drone i respectively. This combined field is visualised in Fig. 1.

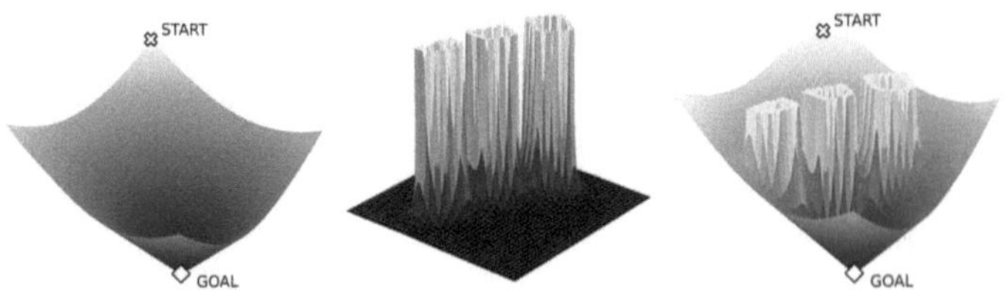

Fig. 1. Attractive potential field that guides a drone to its target (Left). Repulsive potential field generated by three towers (Centre). The overall potential field of the environment (Right). The potential fields were all generated at one altitude.

3.1   Attractive Potential Fields

In both the work presented here and the original, an attractive potential is used to guide drones towards their destinations. This is defined for Drone i at position x_i as a function of its distance to its destination:

$$P_{\text{att},i}(x_i) = k_{\text{att}} ||(x_i - x_{\text{target},i})^{\alpha}||, \tag{2}$$

where $x_{\text{target},i}$ is the target location, and tunable constants $k_{\text{att}} = 1 \times 10^3$ and $\alpha = 2$. This function ensures that the potential field experienced by the drone reduces the closer it is to its destination.

3.2   Repulsive Potential Fields

In addition to the attractive potential field, a repulsive potential field is used to push drones away from obstacles. A Drone i perceives a neighbouring Drone j as a peak in its repulsive field given by:

$$P_{\text{rep},i,j}(x_i) = k_{\text{rep}} \left(\frac{1}{D_{i,j}} - \frac{1}{R} \right)^{\beta} ||(x_i - x_{\text{target},i})||, \tag{3}$$

where $D_{i,j}$ is the distance between Drones i and j, $R = 30$m is the Drone j's influence range, and the tunable constants are $k_{\text{rep}} = 1 \times 10^7$ and $\beta = 2$.

To extend this to obstacles, we use point clouds to delimit restricted areas around buildings. The repulsive fields from the areas are computed by summing the contributions from each point with Eq. 3, where $D_{i,j}$ is the drone's distance to each point. Furthermore, buildings located close to each other are aggregated into one large restricted area to avoid the potential trapping of drones in potential wells generated within narrow corridors between buildings.

3.3   Control Decision Vectors

To optimise drone control at each time-step, the planner seeks a binary solution vector U that minimises the cost function:

$$C(U) = U^T Q(X) U, \tag{4}$$

where $Q(X)$ is the QUBO matrix derived from the potential field values of the controlled drones. The vector U represents all discretised control decisions across UAVs and can be split into smaller vectors u, each corresponding to a single UAV. For instance, a fixed-wing Drone i with a speed range of 10–15 m/s and a maximum turning rate of 25 deg/s may have two binary variables for speed and three for turning rate, giving:

$$u_i^T = [\text{Turning Rate: } -25, 0, 25 \text{ deg/s}; \text{ Speed: } 10, 15 \text{ m/s}]. \tag{5}$$

This formulation essentially turns the path planning problem into a combinatorial optimisation problem, where the optimiser selects the combination of control actions that minimises the cost function. With this scheme, a drone flying at 15 m/s with a turning rate of 0 deg/s is associated with the binary control vector $u_i^T = [0, 1, 0, 1, 0]$. Figure 2 illustrates this setup. While the current approach handles two-dimensional motion, it can be extended to three dimensions by a separate control group for climb and descent rates.

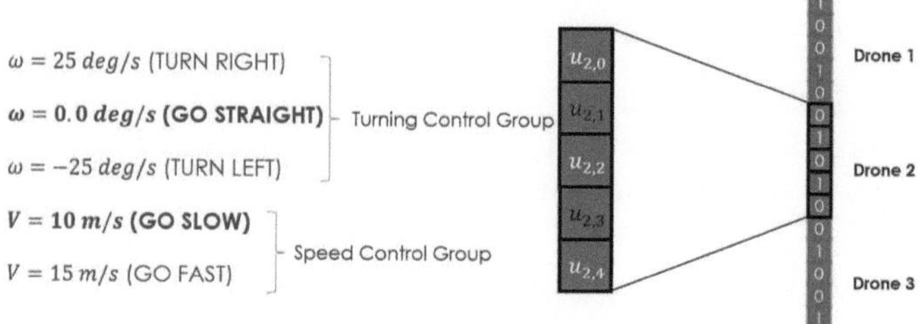

Fig. 2. The overall control vector, U, with the control vector, u_2, for Drone 2 highlighted. The drone is travelling at a speed of 10 m/s and a turning rate of 25 deg/s.

With this formulation, combinations of control decisions results in different positions for the drone at the subsequent time-step, which is used to calculate the future potential value of the drone. These serve as the coefficient values for the QUBO matrix. Formally, for an element q_{ij} in the QUBO matrix, $Q(x)$,

$$q_{ij} := P(x(u_i, u_j)), \tag{6}$$

where $x(u_i, u_j)$ is the future position of the drone when control decisions u_i and u_j are selected. Consequently, the number of logical qubits required to encode the QUBO problem scales as $O(N \times M)$, where N is the number of drones and M is the number of controls per drone.

3.4 Mutual Exclusivity Groups

Some control combinations are invalid. E.g., $u_i = [1, 0, 1, 0, 0]$ instructs the drone to turn both left and right while omitting a speed command. To prevent such conflicts, mutual exclusivity constraints are applied using the following identity:

$$\sum_{g=1}^{G} \left(1 - \sum_{c=1}^{C} u_{gc}\right)^2 = 0, \tag{7}$$

Here, G is the number of exclusivity groups (e.g., speed and turning), and C is the number of control options in each group. In Fig. 2, there are two groups: turning (3 controls) and speed (2 controls), giving $G = 2$ and $C = 3, 2$ respectively. These constraints are incorporated into the optimiser via a weighted minimiser, which penalises the selection of invalid control combinations:

$$K \min_{u} \sum_{g=1}^{G} \left(1 - \sum_{c=1}^{C} u_{gc}\right)^2 = 0, \tag{8}$$

where K is a large penalty constant.

4 QPU Implementation

4.1 Quantum Annealer

Quantum annealing is an approach for solving combinatorial optimisation problems by exploiting the adiabatic theorem: if a quantum system evolves slowly enough from an easy-to-prepare initial Hamiltonian to a final problem Hamiltonian, it remains in its ground state, thereby yielding the optimal solution. In the QUBO-to-Ising mapping used here, the Hamiltonian evolves according to

$$H(t) = A(t)H_i + B(t)H_f, \tag{9}$$

where $H_i = \sum_i \sigma_i^x$ is the initial Hamiltonian, and $H_f = \sum_i h_i \sigma_i^z + \sum_{i,j} J_{i,j} \sigma_i^z \sigma_j^z$ encodes the QUBO problem through local fields h_i and couplings $J_{i,j}$. The schedule functions $A(t)$ and $B(t)$ interpolate between the two.

On D-Wave's Advantage 4.1 quantum annealer, QUBO problems are represented as dictionaries of linear and quadratic coefficients. These logical variables are then mapped to physical qubits arranged in a Pegasus topology. Because the Pegasus graph has sparse, fixed connectivity, a *minor embedding* process is required to map QUBO variables onto physical qubits. D-Wave's SDK provides a heuristic embedding algorithm that automates this step, after which the annealing schedule is executed to sample low-energy solutions.

4.2   Neutral Atom Processor

On Pasqal's neutral atom QPUs, the coefficients of the QUBO matrix, which define the binary variables' interactions, are encoded through the controlled Rydberg interaction between atom pairs [8,11]. The interaction strength is defined as $U_{ij} = C_6/r_{ij}^6$, where C_6 is a constant and r_{ij}^6 is the inter-atomic distance between atoms i and j. To encode a QUBO problem, atoms are arranged in a register to set the desired inter-atomic distances, which then determine the interaction strengths representing the off-diagonal coefficients. Formally, this can be formulated as a minimisation problem:

$$\underset{X}{\text{minimize}}\,\|U(X) - Q\|_F, \tag{10}$$

where Q is the QUBO matrix, X is a vector of the atoms' placements to be optimised, and $U(X)$ is the $n \times n$ matrix with the Rydberg interaction levels between atom pairs, given as:

$$U_{ij} = \begin{cases} \frac{C_6}{\|\mathbf{x_i}-\mathbf{x_j}\|^6}, & \text{if } i \neq j \\ 0, & \text{if } i = j \end{cases}, \tag{11}$$

where $\mathbf{x_i}$ and $\mathbf{x_j}$ are vectors containing the x and y positions of atoms i and j respectively. In our implementation, these were found using the Nelder-Mead method repeated 30 times with random atom initialisations. The configuration with the lowest cost was chosen as the final atomic placements. This qubit placement method effectively maps one QUBO variable to one physical qubit. While this is sufficient for our current implementation, one must consider alternative embedding strategies should large numbers of controls be considered due to the finite level of connectivity for each qubit, including the use of quantum wires to connect physically distant qubits [5,13].

After placing the atoms, we optimise the time evolution of the laser pulse's amplitude and detuning during the adiabatic process, subject to two conditions: (1) the amplitude must start and end at zero, and (2) the detuning must increase monotonically through zero. Using an interpolated waveform, we optimise the pulse at three equally spaced points (six parameter values), which is done based on the expected cost of the sampled bitstrings.

5   Flight Path Planning

Three different solvers were implemented to evaluate the flight planning method. First was D-Wave's Leap software package running on their Advantage-4.1 system, using the quantum annealing process described in Sect. 4.1. The second solver was Pasqal's neutral atom QPU that was emulated via their Pulser library and implemented as described in Sect. 4.2. Finally, a comparison was made between the two quantum approaches against a classical approach, simulated annealing, which was also available via D-Wave's SDK.

For this work, we considered flights taking place in the Marina Bay area of Singapore. As mentioned in Sect. 3, restricted areas were drawn around buildings to prevent drone collisions. In addition, buildings that were located very close to each other were grouped into a larger restricted zone to prevent drones from being trapped between them (see Fig. 3). In this work, for simplicity and to demonstrate its ability to avoid buildings, only fixed-winged drones with two speed controls (10 and 15 m/s) and three turning rate controls (-25, 0, 25 deg/s), as well as a time-step of $dt = 1$s were considered. We also restrict the drones to flying at a fixed altitude, similar to how a controller may assign a specific altitude for drone operations [12]. Despite this, we note that this path planning framework will work for drones with different available controls, including altitude controls, as demonstrated by James and Raheb in clear environments [4].

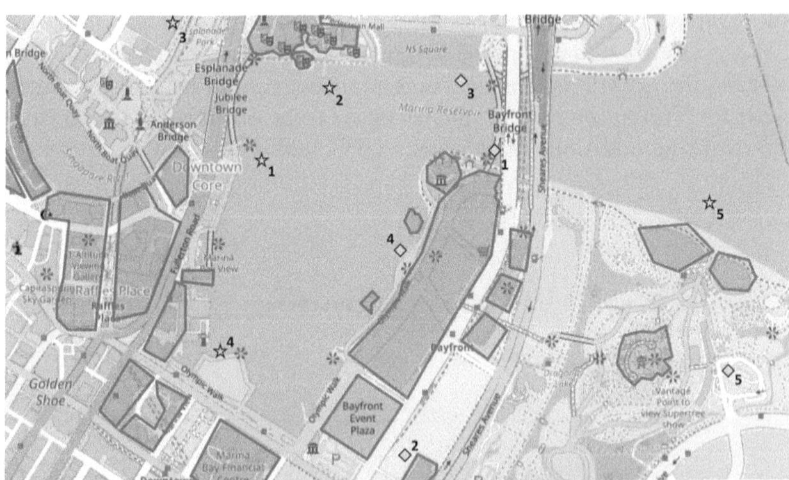

Fig. 3. Flight planning area over Marina Bay, Singapore. Restricted zones are demarcated in blue. Route start points (stars) and destinations (diamonds) are labelled with route numbers. (Color figure online)

6   Results

6.1   Solution Quality

During testing, five flight scenarios were used, each with its own launch and destination point. One example route can be seen in Fig. 4. Table 1 presents the total distance covered by the fixed-wing drone for each test case, as planned by the different solvers. This table shows that the quantum methods produced solution qualities similar to, and in some cases better than, the classical simulated annealing method, with all results falling within 4% of each other.

Table 1. Total distance planned for a fixed-wing drone flying on different routes.

Route No.	D-Wave (m)	Pasqal (m)	Simulated Annealing (m)
1	639	648	**633**
2	**1271**	1295	1290
3	915	900	**898**
4	**535**	555	556
5	**510**	528	520

At this stage, we note that the neutral atom-based solver used a single optimized laser pulse sequence to generate all the paths, including those for flying in both clear airspace and near obstacles (see Fig. 4). This is an advantage over the use of VQAs as although it is possible to run VQAs on neutral atom QPUs, doing so would require additional classical and quantum computational resources. This is because for VQAs, the pulse sequence needs to be optimized at each time-step, thereby requiring additional calls to the QPU and a classical optimiser.

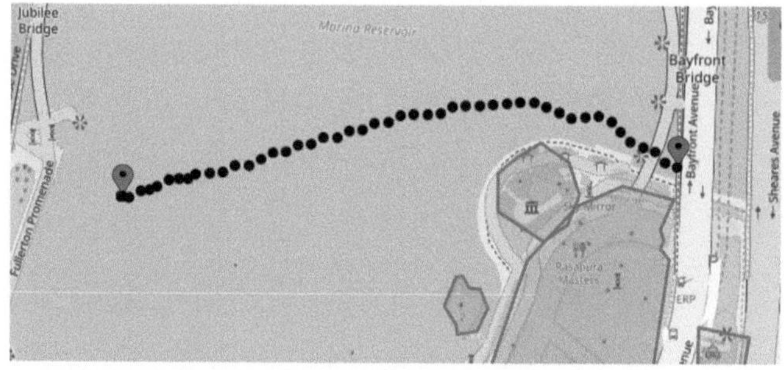

Fig. 4. Flight path for a fixed-wing drone on Route 1, as produced by the simulated Pasqal neutral atom solver.

6.2   Solution Time

Despite the relatively similar result quality, the amount of time required to derive these results from the two QPUs differs. For D-Wave annealers, the time required to return a solution is split into four broad stages[1]: (1) preparation, i.e., the mapping of variables to physical qubits, (2) QPU access time, which includes QPU programming and sampling, (3) post processing, and (4) worker and QPU queues, with steps 2 and 3 occurring in parallel. For our framework, the time required to obtain the solution to each submitted QUBO problem was $\sim$ 6s, with the QPU access time being measured at $\sim$ 550ms per result. This suggests that the majority of the end-to-end latency arises from queueing for both the QPU and the pre-processing worker and pre-processing overhead rather than from quantum computation itself.[1]

In contrast, the preparation phase for Pasqal's QPU occurs locally using the process described in Sect. 4.2. For the 5-control vector single drone case used, this took $\sim$ 60ms on an Intel i5-1145G7 CPU. However, it should be noted that initialising the atomic register using a structured layout that reflects the connectivity of the QUBO problem, rather than a random placement, may reduce the preprocessing time associated with atom positioning and potentially improve solution quality by providing a better starting point for layout optimisation. Although simulating quantum processes to solve the QUBO problem took less than 100ms per instance, this step is expected to take significantly longer on a physical QPU. Due to the qubit preparation processes, these QPUs currently have a base repetition rate of $2 - 3$Hz [11]. As such, solving a QUBO problem for each time-step with 5000 samples on a physical QPU is expected to take around 30mins, while solving for a drone's entire trajectory over 100 time-steps is expected to take 50 h.

7   Discussion

In this paper, we presented an extension to the method proposed by James and Raheb to account for fixed environmental obstacles such as buildings, and its implementation on two QPUs. Unlike waypoint-based systems, this approach permits free UAV flight without predefined routes, assuming obstacles have been pre-mapped. The required number of logical qubits scales with the number of drones and control variables rather than flight area size. Tests with a single fixed-wing UAV demonstrate the feasibility of hybrid quantum–classical frameworks for obstacle-aware urban path planning, with quantum solvers producing path lengths comparable to or better than a classical solver. Further validation and scalability analysis is required for multi-drone operations.

Despite the similar solution qualities, solution times varied significantly depending on the QPU used. D-Wave's quantum annealers returned each time-step's solution in $\sim$ 6s while physical execution on Pasqal's QPU is estimated to take $\sim$ 30mins per QUBO problem. This is mainly due to the time required for atomic register preparation. As such, D-Wave is more suitable for pre-flight or urgent replanning, while Pasqal's QPUs are better suited to generating routes well in advance of the flight. Fundamentally, implementing such a framework in live path planning requires a reduction in solving time for each submitted QUBO

---

[1] D-Wave Operation and System Timing.

problem. This means reducing the base repetition rate for neutral atom QPUs[2] or the preparation and queue time for D-Wave's quantum cloud services.

Scaling the number of drones or control variables produces larger, denser QUBO graphs, increasing both embedding complexity and physical qubit requirements. For instance, moving from one drone with five controls to ten drones increases the QUBO size and logical qubit requirement by a factor of 10. This could quickly exceed the capacity of current annealing and neutral atom systems without qubit-efficient encodings or improved mapping strategies. While D-Wave offers mature minor embedding heuristics and gate-based QPUs support qubit efficient encodings [6,10], no comparable approach exists for neutral atom architectures. Future work should focus on geometry-aware heuristics or variational embedding strategies tailored to neutral atom QPUs to reduce preprocessing overhead and improve scalability.

Acknowledgements. The authors would like to thank Prof. Dimitris Angelakis, Mr. Andreas Stratakis, and Mr. Gordon Ma from AngelQ for the insightful discussions on problem formulation, and Dr. Krisztian Benyo from Pasqal for the discussions on implementing quantum algorithms on their QPUs emulators.

References

1. Bauckhage, C., Brito, E., Cvejoski, K., Ojeda, C., Schücker, J., Sifa, R.: Towards shortest paths via adiabatic quantum computing. In: Proc. Mining Learn. Graphs (2018)
2. Chella, A., et al.: Quantum planning for swarm robotics. Robot. Auton. Syst. **161**, 104362 (2023)
3. Glover, F., Kochenberger, G., Du, Y.: Applications and computational advances for solving the QUBO model. In: The Quadratic Unconstrained Binary Optimization Problem: Theory, Algorithms, and Applications, pp. 39–56. Springer (2022)
4. James, S.H., Raheb, R.N.: Path planning for critical ATM/UTM areas. In: 2019 IEEE/AIAA 38th Digital Avionics Systems Conference (DASC) (2019)
5. Kim, M., Kim, K., Hwang, J., Moon, E.G., Ahn, J.: Rydberg quantum wires for maximum independent set problems. Nat. Phys. **18**(7), 755–759 (2022)
6. Leonidas, I.D., Dukakis, A., Tan, B., Angelakis, D.G.: Qubit efficient quantum algorithms for the vehicle routing problem on noisy intermediate-scale quantum processors. Adv. Quantum Technol. **7**(5), 2300309 (2024)
7. Sanchez-Ibanez, J.R., Pérez-del Pulgar, C.J., García-Cerezo, A.: Path planning for autonomous mobile robots: A review. Sensors **21**(23), 7898 (2021)
8. Silvério, H., et al.: Pulser: an open-source package for the design of pulse sequences in programmable neutral-atom arrays. Quantum **6**, 629 (2022)
9. Symons, B.C., Galvin, D., Sahin, E., Alexandrov, V., Mensa, S.: A practitioner's guide to quantum algorithms for optimisation problems. J. Phys. A: Math. Theor. **56**(45), 453001 (2023)

---

[2] Pasqal's QPU development roadmap states their aim to achieve a base repetition of 100Hz in 2029.

10. Tan, B., Lemonde, M.A., Thanasilp, S., Tangpanitanon, J., Angelakis, D.G.: Qubit-efficient encoding schemes for binary optimisation problems. Quantum **5**, 454 (2021)
11. Tibaldi, S., Leclerc, L., Vodola, D., Tignone, E., Ercolessi, E.: Analog QAOA with Bayesian optimisation on a neutral atom QPU (2025). arxiv:2501.16229
12. Veytia, A.M., et al.: Metropolis II: benefits of centralised separation management in high-density urban airspace. In: 12th SESAR Innovation Days (2022)
13. Wintersperger, K., et al.: Neutral atom quantum computing hardware: performance and end-user perspective. EPJ Quantum Technol. **10**(1), 32 (2023)
14. Yarkoni, S., et al.: Quantum shuttle: traffic navigation with quantum computing. In: Proceedings of the 1st ACM SIGSOFT International Workshop on Architectures and Paradigms for Engineering Quantum Software, pp. 22–30 (2020)

Quantum-Backed Integrity for Industrial Blockchain Events: A Secure Logging Framework Using Qiskit and Ethereum

Benoît Prieur[✉]

Medialoco LLC, Austin, USA
benoit.prieur@protonmail.com

Abstract. In the era of Industry 4.0, where automation, traceability, and secure data flows are foundational to modern industrial ecosystems, blockchain technology has emerged as a powerful tool to ensure transparency, auditability, and resistance to tampering. By offering an immutable ledger of transactions or events, blockchains are increasingly adopted in manufacturing, logistics, healthcare, and energy sectors to track critical operations and ensure trust among stakeholders.

However, the integrity of any blockchain-based system critically depends on the quality of the data it ingests. If the data recorded is predictable, manipulated, or inserted without sufficient entropy or verification, the immutability of the blockchain becomes moot. This paper introduces a proof-of-concept framework that integrates Quantum Random Number Generation (QRNG) and smart contract-based logging on Ethereum to strengthen the cryptographic robustness of industrial event recording. Each event is timestamped, tagged with a quantum-generated number, described via metadata, and hashed using SHA-256 to produce a unique fingerprint. The hash is then immutably stored on a local Ethereum blockchain. The system demonstrates a fusion of quantum physical entropy and decentralized blockchain trust.

Keywords: Quantum Random Number Generation (QRNG) · Blockchain immutability · Ethereum smart contract · Industrial event logging · Cryptographic integrity

1 Motivation and Threat Model

Modern industrial systems increasingly depend on the real-time logging of operational events to guarantee quality control, safety compliance, and system transparency. In domains such as energy, aerospace, pharmaceuticals, and defense, industrial logs serve not only as technical records but as legal and operational anchors for decision-making and accountability.

While blockchain offers immutability once data is recorded, it cannot guarantee the quality or unpredictability of the input itself. Attackers may intercept or forge log entries before they are committed to the chain. Classical sources of entropy, such as pseudorandom number generators (PRNGs), are vulnerable to manipulation if seeds or

F. Barbaresco and F. Gerin (Eds.): QUEST-IS 2025, CCIS 2744, pp. 346–351, 2026.
https://doi.org/10.1007/978-3-032-13855-2_32

initial conditions are exposed. Predictable or replayable data allows adversaries to mimic legitimate events or tamper with historical records.

Quantum Random Number Generation (QRNG), based on the inherent unpredictability of quantum measurement, provides a remedy. By sourcing entropy from quantum states, one ensures the non-reproducibility of each event, enhancing trust at the data insertion layer. The hash of this event, constructed with SHA-256, becomes a tamper-evident commitment to the quantum measurement, timestamp, and metadata.

In our threat model, an attacker may attempt to replay events, inject forged data, or alter metadata before blockchain commitment. The proposed system thwarts such attempts by binding unpredictable quantum entropy with cryptographic sealing and immutable storage.

2   System Design

The proposed framework integrates four components: a quantum entropy source (QRNG with Qiskit), a cryptographic hashing layer (SHA-256), a smart contract for event recording (QSecure.sol), and a blockchain interaction layer (Node.js orchestration).

2.1   Quantum Entropy Extraction

To obtain physically unpredictable values, we construct an 8-qubit quantum circuit using Qiskit. Each qubit is placed in superposition via a Hadamard gate, producing a uniform distribution over all 256 possible bitstrings (see Fig. 1). Upon measurement, the circuit collapses to one classical outcome, interpreted as a random integer.

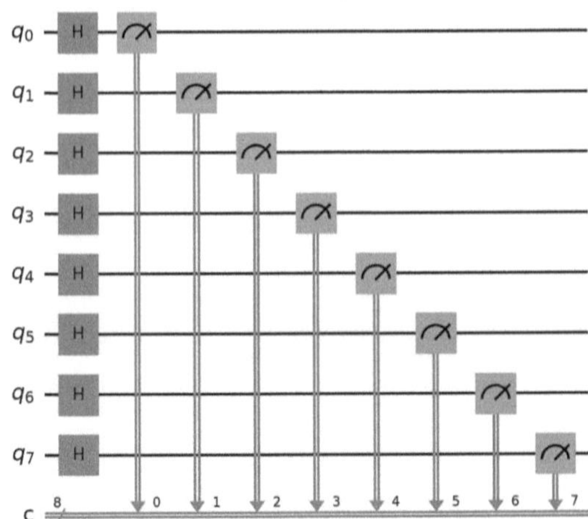

Fig. 1. Diagram of the quantum circuit.

The following Python method (file qrng.py) implements this QRNG pipeline and persists all artifacts, bitstring, integer value, timestamp, and ASCII diagram, in a JSON file for forensic inspection:

```python
def generate_quantum_random_bits(n_bits=8,
save_path="quantum_proof.json"):
    circuit = QuantumCircuit(n_bits, n_bits)
    circuit.h(range(n_bits))
    circuit.measure(range(n_bits), range(n_bits))
    backend = Aer.get_backend("qasm_simulator")
    job = backend.run(circuit, shots=1)
    result = job.result()
    counts = result.get_counts()
    bitstring = list(counts.keys())[0]
    value = int(bitstring, 2)
    proof = {
        "timestamp": datetime.now(timezone.utc).isofor-
mat(),
        "bitstring": bitstring,
        "value": value,
        "n_bits": n_bits,
        "circuit_ascii": circuit.draw(out-
put='text').single_string()
    }
    with open(save_path, "w") as f:
        json.dump(proof, f, indent=2)
    return value
```

This quantum-generated integer serves as the entropy anchor for each logged event.

2.2 Event Fingerprint and Integrity Seal

The quantum value is combined with metadata (e.g., component label, checkpoint name) and a system timestamp to form a unique event fingerprint. Using Node.js's crypto module, this string is hashed with SHA-256 to produce a tamper-evident commitment.

```javascript
const eventData = `${timestamp}-${randomNumber}-${meta-
data}`;
const eventHash = crypto.createHash('sha256').up-
date(eventData).digest('hex');
```

2.3 Smart Contract for Immutable Logging

The integrity seal, together with the quantum proof (the quantum-generated integer) and metadata, is transmitted to the QSecure smart contract, implemented in Solidity and deployed locally via Hardhat. The contract stores a timestamp, the quantum-generated integer, metadata, and the SHA-256 hash as an immutable record:

```
struct Event {
    uint256 timestamp;
    uint256 quantumProof;
    string metadata;
    bytes32 eventHash;
}

function recordEvent(
    uint256 _quantumProof,
    string calldata _metadata,
    bytes32 _eventHash
) external { ... }
```

Access control ensures only the deployer may append new events, while any participant can read back entries to verify provenance.

2.4 Orchestration Pipeline

The orchestration script (scripts/setQuantum.js) automates the full pipeline: it runs the Python QRNG (qrng.py) to obtain the quantum generated integer, computes the SHA-256 hash, invokes recordEvent on the deployed QSecure contract via Web3.js, retrieves the most recent event (getEvent) to confirm successful commitment, and serializes all results to JSON for external archiving.

```
const eventData = `${timestamp}-${randomNumber}-${meta-
data}`;
const eventHash = crypto.createHash('sha256').up-
date(eventData).digest('hex');

await qsecure.methods
    .recordEvent(randomNumber, metadata, '0x' + eventHash)
    .send({ from: signer.address });
```

With this end-to-end pipeline design [qrng.py → setQuantum.js → QSecure.sol] each industrial event is cryptographically bound to unpredictable quantum entropy and immutably anchored on the blockchain.

2.5 Example Execution Output

The following output, captured from a local Hardhat testnet, demonstrates a full end-to-end run. After deploying the QSecure contract, a quantum-generated integer is produced via Qiskit, combined with metadata and timestamp, and then hashed using SHA-256. This hash, along with the raw data, is recorded on-chain. Finally, the event is retrieved from the blockchain to verify the consistency of the stored values.

```
QSecure deployed at:
0xB7f8BC63BbcaD18155201308C8f3540b07f84F5e
Quantum-generated number: 214
SHA-256 event hash:
d4f8665e0bc8a3ffb589eb1e0f48e889cdc2a3af893f0451c994a48f7
9d82ac2
Total events recorded: 1
Event details:
   Timestamp        : 1744230722
   Quantum Proof  : 214
   Metadata         : Component validated - Checkpoint A
   Stored SHA-256 :
0xd4f8665e0bc8a3ffb589eb1e0f48e889cdc2a3af893f0451c994a48
f79d82ac2
```

This console output confirms that all components of the pipeline, quantum entropy generation, cryptographic hashing, smart contract interaction, and blockchain storage, executed successfully. The SHA-256 hash retrieved from the blockchain exactly matches the one derived from the original quantum value, timestamp, and metadata. This confirms that the event was stored immutably and verifiably. Any attempt to replay or forge the event would inevitably produce a hash mismatch.

3   Classical vs Quantum Randomness and the Role of Qubit Count

Pseudorandom number generators (PRNGs) are algorithms whose unpredictability depends on how well a secret seed is protected. Physical noise sources, often used as true random number generators (TRNGs), improve unpredictability but can drift or be influenced unless carefully conditioned. A quantum random number generator (QRNG) is different: each bit comes from a quantum measurement whose outcome cannot be reconstructed from hidden device state, so unpredictability stems from superposition and collapse rather than secrecy. For logging, the key metric is extractable entropy, not raw bit rate. In practice, QRNGs combine shallow per-qubit measurements with vetted randomness extractors to remove bias and correlations and to deliver an auditable minimum-entropy budget.

Adding qubits mostly increases throughput under ideal conditions, where n qubits can yield up to n bits per shot. Quality is limited by noise, calibration drift, and crosstalk, so moving to 16 or 32 qubits raises subtle asymmetries that require the extractor to use larger windows and lower rates to match measured minimum entropy. The secure yield is therefore $\alpha \cdot n$ with α less than 1 and determined during operation. Beyond a few dozen qubits, bottlenecks often shift to input and output (I/O), post-processing, and anchoring costs, so the right target is verified entropy per second rather than raw counts. For secure logging, simple, parallel, well-calibrated measurements generally outperform deeper, entangling circuits.

4 Security Analysis

Security links three properties end to end: unpredictability at the source, binding by hash, and immutability at the ledger. An attacker might tamper with firmware, reorder network traffic, skew clocks, or attempt chain reorganizations. Quantum generation supplies non-derivable entropy at the edge, continuous health tests enforce fail-closed behavior, and periodic minimum-entropy estimates throttle extraction so commitments never overstate entropy. Each event is sealed with the Secure Hash Algorithm 256-bit (SHA-256) over quantum bits, metadata, and a timestamp. Even with Grover's quadratic speedup, margins remain comfortable, and the Secure Hash Algorithm 384-bit (SHA-384) or 512-bit (SHA-512) can be substituted if policy requires. Non-repudiation comes from signing before anchoring, with a migration path to post-quantum cryptography (PQC) for future resilience.

Time is treated as a security parameter by recording two timestamps, an attested local time and the block time at anchoring, which exposes back-dating and replay. Practical immutability is achieved through finality using confirmations on public chains or Byzantine Fault Tolerant (BFT) consensus in consortium settings. Costs and availability are managed by batching event hashes into a Merkle root on chain (the root hash of a Merkle tree) and keeping encrypted logs off chain under hardware security module (HSM) protected keys, revealing only commitments and minimal metadata. Redundant entropy sources mixed cryptographically prevent single-device bias, while telemetry on entropy, health-test failures, anchoring latency, and clock drift ensures the system fails loudly rather than silently.

5 Conclusion

This framework turns operations into verifiable facts by combining quantum-origin entropy, cryptographic commitments, and an immutable ledger. Quantum generation removes reliance on seed secrecy, while disciplined extraction and monitoring convert raw outcomes into audited minimum entropy. Scaling to 16 to 32 qubits mainly boosts throughput; quality remains a calibration issue, so the right objective is secure entropy per second. Strong commitments, signatures with a path to post-quantum cryptography, dual timestamps, and batched anchoring provide an audit trail resilient to device compromise and network manipulation, suitable for industrial compliance at scale.

References

1. IBM Qiskit Documentation. https://qiskit.org/documentation. Accessed 06 Sept 2025
2. Ethereum Smart Contract Best Practices. https://consensys.github.io/smart-contract-best-practices. Accessed 06 Sept 2025
3. NIST Quantum Randomness Initiative, https://www.nist.gov/programs-projects/quantum-randomness. Accessed 06 Sept 2025
4. Nakamoto, S.: Bitcoin: a peer-to-peer electronic cash system (2008)
5. Herrero-Collantes, M., Garcia-Escartin, J.-C.: Quantum random number generators. Rev. Mod. Phys. **89**, 015004 (2017)
6. Boneh, D., Shoup, V.: A Graduate Course in Applied Cryptography (2020)

QUBO Formulation for Fault Location on Power Grid with Sparse Measurements

Eloi Gravot[1,2]([✉])[iD], Beatriz Moya[1], Sergio Torregrosa[3], Nicolas Hascöet[1], Xavier Kestelyn[1,2], Francisco Chinesta[1,4], Fikri Hafid[5], and Paul-Henri Langlois[5]

[1] RTE Chair @ PIMM Lab, Arts et Métiers Institute of Technology, Paris, France
`eloi.gravot@ensam.eu`
[2] L2EP Lab, Centrale Lille, Junia ISEN, Arts et Métiers,
University of Lille, Lille, France
[3] PIMM Lab, Arts et Métiers Institute of Technology, Paris, France
[4] CNRS @ CREATE, CREATE Tower, Singapore, Singapore
[5] RTE, Réseau de Transport d'Électricité, La Défense, France

Abstract. The paper proposes a Quadratic Unconstrained Binary Optimization (QUBO) formulation in a bipartite power grid fault location problem using sparse measurements. The sparse approximation problem consists of solving an underdetermined system of complex-valued equations. Through enforcement of the grid-depending sparsity of the problem, the amount of quantum bits necessary to solve the system is reduced, opening possibilities for industrial, large grid applications with lowered computational cost.

Keywords: QUBO · Fault Location · Power Grid

1 Introduction

Transmission operators often face electrical faults on their lines due to various external events leading to an isolation of the line through circuit breakers. Some faults are considered fugitive; the short circuit vanishes in few milliseconds after extinguishment of the arc by circuit breakers or end of the physical contact with the line. These faults are mostly due to lightning strikes (+90%) and do not require any sort of intervention by field teams. However, electric flash can cause serious dielectric and thermal damage to conductors and isolators, reducing their respective reliability and performance. Locating these faults has an added value for implementing preventive maintenance strategies. Permanent faults are the consequence of persistent contact with a conductor (vehicle, farming equipment, trees, severed line, etc.). The line is therefore isolated until the contact is removed by field teams. To ensure fast recovery of the line and ensure system stability, various methods exist to locate the fault. Two main fault location families arise: traveling wave based, and impedance based.

The traveling wave method determines the fault location by measuring the time delay between the disturbance's arrival at both terminals of the transmission line [1]. As the disturbance travels around the speed of light, the location estimation leads to good results. This method can also be adapted to a single terminal using the disturbance echoes. The latter may prove less accurate, as it becomes difficult to distinguish echoes from noise. Despite good localization results, traveling wave methods require extremely precise measuring equipment.

Impedance-based methods rely on the steady-state phasors of the network and impedance data of the lines. Depending on the amount of data, there are many variations. One-ended method use phasor from one end of the faulted line and two-ended from both ends, synchronized or not. One-ended method, due to many unknowns, are based on various hypotheses and lack robustness. Two-ended methods are proven to be accurate as they do not depend on the hypothesis of the network. However, they require IEDs (Intelligent Electronic Device) at each bus/node of the system [2].

The last category of impedance methods is the so-called wide area measurement. Using synchronized PMUs (Phasor Measurement Unit) scattered around the network, an underdetermined system of linear complex-value equations with a sparse solution is solved. Different approaches can be found such as LARS (Least Angle Regression) to solve the Lasso problem [3,4] or Compressive Sensing [5–7]. These methods bring a compromise of having accurate fault location while minimizing the number of recorders on the grid. To enforce the sparsity of the solution and convergence of the optimization problem, the mentioned methods use ℓ_1-norm minimization as followed: given a matrix $A \in \mathbb{R}^{m \times n}$ and a vector $b \in \mathbb{R}^m$ with $m \ll n$, we seek a vector $x \in \mathbb{R}^n$ that minimizes :

$$\min_{x} \|Ax - b\|_2^2 + \lambda \|x\|_1. \tag{1}$$

However, depending on topology of the network in fault location, the number of k nonzero values in the solution can be known a priori. Yet the ℓ_1 norm relaxation does not strictly enforce this constraint and is sensitive to the choice of the regularization parameter λ, often requiring extensive tuning to approximate the desired sparsity. By contrast, formulating the problem with the ℓ_0 norm as a combinatorial minimization would directly enforce k nonzero values, addressing both the known sparsity and the sensitivity to regularization. Although, such a problem can not be handled by iterative approach on classical computers due to the discontinuity of the ℓ_0 norm.

In recent years, Quadratic Unconstrained Binary Optimization (QUBO) formulations have gained attention for their application in solving linear problems [8,9] due to the growing availability of adiabatic quantum computer (AQC) or GPU-based Annealing Engines (AE). QUBO problems map naturally to quantum annealing computing frameworks, which are designed to handle binary quadratic formulations directly. This approach enables true ℓ_0 minimization, directly seeking the sparsest solution unlike LARS or CS. However, one of the downsides of solving such system with QUBO is the need of converting real-valued variables to binary ones which can become extremely costly in terms of qubits, and therefore not suited for industrial application on AQCs, or even on AEs that allow higher ranked QUBO matrices. Nevertheless, recent studies have investigated the application of quantum annealing for solving industrial-scale problems, such as fault localization in graph-based network structures like electrical distribution grids [10].

This paper proposes a QUBO formulation for fault location on a power bipartite grid with sparse measurements. By optimizing PMUs placement, the sparsity of the solution is enforced according to the topology of a bipartite graph/grid. This formulation offers scalability advantages by reducing the amount of qubits to solve the system.

2   Problem Formulation

Consider a N-bus power grid. The nodal equation in steady-state at each bus is the following:

$$ZI = V \tag{2}$$

With Z the impedance matrix of the system, I and V the vector of buses' respective current and voltage. When a fault occurs on a power system, voltage variations ΔV can be observed on the nodes of the system. The nodal equations under a steady-state fault becomes :

$$Z\Delta I = \Delta V \tag{3}$$

The resulting ΔI is referred as the sparse virtual fault current injection vector [3]. When studying the positive sequence of the system (for asymmetrical fault), ΔI is $N \times 1$ with two non-zero complex values I_g and I_h corresponding to the two buses g and h surrounding the faulted line. The fault location on the line (from g) can be computed as follows [7] :

$$Fault\ location = \frac{I_h}{I_g + I_h} * 100\% \tag{4}$$

However, solving (3) requires having PMUs on all buses of the networks. When PMUs are spread out, some buses aren't measured. The number of these unmeasured buses is denoted by u, and it's usually larger than the number of measured ones. The whole system is therefore not observable, and the nodal equation becomes underdetermined. To apply sparsity recovery algorithms such as LARS and CS, the complex-value system $Ax = b$ is transformed into a real one:

$$A_r = \begin{bmatrix} real(A) & -imag(A) \\ imag(A) & real(A) \end{bmatrix} \tag{5}$$

$$x_r = \begin{bmatrix} real(x) \\ imag(x) \end{bmatrix} \tag{6}$$

$$b_r = \begin{bmatrix} real(b) \\ imag(b) \end{bmatrix} \tag{7}$$

$$A_r x_r = b_r \tag{8}$$

3 QUBO Formulation

3.1 Ising Model

The Ising model is a mathematical model from statistical mechanics [11]. It represents a system of n discrete bipolar variables called spins. Each spin s_i can take values in $\{-1, +1\}$. Spins interact with their neighbors and with an external field. The energy of the system is defined by the Hamiltonian:

$$H(s) = -\sum_{i<j} J_{ij} s_i s_j - \sum_i h_i s_i \tag{9}$$

The coefficients J_{ij} define the interaction strength between spins i and j. The values h_i represent the effect of the local field. The objective is to find the configuration of spins that minimizes the energy. In quantum computing, optimization problems are often written in QUBO form. In this model, variables are binary: $x_i \in \{0, 1\}$. The QUBO objective function is:

$$Q(x) = \sum_{i<j} Q_{ij} x_i x_j + \sum_i Q_{ii} x_i \tag{10}$$

To map the Ising model to QUBO, we use a change of variables. Each spin s_i is rewritten as $s_i = 2x_i - 1$. This transformation converts Ising variables into binary variables. It preserves the structure of optimization. This equivalence allows solving Ising problems on quantum annealers. Quantum processors like D-Wave natively work with Ising or QUBO models, as well as AEs.

3.2 Solving Linear Least Squares

In QUBO formulation, variables are binary, which is not the case in equation (8). We approximate each real-valued variable x_j using binary variables $q_{j,l} \in \{0, 1\}$ as follows:

$$x_j \approx \sum_{l=0}^{m} 2^l q_{j,l} - 2^{m+1} q_{j,\emptyset}, \tag{11}$$

where $q_{j,\emptyset} \in \{0, 1\}$ acts as a sign bit (0 for non-negative values, 1 for negative values). Substituting this binary expansion into the original system leads to a reformulation of each equation. Specifically, each row i of the expression $A\mathbf{x}$ becomes:

$$\sum_{j=1}^{n} a_{ij} x_j \longrightarrow \sum_{j=1}^{n} a_{ij} \left(\sum_{l=0}^{m} 2^l q_{j,l} - 2^{m+1} q_{j,\emptyset} \right), \tag{12}$$

resulting in a quadratic objective in terms of binary variables, suitable for QUBO formulation. We minimize a least-squares objective function:

$$min \, \|Ax - b\|_2^2 \tag{13}$$

$$\|Ax - b\|_2^2 = x^T A^T A x - 2b^T A x + b^T b \tag{14}$$

where $b^T b$ is a constant. It can be rewritten as a QUBO formulation :

$$min \, y = x^T Q x \tag{15}$$

$$Q = A^T A + diag(-2b^T A) \tag{16}$$

The major limitation of such an approach is that, even if we enforce a sparsity constraint, there will be $m+2$ qubits to express each variable. As m increases for higher accuracy, the size of the QUBO matrix scales linearly with the number of unknown u and m.

3.3 Enforcing a Graph-Dependent Sparsity

We consider a power grid with a bipartite graph topology: vertices can be divided into two distinct and independent sets U and V. The IEEE 9-bus satisfies this conditions (Fig. 1) where $V = \{v_1, v_2, v_3\}$ and $U = \{u_1, u_2, u_3, u_4, u_5, u_6\}$.

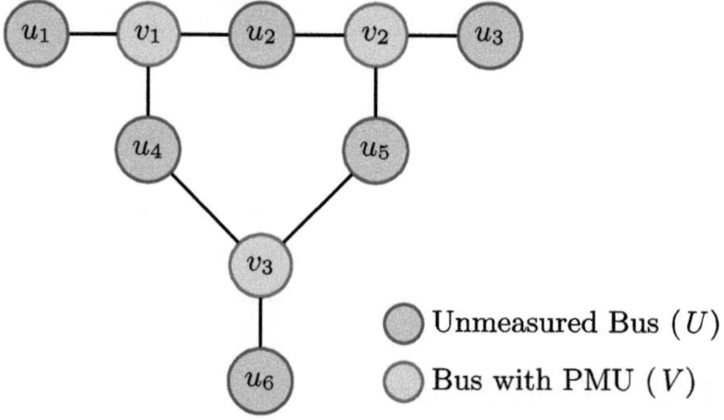

Fig. 1. IEEE 9-bus system consisting of 9 buses and 9 lines. By putting the PMUs on set V, every line has exactly one measured end, and one unmeasured.

If PMUs are placed on all the buses belonging to one set (V in Fig. 1), strictly one bus of the unmeasured ones (U in Fig. 1) will have a non-zero injected current value during a fault on the grid. Therefore, $real(x)$ and $imag(x)$ (6) both have a single non-zero entry at the same index. This implies that the two vectors share exactly the same support. To make the structure explicit, we introduce a additional qubits $q_{bus,i} \in \{0, 1\}$, one for each bus $i \in \{1, ..., |U|\}$, that takes 1 if the bus is at one end of the faulted line, 0 otherwise. Using this notation, the original expression of x_r (17) can be transformed to (18) where X_1 and

X_2 are single real-valued approximations obtained from the underlying binary variables using (11). That is, only one approximation per vector is needed, and it is shared across all active entries, with the binary coefficients $q_{bus,i}$ determining which position is active. Moreover, to enforce sparsity, the sum of activation bus qubits is equal to one: The shunt fault happens on only one line (19).

$$x_r = \begin{bmatrix} \sum_{l=0}^{m} 2^l q_{1,l} - 2^{m+1} q_{1,\emptyset} \\ \sum_{l=0}^{m} 2^l q_{2,l} - 2^{m+1} q_{2,\emptyset} \\ \sum_{l=0}^{m} 2^l q_{3,l} - 2^{m+1} q_{3,\emptyset} \\ \cdots \end{bmatrix} \tag{17}$$

$$\begin{cases} real(x) = \begin{bmatrix} q_{bus,1} * X_1 \\ q_{bus,2} * X_1 \\ q_{bus,3} * X_1 \\ \cdots \end{bmatrix} & \text{with } X_1 = \sum_{l=0}^{m} 2^l q_{1,l} - 2^{m+1} q_{1,\emptyset} \\[2em] imag(x) = \begin{bmatrix} q_{bus,1} * X_2 \\ q_{bus,2} * X_2 \\ q_{bus,3} * X_2 \\ \cdots \end{bmatrix} & \text{with } X_2 = \sum_{l=0}^{m} 2^l q_{2,l} - 2^{m+1} q_{2,\emptyset} \end{cases} \tag{18}$$

$$\left(\sum_{i=1}^{|U|} q_{\text{bus},i} = 1 \right) \quad \Rightarrow \quad \forall i \neq j, \ q_{\text{bus},i} \cdot q_{\text{bus},j} = 0 \tag{19}$$

The variable change from (17) to (18) leads to, once (15) is expanded, some quadratic, cubic and quartic terms. The quartic terms necessary include bus activation qubits products such as in (19) and are therefore removed. To enforce sparsity of (19), a penalty is added to the expression (20, 21) [12].

$$P \left(\sum_{i=1}^{|U|} q_{\text{bus},i} - 1 \right)^2 \tag{20}$$

$$\Leftrightarrow P \left(2 \sum_{i<j} q_{\text{bus},i} \cdot q_{\text{bus},j} - \sum_{i=1}^{|U|} q_{\text{bus},i} + 1 \right) \tag{21}$$

The cubic terms are due to products of X_1 and X_2 qubits. These products are replaced by an additional binary variable y_i [12] and an added penalty as in the following example :

$$q_{\text{bus},i}\, q_{1,l}\, q_{2,l} \Rightarrow q_{\text{bus}}\, y_1 + P(q_{1,l}\, q_{2,l} - 2q_{1,l}\, y_1 - 2q_{2,l}\, y_1 + 3y_1) \tag{22}$$

By considering those added variables, the total amount of qubits necessary for fault location on a bipartite graph where PMUs are placed on set V and where set U contains u unmeasured bus is :

$$nb_{qubits} = \frac{(2\,m+4)(2\,m+3)}{2} + 2(m+2) + u \tag{23}$$

The first term corresponds to the added qubits due to the cubic terms of X_1 and X_2 that each have $m+2$ qubits (second term) plus the u unmeasured buses with their respective activation qubits. This formulation offers a significant scalability advantage over the classic one (17), which requires $2(m+1) \times u$ qubits. While the proposed formulation introduces a quadratic term in m, the number of qubits needed grows linearly with the system size u, whereas formulation (16) scales as $\mathcal{O}(mu)$. It makes the approach much more efficient for large systems.

4   Example

The formulation is tested on complex-value system inspired by the IEEE 9-bus and solved on the Fixstars Amplify AE [13]. We consider the following system with A of size 3×6 (3 measured buses, 6 unknowns) :

$$\begin{cases} A = \begin{bmatrix} 35-36j & 36-28j & -20+32j & -17-28j & -24-39j & -1+36j \\ 36-27j & -34+22j & -13+10j & -3+11j & 26+7j & -22-20j \\ -24+38j & -11-25j & -20-26j & -36-9j & 28+2j & -7+19j \end{bmatrix} \\ b = \begin{bmatrix} -656+2960j \\ 1938+1504j \\ -840+1904j \end{bmatrix} \end{cases}$$

$$\tag{24}$$

The sparse solution of the system is $-98+38j$ for x_5. Using the proposed method, the size of Q is 142×142 with accuracy set at $m = 6$. The output of the AE is displayed in Fig 2. The computation time, including the server connection, is 2.6 s. Once the binary values of X_1 and X_2 are transformed back to integers with (11), the computed solution is $-98 + 38j$, obtained on bus 5. Although based on synthetic data, this example demonstrates how the resulting complex output could be used to localize the fault along the line in a real-world electrical network using (4).

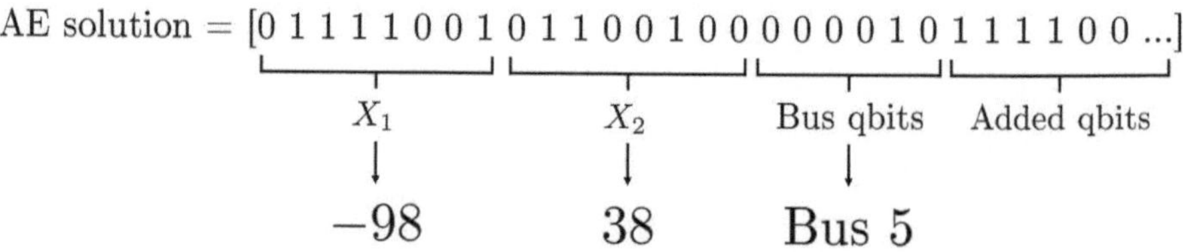

Fig. 2. First qubits of the 142 long AE output. First 8 corresponds to the real part of the solution, 8 following to the imaginary part, then the 6 bus qubits, followed by 120 added qubits due to cubic terms.

5   Conclusion

This paper proposed a QUBO formulation of an underdetermined sparse complex-value linear system to solve a fault location problem on a bipartite power grid with sparse measurements. Through enforcement of sparsity based on the topology of the graph, the formulation proposed demonstrates scalability advantages and opens doors to application on larger grids. Future work aims to adapt the method to tripartite graph that are more representative of transmission grids with also less PMUs: the proposed method relies on having a PMU at least one end of each line. Their optimal placement on the grid could be studied as a constrained graph coloring problem. Also, the data considered in this study is purely analytical, and future work will include simulations, incorporating tests with noise, to assess the performance of the proposed formulation under more realistic system conditions.

Acknowledgments. The authors acknowledge the support of the RTE research chair at Arts et Métiers Institute of Technology.

Disclosure of Interests. The authors have no competing interests to declare that are relevant to the content of this article.

References

1. Lee, H., Mousa, A.M.: GPS travelling wave fault locator systems: investigation into the anomalous measurements related to lightning strikes. IEEE Trans. Power Delivery **11**(3), 1214–1223 (1996)
2. Das, S., Santoso, S., Gaikwad, A., Patel, M.: Impedance-based fault location in transmission networks: theory and application. IEEE Access **2**, 537–557 (2014)
3. Feng, G., Abur, A.: Identification of faults using sparse optimization. In: 2014 52nd Annual Allerton Conference on Communication. Control, and Computing (Allerton), pp. 1040–1045. IEEE, Urbana (2014)
4. Mouco, A., Abur, A.: Improving the wide-area PMU-based fault location method using ordinary least squares estimation. Electr. Power Syst. Res. **189**, 106620 (2020)
5. Jia, K., Yang, B., Bi, T., Zheng, L.: An improved sparse-measurement based fault location technology for distribution networks. IEEE Trans. Industr. Inf. **16**(1), 1–1 (2020)
6. Majidi, M., Etezadi-Amoli, M., Fadali, M.S.: A sparse-data-driven approach for fault location in transmission networks. IEEE Trans. Smart Grid **8**(2), 548–556 (2017)
7. Shan, H., Zhang, L., Wu, Q.H., Li, M.: Location of asymmetric ground fault using virtual injected current ratio and two-stage recovery strategy in distribution networks. CSEE J. Power Energy Syst. **10**(1), 151–161 (2024)
8. Borle, A., Lomonaco, S. J.: Analyzing the quantum annealing approach for solving linear least squares problems, pp. 1–5 (2018). arXiv:1809.07649
9. Jun, K.: QUBO formulations for a system of linear equations. Results Control Optim. **14**, 100380 (2024)

10. Bi, Z., Yang, X., Wang, B., Zhang, W., Dong, Z., Zhang, D.: Quantum annealing algorithm for fault section location in distribution networks. Appl. Soft Comput. **149**, 110973 (2023)
11. Brush, S.G.: History of the Lenz-Ising model. Rev. Mod. Phys. **39**(4), 1–10 (1967)
12. Glover, F., Kochenberger, G., Du, Y.: Quantum bridge analytics i: a tutorial on formulating and using QUBO models. Ann. Oper. Res. (2018)
13. Fixstars: Fixstars Amplify. Online documentation (2024). https://amplify.fixstars.com/en/docs/amplify/v1/index.html

Author Index

F. Barbaresco and F. Gerin (Eds.): QUEST-IS 2025, CCIS 2744, pp. 361–363, 2026.
https://doi.org/10.1007/978-3-032-13855-2

MIX
Papier aus verantwortungsvollen Quellen
Paper from responsible sources
FSC® C105338

If you have any concerns about our products,
you can contact us on
ProductSafety@springernature.com

In case Publisher is established outside the EU,
the EU authorized representative is:
Springer Nature Customer Service Center GmbH
Europaplatz 3, 69115 Heidelberg, Germany

Printed by Libri Plureos GmbH
in Hamburg, Germany